AF595564

Visible Light Communication (VLC)

Visible Light Communication (VLC)

Editor

Chen Chen

MDPI • Basel • Beijing • Wuhan • Barcelona • Belgrade • Manchester • Tokyo • Cluj • Tianjin

Editor
Chen Chen
Chongqing University
China

Editorial Office
MDPI
St. Alban-Anlage 66
4052 Basel, Switzerland

This is a reprint of articles from the Special Issue published online in the open access journal *Photonics* (ISSN 2304-6732) (available at: https://www.mdpi.com/journal/photonics/special_issues/VLC).

For citation purposes, cite each article independently as indicated on the article page online and as indicated below:

LastName, A.A.; LastName, B.B.; LastName, C.C. Article Title. *Journal Name* **Year**, *Volume Number*, Page Range.

ISBN 978-3-0365-4087-0 (Hbk)
ISBN 978-3-0365-4088-7 (PDF)

Contents

About the Editor

Chen Chen (Research Professor) received B.S. and M.Eng. degrees from the University of Electronic Science and Technology of China, Chengdu, China, in 2010 and 2013, respectively, and a Ph.D. from Nanyang Technological University, Singapore, in 2017. He was a Post-Doctoral Researcher at the School of Electrical and Electronic Engineering, Nanyang Technological University, from 2017 to 2019. He is currently a Research Professor with the School of Microelectronics and Communication Engineering, Chongqing University, Chongqing, China. His research interests include optical wireless communication, optical access networks, Internet of Things, and machine learning. He currently serves as Section Board Member for "Optical Communication and Network" of the MDPI journal *Photonics*.

Editorial

Special Issue on "Visible Light Communication (VLC)"

Chen Chen

School of Microelectronics and Communication Engineering, Chongqing University, Chongqing 400044, China; c.chen@cqu.edu.cn

Due to its appealing advantages, including abundant and unregulated spectrum resources, no electromagnetic interference (EMI) radiation and high security, visible light communication (VLC) using light-emitting diodes (LEDs) or laser diodes (LDs) has been envisioned as one of the key enabling technologies for 6G and Internet of Things (IoT) systems [1–3]. Despite its many advantages, VLC faces several key technical challenges such as a limited bandwidth and severe nonlinearity of opto-electronic devices, link blockage, user mobility, etc. [4]. Therefore, significant efforts are needed from the global VLC community to further develop the VLC technology.

This Special Issue aims to provide an opportunity for global researchers to share their new ideas and cutting-edge techniques to address the above-mentioned challenges. A total of 16 selected papers are published in this Special Issue, representing the fascinating progress of VLC in various scenarios, including general indoor scenarios [5–8], underwater scenarios [9–12], and other potential scenarios [13–16], and the emerging application of machine learning/artificial intelligence (ML/AI) techniques in VLC [17–20].

VLC for general indoor scenarios: In order to improve the spectral efficiency of bandlimited VLC systems, Hong, H. et al. proposed a hybrid, adaptive bias, optical orthogonal frequency division multiplexing (HABO-OFDM) scheme, which has been shown to outperform benchmark schemes in terms of peak-to-average-power ratio (PAPR) and power efficiency [5]. Moreover, Nie, Y. et al. further proposed a pairwise coding (PWC)-based, multiband, carrierless amplitude and phase (mCAP) modulation with chaotic, dual-mode index modulation (DM) for secure bandlimited VLC systems [6]. The spectral efficiency of the system can be greatly improved by combining mCAP with DM and the signal-to-noise ratio (SNR) imbalance caused by the low-pass frequency response can be mitigated by applying PWC coding. As VLC systems built upon lighting LEDs need to fulfil the dual functions of lighting and communication, dimming control should be taken into consideration. To achieve constant transmission efficiency, Guo, J.-N. et al. proposed a dimming control scheme based on extensional constant weight codeword sets and provided a low implementation complexity decoding algorithm for the scheme [7]. A practical VLC system usually needs to serve multiple users; hence, efficient multiple access schemes should be employed in multi-user VLC systems. Since the BER performance of VLC systems using non-orthogonal multiple access (NOMA) is poor, due to the unequal distances between adjacent points in the superposition constellation (SC), Wu, T. et al. proposed a novel scheme to improve the BER performance by adjusting the parameters to change the shape of SC at the transmitter and by simultaneously adjusting the parameters of successive interference cancellation (SIC) decoding at the receiver [8].

VLC for underwater scenarios: For the underwater VLC system using on–off-keying (OOK) modulation, Zhang, J. et al. systematically studied the bandwidth limitation due to the transceiver and underwater channel through experiments and simulations [9]. The obtained results show that the maximum likelihood sequence estimation (MLSE) detection has great potential to improve the performance of bandwidth-limited communication systems. Moreover, Li, C. et al. proposed and demonstrated a high-sensitivity, long-reach, underwater VLC system using a commercial blue LED source, a photon counting receiver

Citation: Chen, C. Special Issue on "Visible Light Communication (VLC)". *Photonics* **2022**, *9*, 284. https://doi.org/10.3390/photonics9050284

Received: 6 April 2022
Accepted: 20 April 2022
Published: 21 April 2022

and OOK modulation [10]. Receiver sensitivities of −76 dBm, −74 dBm, and −70 dBm were achieved for 1 Mbps, 2 Mbps, and 5 Mbps data rates over a 10 m underwater channel, respectively, and a more than 100 m distance can be transmitted for a 2 Mbps data rate in pure seawater at 1 W transmitted power. For the underwater VLC system using OFDM modulation, Wei, Z. et al. proposed a dual-branch, pre-distorted, enhanced, asymmetrically clipped, direct current biased, optical orthogonal frequency division multiplexing (PEADO-OFDM) scheme for underwater VLC systems [11]. The simulation results show that PEADO-OFDM can increase the spectral efficiency, enhance the tolerance against ISI, and improve the bit error rate (BER) performance. Furthermore, Li, H. et al. proposed and experimentally demonstrated a simple sampling frequency offset (SFO) estimation and compensation scheme for OFDM-based underwater VLC systems [12]. The experimental results show that the proposed scheme can achieve a high estimation accuracy with low computational complexity.

VLC for other potential scenarios: Besides, general indoor scenarios and underwater scenarios, VLC is also applicable to many other potential scenarios. Specifically, VLC can be used to realize high-accuracy positioning and navigation in indoor environments. Martínez-Ciro, R.A. et al. presented an indoor visible-light positioning (VLP) system based on red–green–blue (RGB) LEDs and a frequency division multiplexing (FDM) scheme, where the received signal strength (RSS) technique is adopted to estimate the receiver position for multi-cell networks [13]. Simulation results demonstrated an average positioning error of less than 2.2 cm for all chromatic points. Moreover, VLC can also be applied in vehicular communications by utilizing the headlamps or taillights of vehicles to transmit data. Zhan, L. et al. investigated the performance of vehicular VLC employing mirror array-based intelligent reflecting surface (IRS) [14]. By optimizing the number of mirrors in the IRS, the energy efficiency (EE) can be successfully maximized. In addition, VLC can be integrated with other communication technologies to establish a hybrid system. Particularly, Liu, B. et al. proposed a hybrid millimeter-wave/VLC system and compared the propagation characteristics, including path loss, root mean square (RMS) delay spread (DS), K-factor, and cluster characteristics, between mmWave and VLC bands based on a measurement campaign and ray tracing simulation in a conference room [15], while Le-Tran, M. et al. demonstrated a 294-Mb/s hybrid fiber/wireless link based on a single visible LED over a 1.5-m polymer optical fiber (POF) and 1.5-m free-space distance [16].

The application of ML/AI in VLC: Aiming to mitigate the adverse effect of nonlinearity in VLC systems, Park, Y.-J. et al. proposed a predistortion approach using coefficient approximation without sampling the LED transfer function, and further utilized the bidirectional long short-term memory (BLSTM) approach to train the LED distortion correction without knowing the LED modeling at the transmitter side [17], while Cao, B. et al. proposed a pilot-assisted reservoir computing (PA-RC) nonlinear equalization algorithm for nonlinearity mitigation at the receiver side [18]. In addition to bandwidth limitation and nonlinearity, inter-symbol-interference (ISI) due to multipath reflection is another issue that needs to be addressed in high-speed VLC systems. Li, L. et al. constructed a neural network (NN)-based transceiver to compensate for the varying ISI effect in VLC systems [19]. The application of multiple-input multiple-output (MIMO) transmission in VLC systems is a natural and efficient way to enhance the system capacity, and the combination of MIMO and index modulation can convey additional bits. To improve the BER performance of the OFDM-based generalized LED index modulation (GLIM) system, Le-Tran, M. et al. proposed a deep neural network (DNN)-aided active LED index estimator (IE) for improving the GLIM-OFDM detector performance [20].

Acknowledgments: We would like to thank all authors, the many dedicated reviewers, and the editorial team of *Photonics*, for their valuable contributions, making this Special Issue possible and successful.

Conflicts of Interest: The authors declare no conflict of interest.

References

1. Chi, N.; Zhou, Y.; Wei, Y.; Hu, F. Visible Light Communication in 6G: Advances, Challenges, and Prospects. *IEEE Veh. Technol. Mag.* **2020**, *15*, 93–102. [CrossRef]
2. Demirkol, I.; Camps-Mur, D.; Paradells, J.; Combalia, M.; Popoola, W.; Haas, H. Powering the Internet of Things Through Light Communication. *IEEE Commun. Mag.* **2019**, *57*, 107–113. [CrossRef]
3. Chen, C.; Fu, S.; Jian, X.; Liu, M.; Deng, X.; Ding, Z.G. NOMA for Energy-Efficient LiFi-Enabled Bidirectional IoT Communication. *IEEE Trans. Commun.* **2021**, *69*, 1693–1706. [CrossRef]
4. Matheus, L.E.M.; Vieira, A.B.; Vieira, L.F.M.; Vieira, M.A.M.; Gnawali, O. Visible Light Communication: Concepts, Applications and Challenges. *IEEE Commun. Surv. Tutor.* **2019**, *21*, 3204–3237. [CrossRef]
5. Hong, H.; Li, Z. Hybrid Adaptive Bias OFDM-Based IM/DD Visible Light Communication System. *Photonics* **2021**, *8*, 257. [CrossRef]
6. Nie, Y.; Zhang, W.; Yang, Y.; Deng, X.; Liu, M.; Chen, C. Pairwise Coded mCAP with Chaotic Dual-Mode Index Modulation for Secure Bandlimited VLC Systems. *Photonics* **2022**, *9*, 141. [CrossRef]
7. Guo, J.-N.; Zhang, J.; Xin, G.; Li, L. Constant Transmission Efficiency Dimming Control Scheme for VLC Systems. *Photonics* **2021**, *8*, 7. [CrossRef]
8. Wu, T.; Wang, Z.; Han, S.; Yu, J.; Jiang, Y. Demonstration of Performance Improvement in Multi-User NOMA VLC System Using Joint Transceiver Optimization. *Photonics* **2022**, *9*, 168. [CrossRef]
9. Zhang, J.; Gao, G.; Li, J.; Ma, Z.; Guo, Y. Experimental Demonstration and Simulation of Bandwidth-Limited Underwater Wireless Optical Communication with MLSE. *Photonics* **2022**, *9*, 182. [CrossRef]
10. Li, C.; Liu, Z.; Chen, D.; Deng, X.; Yan, F.; Li, S.; Hu, Z. Experimental Demonstration of High-Sensitivity Underwater Optical Wireless Communication Based on Photocounting Receiver. *Photonics* **2021**, *8*, 467. [CrossRef]
11. Wei, Z.; Li, Y.; Wang, Z.; Fang, J.; Fu, H. Dual-Branch Pre-Distorted Enhanced ADO-OFDM for Full-Duplex Underwater Optical Wireless Communication System. *Photonics* **2021**, *8*, 368. [CrossRef]
12. Li, H.; Chen, T.; Wang, Z.; Cao, B.; Li, Y.; Zhang, J. Low-Complexity Sampling Frequency Offset Estimation and Compensation Scheme for OFDM-Based UWOC System. *Photonics* **2022**, *9*, 216. [CrossRef]
13. Martínez-Ciro, R.A.; López-Giraldo, F.E.; Luna-Rivera, J.M.; Ramírez-Aguilera, A.M. An Indoor Visible Light Positioning System for Multi-Cell Networks. *Photonics* **2022**, *9*, 146. [CrossRef]
14. Zhan, L.; Zhao, H.; Zhang, W.; Lin, J. An Optimal Scheme for the Number of Mirrors in Vehicular Visible Light Communication via Mirror Array-Based Intelligent Reflecting Surfaces. *Photonics* **2022**, *9*, 129. [CrossRef]
15. Liu, B.; Tang, P.; Zhang, J.; Yin, Y.; Liu, G.; Xia, L. Propagation Characteristics Comparisons between mmWave and Visible Light Bands in the Conference Scenario. *Photonics* **2022**, *9*, 228. [CrossRef]
16. Apolo, J.A.; Ortega, B.; Almenar, V. Hybrid POF/VLC Links Based on a Single LED for Indoor Communications. *Photonics* **2021**, *8*, 254. [CrossRef]
17. Park, Y.-J.; Kim, J.-Y.; Jung, J.-I. Predistortion Approaches Using Coefficient Approximation and Bidirectional LSTM for Nonlinearity Compensation in Visible Light Communication. *Photonics* **2022**, *9*, 198. [CrossRef]
18. Cao, B.; Yuan, K.; Li, H.; Duan, S.; Li, Y.; Ouyang, Y. The Performance Improvement of VLC-OFDM System Based on Reservoir Computing. *Photonics* **2022**, *9*, 185. [CrossRef]
19. Li, L.; Zhu, Z.; Zhang, J. Neural Network-Based Transceiver Design for VLC System over ISI Channel. *Photonics* **2022**, *9*, 190. [CrossRef]
20. Le-Tran, M.; Kim, S. Deep Learning-Assisted Index Estimator for Generalized LED Index Modulation OFDM in Visible Light Communication. *Photonics* **2021**, *8*, 168. [CrossRef]

Article

Hybrid Adaptive Bias OFDM-Based IM/DD Visible Light Communication System

Huandong Hong † and Zhengquan Li *,†

Jiangsu Provincial Engineering Laboratory of Pattern Recognition and Computational Intelligence, Jiangnan University, Wuxi 214122, China; 6191918015@stu.jiangnan.edu.cn
* Correspondence: lzq722@jiangnan.edu.cn; Tel.: +86-138-6183-7081
† These authors contributed equally to this work.

Abstract: Conventional optical orthogonal frequency division multiplexing (OFDM) schemes, such as adaptively biased optical OFDM (ABO-OFDM) and hybrid asymmetrically clipped optical OFDM (HACO-OFDM), are unable to tap all the resources of the subcarriers and only achieve relatively high power efficiency. In this paper, a hybrid adaptive bias optical OFDM (HABO-OFDM) scheme for visible light communication (VLC) is proposed to improve spectral efficiency and power efficiency. In the proposed HABO-OFDM scheme, different optical OFDM components are combined for transmission at the same time, and the adaptive bias is designed to ensure the non-negativity, as well as obtaining significantly high power efficiency. Meanwhile, the implementation complexity of the HABO-OFDM receiver is notably lower than the conventional superimposed optical OFDM schemes. Simulation results show that the proposed HABO-OFDM scheme outperforms ABO-OFDM and HACO-OFDM in terms of both peak-to-average-power ratio (PAPR) and power efficiency. The PAPR performance of HABO-OFDM is about 3.2 dB lower than that of HACO-OFDM and 1.7 dB lower than that of ABO-OFDM. Moreover, we can see that the $E_{b(elec)}/N_0$ required for HABO-OFDM to reach the BER target is lower than the other two schemes at the Bit rate/Normalized bandwidth range of 3.5 to 8.75, which means that the power efficiency of HABO-OFDM is higher in this range.

Keywords: orthogonal frequency division multiplexing (OFDM); visible light communication (VLC); power efficiency; peak-to-average-power ratio (PAPR)

Citation: Hong, H.; Li, Z. Hybrid Adaptive Bias OFDM-Based IM/DD Visible Light Communication System. *Photonics* **2021**, *8*, 257. https://doi.org/10.3390/photonics8070257

Received: 9 June 2021
Accepted: 30 June 2021
Published: 5 July 2021

Publisher's Note: MDPI stays neutral with regard to jurisdictional claims in published maps and institutional affiliations.

1. Introduction

With the rapid increase in wireless mobile devices, the continuous increase of wireless data traffic has brought challenges to the continuous reduction of radio frequency (RF) spectrum, which has also driven the demand for alternative technologies [1,2]. In order to solve the contradiction between the explosive growth of data and the consumption of spectrum resources, visible light communication (VLC) has become the development direction of the next generation communication network with its huge spectrum resources, high security, low cost, and so on [3–5]. The orthogonal frequency division multiplexing (OFDM) technology has the advantages of high spectrum utilization and strong ability of resisting inter-symbol interference (ISI) [6,7]. Hence, the OFDM technology is also an excellent choice for VLC.

Considerable research has applied OFDM technology to VLC. As we all know, the VLC requires non-negative and real OFDM signals, due to intensity modulated direct detection (IM/DD) is widely used in VLC systems [8–10]. In order to meet the requirement, several optical OFDM schemes have been proposed for IM/DD VLC systems. Asymmetrically clipped optical OFDM (ACO-OFDM), where only the odd subcarriers are modulated to transmit the information data, achieves high power efficiency [11]. Pulse amplitude modulated discrete multitone (PAM-DMT) is one of clipping-based schemes, where only the imaginary parts of the subcarriers are used, while the real parts are set to zero [12].

However, due to half of the subcarrier resources not being utilized, the ACO-OFDM and PAM-DMT schemes lead to a significant loss of spectral efficiency [13].

Hence, some advanced schemes superimposing various optical OFDM components have been proposed to improve spectral efficiency, including hybrid asymmetrically clipped optical OFDM (HACO-OFDM) [14] with pulse-amplitude-modulated based hybrid optical OFDM (PHO-OFDM) [15] and adaptively biased optical OFDM (ABO-OFDM) [16]. The ACO-OFDM signals and PAM-DMT signals are transmitted simultaneously in HACO-OFDM scheme, where high spectrum efficiency is achieved compared with the ACO-OFDM and PAM-DMT schemes. However, the interference of the clipping of the bipolar signal falls on the superimposed components. The operation of mitigating the interference at the receiver leads to a significantly high receiver complexity [17]. Similarly, the PAM signals and the QAM signals are transmitted on odd and even subcarriers in the PHO-OFDM scheme, respectively. Unfortunately, the clipping noise of the PAM signals affects the demodulation of the QAM signals that are modulated on the even subcarriers, which significantly increase the complexity of the receiver. Another advanced scheme, known as ABO-OFDM, uses the odd and a part of the even subcarriers to convey the information data, which enhances the spectral efficiency compared to ACO-OFDM scheme. Nevertheless, the reserved subcarriers are not utilized, which incurs a bandwidth penalty.

In this paper, a novel hybrid adaptive bias optical OFDM (HABO-OFDM) scheme is conceived to enhance spectral efficiency with power efficiency, and achieve low-complexity implementation of the receiver. Our main contributions are as follows:

- A novel structure combining different optical OFDM components to transmit the signals simultaneously is proposed for VLC. This structure occupies all subcarriers to convey the signals, thus significantly enhancing the spectral efficiency in comparison with ABO-OFDM.
- The adaptive bias is carefully designed to guarantee the non-negativity of the transmitted signal and improve power efficiency. Furthermore, since the interference caused by the adaptive bias does not disturb the estimation of the superimposed signals at the receiver, a relatively simple receiver can be employed in HABO-OFDM.
- Different performances analysis of our proposed system is presented. We show its performance and perform analysis by comparing the proposed scheme with HACO-OFDM and ABO-OFDM in terms of bit error rate (BER), peak-to-average-power ratio (PAPR), and power efficiency performances. In addition, the BER performance under nonlinear conditions is also demonstrated.

This paper is organized as follows. Section 2 describes the transmitter and receiver in the HABO-OFDM scheme in detail. Additionally, it presents the design of the adaptive bias and the proof that the noise introduced by the adaptive bias does not affect the demodulation of the superimposed signals. Section 3 analyzes the complexity of HABO-OFDM and conventional superimposed schemes, including HACO-OFDM and PHO-OFDM. The results of the different schemes performance are shown in Section 4. We provide the BER performance of these schemes in the additive white Gaussian noise (AWGN) channel. We also provide a comparison of the performances of the PAPR and power efficiency for these schemes. In Section 5, we draw conclusions for the proposed scheme.

2. System Model

In this section, the transmitter and receiver of the proposed HABO-OFDM scheme are investigated concretely. In addition, the adaptive bias is also introduced specifically.

2.1. Transmitter

The block diagram structure of the transmitter of the proposed HABO-OFDM system is presented in Figure 1.

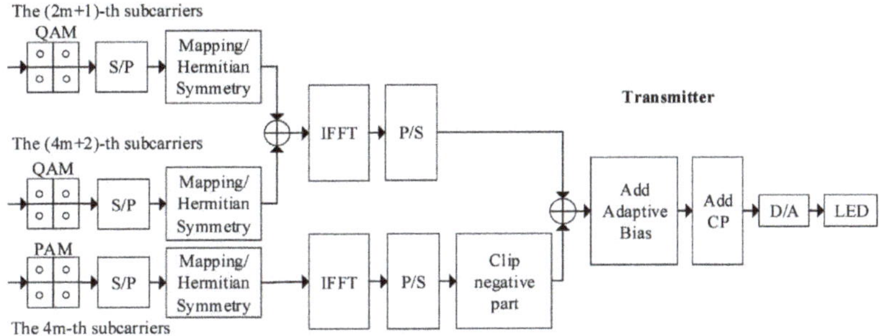

Figure 1. Block diagram structure of the transmitter of the proposed HABO-OFDM system.

In the proposed scheme, for the $(2m + 1)$-th $(m = 0, 1, \cdots, N/8 - 1)$ subcarriers, the transmitted signal is mapped in line with the Hermitian symmetry [11]. Thus, the frequency-domain signal of this component is expressed as

$$\mathbf{X}_1 = [0, Q_1, 0, Q_2, 0, \cdots, Q_{N/4}, 0, Q^*_{N/4}, 0, \cdots, Q^*_2, 0, Q^*_1], \tag{1}$$

where N is the number of subcarriers, and Q_i is the quadrature amplitude modulation (QAM) symbol. For the $(4m + 2)$-th $(m = 0, 1, \cdots, N/8 - 1)$ subcarriers, the modulation of the transmitted signal is QAM. Hence, the frequency-domain signal is given as follows:

$$\mathbf{X}_2 = [0, 0, Q_{N/4+1}, 0, \cdots, Q_{3N/8}, 0, 0, 0, Q^*_{3N/8}, \cdots, 0, Q^*_{N/4+1}, 0], \tag{2}$$

where Q_i denotes the QAM symbol. Furthermore, the modulated transmission signals in $\mathbf{X}_1$ and $\mathbf{X}_2$ are allocated on different subcarriers. Then, the frequency-domain signal of the superimposition of two different component is given by

$$\mathbf{X} = [0, Q_1, Q_{N/4+1}, Q_2, 0, \cdots, Q_{3N/8}, Q_{N/4}, 0, Q^*_{N/4}, Q^*_{3N/8}, \cdots, 0, Q^*_{N/4+1}, 0], \tag{3}$$

where Q_i is the QAM symbol.

The remaining component in HABO-OFDM only modulates the imaginary parts of the $4m$-th $(m = 0, 1, \cdots, N/8 - 1)$ subcarriers, which can be presented as

$$\mathbf{Y} = j[0, 0, 0, 0, P_1, 0, \cdots, P_{N/8-1}, 0, 0, 0, 0, 0, 0, 0, P^*_{N/8-1}, 0, \cdots, 0, P^*_1, 0, 0, 0], \tag{4}$$

where $Y_{4k} = jP_k$ and P_k represents the real PAM symbol, $k = 1, \cdots, N/2 - 1$. Additionally, the time-domain signal y_n holds anti-symmetry and periodic property [18],

$$\begin{cases} y_0 = y_{N/4} = y_{N/2} = y_{3N/4} = 0, \\ y_n = y_{n+N/2} = -y_{N/2-n} = -y_{N-n}, \quad n = 1, 2, \cdots, N/4 - 1 \end{cases}. \tag{5}$$

Therefore, the combined time-domain signal w_n of the proposed scheme, is as follows:

$$w_n = x_n + \lfloor y_n \rfloor, \quad n = 0, 1, \cdots, N - 1, \tag{6}$$

where x_n denotes the time-domain signal of $\mathbf{X}$, and $\lfloor y_n \rfloor_c = \max\{y_n, 0\}$ presents the clipped value of y_n. Thus, x_n does not interfere with the $4m$-th subcarriers of the PAM component due to x_n is unclipped. Moreover, the interference generated by $\lfloor y_n \rfloor$ only falls on the real parts of the $4m$-th subcarriers, which implies that the $(2m + 1)$-th and the $(4m + 2)$-th subcarriers with the imaginary parts of the $4m$-th subcarriers are not interfered with the clipping operation of y_n. The combination of QAM and PAM achieves the advantage of high spectral efficiency. However, the hybrid optical OFDM signal of w_n is a bipolar signal, and it is necessary to ensure the non-negativity of the signal in the optical OFDM scheme [19]. In order to guarantee the non-negativity of w_n, we design the adaptive bias according to the characteristics of the superimposed signal. Furthermore, the properties which are satisfied with the adaptive bias s_n can be expressed as

$$s_n = s_{n+N/4} = s_{n+N/2} = s_{n+3N/4} = -\min\{w_n, w_{n+N/4}, w_{n+N/2}, w_{n+3N/4}\}, \quad n = 0, N/8, \tag{7}$$

$$\begin{aligned} s_n &= s_{N/4-n} = s_{N/4+n} = s_{N/2-n} = s_{N/2+n} = s_{3N/4-n} = s_{3N/4+n} = s_{N-n} \\ &= -\min\{w_n, w_{N/4-n}, w_{N/4+n}, w_{N/2-n}, w_{N/2+n}, w_{3N/4-n}, w_{3N/4+n}, w_{N-n}\}, \; . \\ n &= 1, \cdots, N/8-1 \end{aligned} \tag{8}$$

According to (7), it is obvious that, when at least one of the four signals, w_n, $w_{n+N/4}$, $w_{n+N/2}$, $w_{n+3N/4}$, for $n = 0, N/8$ is negative, the inverse polarity of the minimum of $\{w_n, w_{n+N/4}, w_{n+N/2}, w_{n+3N/4}\}$ can be selected to produce the adaptive bias. Additionally, if the four signals are all non-negative, no bias is required to ensure the non-negativity of the four signals.

From (8), if at least one signal in $\{w_n, w_{N/4-n}, w_{N/4+n}, w_{N/2-n}, w_{N/2+n}, w_{3N/4-n}, w_{3N/4+n}, w_{N-n}\}$ for $n = 1, \cdots, N/8-1$ is negative, the adaptive bias can be generated with the inverse polarity of the minimum of $\{w_n, w_{N/4-n}, w_{N/4+n}, w_{N/2-n}, w_{N/2+n}, w_{3N/4-n}, w_{3N/4+n}, w_{N-n}\}$. In addition, if the eight signals are all non-negative, there is no need to produce the adaptive bias. Furthermore, it can also be observed that adaptive bias is selected according to the different amplitude of the signal, which can decrease the power consumption, and still ensure the non-negativity of the signal.

As a result, the superimposed time-domain signal of the proposed scheme is given by

$$z_n = w_n + s_n, \quad n = 0, 1, \cdots, N-1. \tag{9}$$

Through the introduction of the adaptive bias, the non-negativity of the transmission signal is dynamically adjusted to realize the simultaneous transmission of different optical OFDM signals, while maintaining high spectrum utilization.

2.2. Receiver

Now, we discuss the demodulation process of the proposed HABO-OFDM scheme. The block diagram structure of the receiver of the proposed HABO-OFDM system is shown in Figure 2.

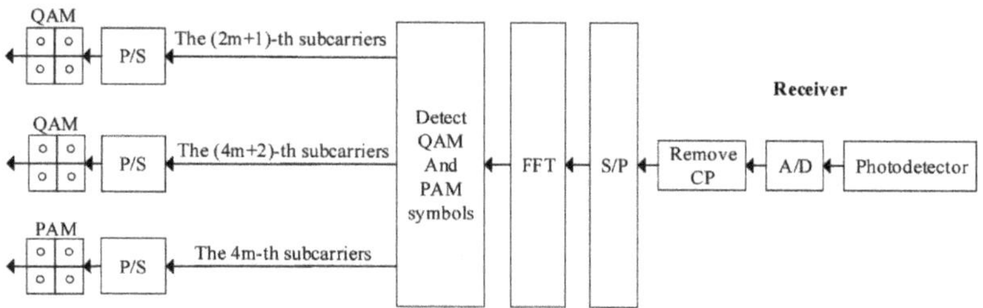

Figure 2. Block diagram structure of the receiver of the proposed HABO-OFDM system.

It can be obtained that x_n occupies the $(2m+1)$-th and $(4m+2)$-th subcarriers, and $\lfloor y_n \rfloor_c$ occupies the imaginary parts of the $4m$ subcarriers in HABO-OFDM scheme. The clipping noise of y_n only falls on the real parts of the $4m$ subcarriers, and there is no interference to x_n. Meanwhile, no clipping noise generated by x_n is superimposed on the imaginary parts of the $4m$ subcarrier of $\lfloor y_n \rfloor_c$ owing to no clipping operation for x_n. Therefore, the two sets of signals have no influence on each other during demodulation. Furthermore, the following will prove that the adaptive bias will not cause interference to the transmission signals x_n and $\lfloor y_n \rfloor_c$.

Specifically, the equation obtained by expanding the Fast Fourier transform (FFT) of s_n can be expressed as

$$\begin{aligned} S_k = & \frac{1}{\sqrt{N}}\sum_{n=0}^{N-1} s_n e^{-j\frac{2\pi nk}{N}} = s_0(1+e^{-j\frac{\pi k}{2}}+e^{-j\pi k}+e^{-j\frac{3\pi k}{2}}) \\ & + s_{\frac{N}{8}} e^{-j\frac{\pi k}{4}}(1+e^{-j\frac{\pi k}{2}}+e^{-j\pi k}+e^{-j\frac{3\pi k}{2}}) \\ & + \frac{1}{\sqrt{N}}\sum_{n=0}^{\frac{N}{8}-1} s_n[(1+e^{-j\frac{\pi k}{2}}+e^{-j\pi k}+e^{-j\frac{3\pi k}{2}})e^{-j\frac{2\pi nk}{N}} \\ & + (1+e^{-j\frac{\pi k}{2}}+e^{-j\pi k}+e^{-j\frac{3\pi k}{2}})e^{j\frac{2\pi nk}{N}}] \\ k = & 0,1,\cdots,N-1 \end{aligned} \tag{10}$$

When k is odd, that is, when $k = 2m+1$, we can obtain $e^{-j\pi k} = -1$. Then, S_k can be given by

$$\begin{aligned} S_k = & s_0 e^{-j\frac{\pi k}{2}}(1+e^{-j\pi k}) + s_{\frac{N}{8}} e^{-j\frac{\pi k}{4}} e^{-j\frac{\pi k}{2}}(1+e^{-j\pi k}) \\ & + \frac{1}{\sqrt{N}}\sum_{n=1}^{\frac{N}{8}-1} s_n[(1+e^{-j\pi k})e^{-j\frac{2\pi nk}{N}}e^{-j\frac{\pi k}{2}} + (1+e^{-j\pi k})e^{j\frac{2\pi nk}{N}}e^{-j\frac{\pi k}{2}}]. \\ = & 0 \end{aligned} \tag{11}$$

When k is even, we can obtain $e^{-j\pi k} = 1$. Thus, S_k can be expressed as the following two cases.

case 1 when $k = 4m+2$,

$$\begin{aligned} S_k = & s_0(1+e^{-j(2m+1)\pi}+1+e^{-j3(2m+1)\pi}) \\ & + s_{\frac{N}{8}} e^{-j\frac{\pi}{2}(2m+1)}(1+e^{-j(2m+1)\pi}+1+e^{-j3(2m+1)\pi}) \\ & + \frac{1}{\sqrt{N}}\sum_{n=1}^{\frac{N}{8}-1} s_n[(1+e^{-j(2m+1)\pi}+1+e^{-j3(2m+1)\pi})e^{-j\frac{4(2m+1)n\pi}{N}} \\ & + (1+e^{-j(2m+1)\pi}+1+e^{-j3(2m+1)\pi})e^{j\frac{4(2m+1)n\pi}{N}}] \\ = & 0, \end{aligned} \tag{12}$$

case 2 when $k = 4m$,

$$\begin{aligned} S_k = & s_0(1+e^{-j2m\pi}+e^{-j4m\pi}+e^{-j6m\pi}) \\ & + s_{\frac{N}{8}} e^{-jm\pi}(1+e^{-j2m\pi}+e^{-j4m\pi}+e^{-j6m\pi}) \\ & + \frac{1}{\sqrt{N}}\sum_{n=1}^{\frac{N}{8}-1} s_n(1+e^{-j2m\pi}+e^{-j4m\pi}+e^{-j6m\pi})e^{-j\frac{8mn\pi}{N}} \\ & + \frac{1}{\sqrt{N}}\sum_{n=1}^{\frac{N}{8}-1} s_n(1+e^{-j2m\pi}+e^{-j4m\pi}+e^{-j6m\pi})e^{j\frac{8mn\pi}{N}} \\ = & 4s_0 + 4s_{\frac{N}{8}} e^{-jm\pi} + \frac{4}{\sqrt{N}}\sum_{n=1}^{\frac{N}{8}-1} s_n(e^{-j\frac{8mn\pi}{N}}+e^{j\frac{8mn\pi}{N}}) \\ = & 4s_0 + 4s_{\frac{N}{8}}\cos(m\pi) + \frac{8}{\sqrt{N}}\sum_{n=1}^{\frac{N}{8}-1} s_n\cos(\frac{8mn\pi}{N}) \end{aligned} \tag{13}$$

Consequently, S_k can be given by

$$S_k = \begin{cases} 0, & k = 2m+1, \\ 0, & k = 4m+2,, \\ 4s_0 + 4s_{\frac{N}{8}}\cos(m\pi) + \frac{8}{\sqrt{N}}\sum_{n=1}^{\frac{N}{8}-1} s_n \cos(\frac{8mn\pi}{N}), & k = 4m \end{cases} \quad (14)$$

where $m = 0, 1, \cdots, N/8 - 1$. It can be easily observed that S_k only falls on the real parts of the $4m$-th subcarriers, which explains that the adaptive bias s_n does not disturb x_n and $\lfloor y_n \rfloor_c$. Therefore, a relatively simple OFDM receiver structure can be applied to HABO-OFDM scheme.

3. Complexity Analysis

We will compare the complexity of HABO-OFDM with HACO-OFDM and PHO-OFDM in this section. It is known that the complexity of the transmitter in the conventional optical OFDM scheme is determined by the number of Inverse Fast Fourier Transform (IFFT) operations [20]. In this way, we can evaluate the system complexity by using the number of the real multiplication operation in IFFT [21]. As for HABO-OFDM, the transmitter of it requires an N-point IFFT and another N-point IFFT used on the imaginary value, which yields $3N\log_2 N - 3N + 4$ times of the real multiplication operation. Similarly, $3N\log_2 N - 3N + 4$ times of the real multiplication operation are required for the HACO-OFDM transmitter, which is the same as PHO-OFDM.

As far as the complexity of the receiver, only one N-point FFT and one N-point real-valued FFT are included in HABO-OFDM. Thus, it results in $3N\log_2 N - 3N + 4$ times of the real multiplication operation for the HABO-OFDM receiver. By contrast, the HACO-OFDM receiver needs an N-point FFT and two N-point real-valued FFT, which causes $4N\log_2 N - 6N + 8$ times of the real multiplication operation. In addition, two N-point FFT and one N-point real-valued FFT are needed for the receiver of PHO-OFDM, which leads to $5N\log_2 N - 3N + 4$ times of the real multiplication operation. The comparison of the complexity of the superimposed optical OFDM schemes is shown in Table 1. It can be clearly observed that the complexity of the receiver of the proposed HABO-OFDM scheme is relatively reduced compared with the other two superimposed schemes. Compared with HACO-OFDM and PHO-OFDM, the complexity of the receiver of the proposed scheme is reduced by $N\log_2 N - 3N + 4$ and $2N\log_2 N$, respectively.

Table 1. Comparison of the complexity of the transmitter and receiver of different optical OFDM schemes.

Modulation	Transmitter	Receiver
HABO-OFDM	$3N\log_2 N - 3N + 4$	$3N\log_2 N - 3N + 4$
HACO-OFDM	$3N\log_2 N - 3N + 4$	$4N\log_2 N - 6N + 8$
PHO-OFDM	$3N\log_2 N - 3N + 4$	$5N\log_2 N - 3N + 4$

4. Simulation Results

In this section, we evaluate the performance of the proposed HABO-OFDM scheme through simulation results, and compare it with HACO-OFDM and ABO-OFDM. In the simulation, the number of subcarriers is 512 for these schemes, while the constellation sizes of QAM and PAM symbols are set to M and $\sqrt{M}$, respectively. For HACO-OFDM, we choose 16QAM-4PAM, 64QAM-8PAM, 256QM-16QAM, 1024QAM-32PAM, and 4096QAM-64PAM as simulation conditions. In ABO-OFDM, 16QAM, 64QAM, 256QM, 1024QAM, and 4096QAM are selected as simulation conditions. For HABO-OFDM, we generate signals from M-QAM and $\sqrt{M}$-PAM constellations, such as 16QAM-4PAM, 64QAM-8PAM, 256QM-16PAM, and 1024QAM-32PAM. Additionally, the total average electric power is normalized in different schemes. The system specifications is shown in Table 2. The bit error rate (BER) and peak-to-average-power ratio (PAPR) of the proposed HABO-OFDM scheme are calculated and compared with HACO-OFDM and ABO-OFDM. We also present the BER performance of the three schemes in the nonlinearity limit. Moreover, calculation results in the electric power efficiency of the proposed HABO-OFDM scheme with HACO-OFDM and ABO-OFDM are also shown.

Table 2. System specifications.

Name of Parameters	Values
The number of subcarriers	512
The constellation sizes of QAM	16, 64, 256, 1024, 4096
The constellation sizes of PAM	4, 8, 16, 32, 64
The clipping ratio τ	9 dB
The BER target	10^{-3}

Figure 3 shows the BER curves of QAM and PAM with different constellation sizes in HACO-OFDM, ABO-OFDM and the proposed HACO-OFDM scheme. It can be observed that HACO-OFDM and ABO-OFDM nearly present the same BER performance. Simultaneously, we also see that the BER performance of the proposed HABO-OFDM scheme is slightly degraded compared with the other two schemes. This is on account of the proposed HABO-OFDM scheme, where part of the power is allocated to the reserved subcarriers that are modulated at the same time, which in the ABO-OFDM scheme are not utilized. In addition, the proposed HABO-OFDM scheme can transmit more information data than HACO-OFDM under the same simulation conditions, which explains that the BER performance of HABO-OFDM is slightly worse than that of HACO-OFDM. In fact, for 256QAM-16PAM, the spectral efficiency of HABO-OFDM is improved to 7 bit/s/Hz, while the other two schemes are 6 bit/s/Hz. Specifically, this is about a 16.6% enhancement. Therefore, the spectral efficiency of the proposed HABO-OFDM scheme has been relatively improved compared with the other two systems.

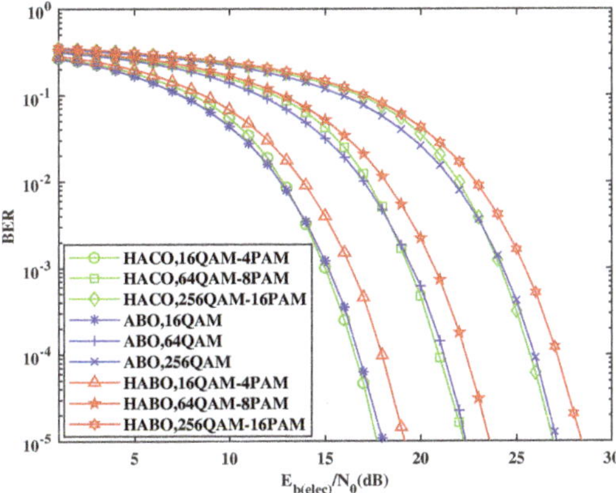

Figure 3. Comparison of the BER performance for HACO-OFDM, ABO-OFDM, and HABO-OFDM.

The complementary cumulative distribution function (CCDF) curves of the PAPR for different schemes are shown in Figure 4. We can clearly observe that the PAPR of HABO-OFDM has a significant reduction compared with HACO-OFDM and ABO-OFDM, which means that the influence of nonlinear distortion is greatly alleviated [22]. Specifically, there is about a 3.2 dB reduction in the PAPR compared to HACO-OFDM, and it is reduced by about 1.7 dB in comparison to ABO-OFDM.

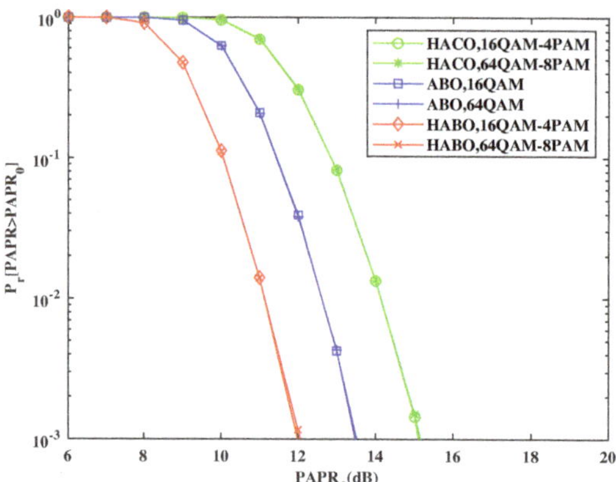

Figure 4. CCDF curves of the PAPR for HACO-OFDM, ABO-OFDM, and HABO-OFDM.

Figure 5 presents the BER performance of HACO-OFDM, ABO-OFDM, and HABO-OFDM under the nonlinear constraints, where the clipping ratio τ is set to 9 dB [23]. It is clear that the anti-nonlinearity ability of HABO-OFDM is noticeably improved compared with HACO-OFDM and ABO-OFDM. Thanks to the fact that PAPR of HABO-OFDM is superior to HACO-OFDM and ABO-OFDM, HABO-OFDM obtains significantly better BER performance.

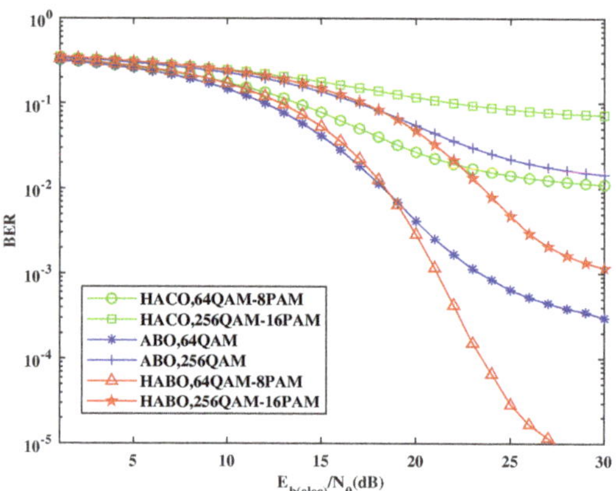

Figure 5. BER of HACO-OFDM, ABO-OFDM, and HABO-OFDM versus $E_{b(elec)}/N_0$ when the clipping ratio is set to 9 dB.

The variation of $\langle E_{b(elec)}/N_0 \rangle_{BER}$ when the proportion of electric energy allocated to subcarriers using QAM ranges from 0.1 to 0.9 is shown in Figure 6. We choose the size of QAM constellation and PAM constellation to be M and $\sqrt{M}$. In this way, it is shown that the minimum $\langle E_{b(elec)}/N_0 \rangle_{BER}$ is obtained when the proportion of electric power allocated to subcarriers using QAM is 0.6 for all cases.

In Figure 7, the $\langle E_{b(elec)}/N_0 \rangle_{BER}$ versus the proportion of electric power allocated to subcarriers using QAM is presented. The size of QAM constellation and PAM constellation is defined as M and $\sqrt{M}$. The minimum $\langle E_{b(elec)}/N_0 \rangle_{BER}$ is achieved when the proportion of electric power on subcarriers using QAM is 0.7 in all cases. This is because, in

HABO-OFDM, the subcarriers that use QAM to modulate the transmitted signal occupy the majority.

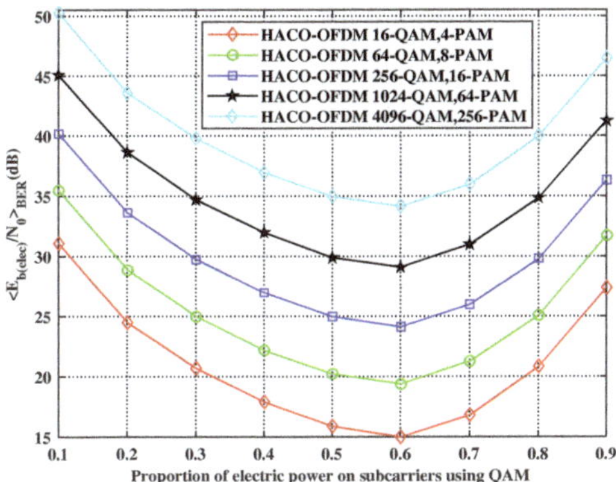

Figure 6. Comparison of $\langle E_{b(elec)}/N_0\rangle_{BER}$ for HACO-OFDM for different proportions of electric power.

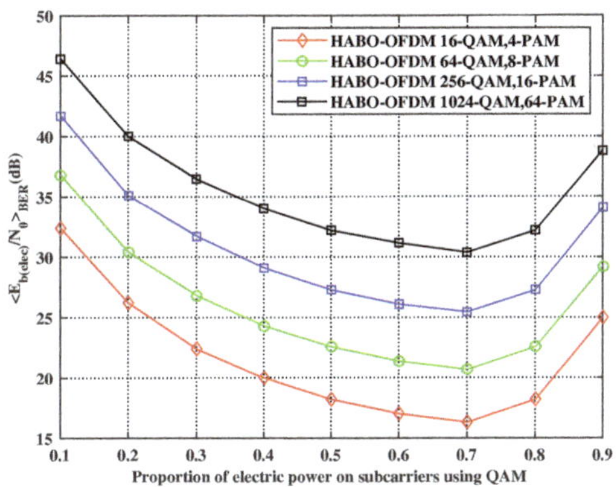

Figure 7. Comparison of $\langle E_{b(elec)}/N_0\rangle_{BER}$ for HABO-OFDM for different proportions of electric power.

In Figure 8, the required $E_{b(elec)}/N_0$ for the BER target of 10^{-3} versus Bit rate/Normalized bandwidth [24] is presented. In order to make a fair comparison, the average electric power of each of the three schemes is set to be uniform. Since HACO-OFDM and HABO-OFDM are combined modulations of QAM and PAM, the distribution of electric power needs to be considered. The $\langle E_{b(elec)}/N_0\rangle_{BER}$ versus the proportion of electric power allocated to subcarriers using QAM for HACO-OFDM and HABO-OFDM is shown in Figures 6 and 7, respectively. Therefore, the lowest values of $\langle E_{b(elec)}/N_0\rangle_{BER}$ for HACO-OFDM and HABO-OFDM can be selected.

From Figure 8, we see that the Bit rate/Normalized bandwidth of 3, 4.5, 6, 7.5, 9 for HACO-OFDM and ABO-OFDM and the Bit rate/Normalized bandwidth of 3.5, 5.25, 7, 8.75 for HABO-OFDM are shown. We can clearly obtain that when the Bit rate/Normalized bandwidth is in the range of 3 to 9, the $E_{b(elec)}/N_0$ required for HACO-OFDM and ABO-OFDM to achieve the BER target of 10^{-3} is almost the same, which means that their power efficiency is similar. Moreover, it is obvious that as the Bit rate/Normalized bandwidth increases, the $E_{b(elec)}/N_0$ required for HABO-OFDM to achieve the BER target of 10^{-3} will be significantly lower than the other two schemes, which shows that in a wide range of the Bit rate/Normalized bandwidth, HABO-OFDM exhibits higher power efficiency in comparison with HACO-OFDM and ABO-OFDM.

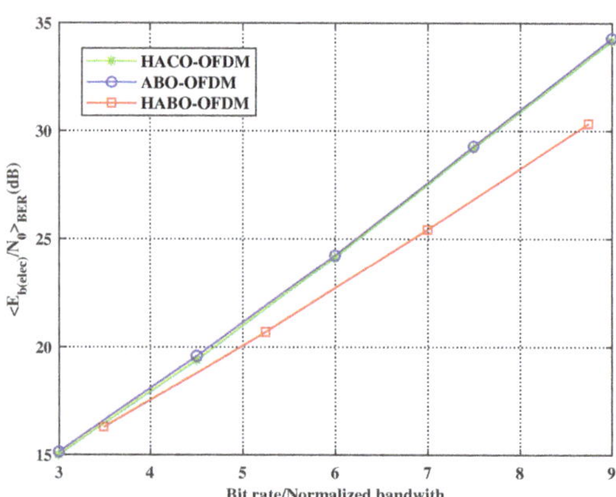

Figure 8. $\langle E_{b(elec)}/N_0 \rangle_{BER}$ required for the BER target of 10^{-3} versus Bit rate/Normalized bandwidth.

5. Conclusions

In this paper, a novel HABO-OFDM scheme was proposed for VLC to achieve the high spectrum efficiency and power efficiency. In terms of 256QAM-16PAM, we saw about a 16.6% enhancement in spectrum efficiency compared with the other two schemes. Compared with the conventional superimposed systems, the complexity of the HABO-OFDM receiver was significantly reduced due to the introduction of adaptive bias, which not only had no interference to the transmitted signal but also guaranteed the non-negativity of the signal. Specifically, the complexity of the HABO-OFDM receiver was reduced by $N\log_2 N - 3N + 4$ and $2N\log_2 N$ compared to HACO-OFDM and PHO-OFDM. Meanwhile, HABO-OFDM had a superior PAPR performance in comparison with HACO-OFDM and ABO-OFDM. There was about a 3.2 dB reduction and 1.7 dB reduction in PAPR compared with HACO-OFDM and ABO-OFDM. Additionally, HABO-OFDM also maintained a better power efficiency performance than HACO-OFDM and ABO-OFDM in a wide range of the Bit rate/Normalized bandwidth. However, in the proposed scheme, the real parts of some subcarriers were still not used. We will continue to exploit the reserved parts in future researches. Furthermore, the proposed HABO-OFDM scheme can be applied in high-speed VLC in future due to its high spectral efficiency, low receiver complexity, low PAPR, and high power efficiency.

Author Contributions: Conceptualization, H.H.; methodology, H.H.; software, H.H.; validation, H.H., Z.L.; formal analysis, H.H.; investigation, H.H.; resources, H.H.; data curation, H.H.; writing—original draft preparation, H.H.; writing—review and editing, Z.L.; visualization, H.H.; supervision, Z.L.; project administration, Z.L.; funding acquisition, Z.L. All authors have read and agreed to the published version of the manuscript.

Funding: This research was funded by the National Natural Science Foundation of China under Grant 61571108; in part by the Wuxi Science and Technology Development Fund(No.H20191001, No.G20192010); in part by the Natural Science Foundation of Jiangsu Province under Grants BK20190582.

Institutional Review Board Statement: Not applicable.

Informed Consent Statement: Not applicable.

Data Availability Statement: Not applicable.

Conflicts of Interest: The authors declare no conflict of interest.

References

1. Zafar, F.; Karunatilaka, D.; Parthiban, R. Dimming Schemes for Visible Light Communication: The State of Research. *IEEE Wireless Commun.* **2015**, *22*, 29–35. [CrossRef]
2. Khan, L.U. Visible Light Communication: Applications, Architecture, Standardization and Research Challenges. *Digit. Commun. Netw.* **2017**, *3*, 78–88. [CrossRef]
3. Wu, S.; Wang, H.; Youn, C. Visible Light Communications for 5G Wireless Networking Systems: From Fixed to Mobile Communications. *IEEE Netw.* **2014**, *28*, 41–45. [CrossRef]
4. Khalid, A.; Asif, H.M.; Kostromitin, K.I.; Al-Otaibi, S.; Huq, K.M.S.; Rodriguez, J. Doubly Orthogonal Wavelet Packets for Multi-Users Indoor Visible Light Communication Systems. *Photonics* **2019**, *6*, 85. [CrossRef]
5. Haas, H.; Yin, L. Physical-Layer Security in Multiuser Visible Light Communication Networks. *IEEE J. Sel. Areas Commun.* **2018**, *36*, 162–174.
6. Islim, M.S.; Ferreira, R.X.; He, X.Y.; Xie, E.Y.; Videv, S.; Viola, S.; Watson, S.; Bamiedakis, N.; Penty, R.V.; White, I.H.; et al. Towards 10Gb/s Orthogonal Frequency Division Multiplexing-Based Visible Light Communication Using A GaN Violet Micro-LED. *Photon. Res.* **2017**, *5*, A35–A43. [CrossRef]
7. Başar, E.; Aygölü, Ü.; Panayırcı, E.; Poor, H.V. Orthogonal Frequency Division Multiplexing with Index Modulation. *IEEE Trans. Signal Process.* **2013**, *61*, 5536–5549. [CrossRef]
8. Armstrong, J.; Schmidt, B.J.C. Comparison of Asymmetrically Clipped Optical OFDM and DC-Biased Optical OFDM in AWGN. *IEEE Commun. Lett.* **2008**, *12*, 343–345. [CrossRef]
9. Armstrong, J.; Lowery, A.J. Power Efficient Optical OFDM. *Electron. Lett.* **2006**, *42*, 370–372. [CrossRef]
10. Le-Tran, M.; Kim, S. Deep Learning-Assisted Index Estimator for Generalized LED Index Modulation OFDM in Visible Light Communication. *Photonics* **2021**, *8*, 168. [CrossRef]
11. Xu, W.; Wu, M.; Zhang, H.; You, X.; Zhao, C. ACO-OFDM-Specified Recoverable Upper Clipping with Efficient Detection for Optical Wireless Communications. *IEEE Photon. J.* **2014**, *6*, 1–17.
12. Lee, S.C.J.; Randel, S.; Breyer, F.; Koonen, A.M.J. PAM-DMT for Intensity-Modulated and Direct-Detection Optical Communication Systems. *IEEE Photon. Technol. Lett.* **2009**, *21*, 1745–1749. [CrossRef]
13. Li, B.L.; Feng, S.; Xu, W. Spectrum-Efficient Hybrid PAM-DMT for Intensity-Modulated Optical Wireless Communication. *Opt. Exp.* **2020**, *28*, 12621–12637. [CrossRef]
14. Ranjha, B.; Kavehrad, M. Hybrid Asymmetrically Clipped OFDM-Based IM/DD Optical Wireless System. *IEEE/OSA J. Opt. Commun. Netw.* **2014**, *6*, 387–396. [CrossRef]
15. Zhang, T.; Zou, Y.; Sun, J.; Qiao, S. Design of PAM-DMT-Based Hybrid Optical OFDM for Visible Light Communications. *IEEE Wirel. Commun. Lett.* **2019**, *8*, 265–268. [CrossRef]
16. Li, B.L.; Xu, W.; Li Z.; Zhou Y. Adaptively Biased OFDM for IM/DD-Aided Optical Wireless Communication Systems. *IEEE Wireless Commun. Lett.* **2020**, *9*, 698–701. [CrossRef]
17. Li, B.L.; Xue, X.; Feng, S.; Xu, W. Layered Optical OFDM with Adaptive Bias for Dimming Compatible Visible Light Communications. *J. Lightw. Technol.* **2021**, *39*, 3434–3444. [CrossRef]
18. Wang, T.; Hou, Y.; Ma, M. A Novel Receiver Design for HACO-OFDM by Time-Domain Clipping Noise Elimination. *IEEE Commun. Lett.* **2018**, *22*, 1862–1865. [CrossRef]
19. Yesilkaya, A.; Bian, R.; Tavakkolnia, I.; Haas, H. OFDM-Based Optical Spatial Modulation. *IEEE J. Sel. Top. Signal Process.* **2019**, *13*, 1433–1444. [CrossRef]
20. Wang, Q.; Qian, C.; Guo, X.; Wang, Z.; Cunningham, D.G.; White, I.H. Layered ACO-OFDM for Intensity-Modulated Direct-Detection Optical Wireless Transmission. *Opt. Exp.* **2015**, *23*, 12382–12393. [CrossRef]
21. Liu, C.; Chan, C.; Cheng, P.; Lin, H. FFT-Based Multi-Rate Signal Processing for 18-Band Quasi-ANSI S1.11 1/3-Octave Filter Bank. *IEEE Trans. Circuits Syst. II Exp. Briefs* **2019**, *66*, 878–882. [CrossRef]
22. Mestdagh, D.J.G.; Monsalve, J.L.; Brossier, J.M. GreenOFDM: A New Selected Mapping Method for OFDM PAPR Reduction. *Electron. Lett.* **2018**, *54*, 449–450. [CrossRef]
23. Li, B.; Xu, W.; Zhang, H.; Zhao, C.; Hanzo, L. PAPR Reduction for Hybrid ACO-OFDM Aided IM/DD Optical Wireless Vehicular Communications. *IEEE Trans. Veh. Technol.* **2017**, *66*, 9561–9566. [CrossRef]
24. Armstrong, J.; Dissanayake, S.D. Comparison of ACO-OFDM, DCO-OFDM and ADO-OFDM in IM/DD Systems. *J. Lightw. Technol.* **2013**, *31*, 1063–1072.

photonics

MDPI

Article

Pairwise Coded *m*CAP with Chaotic Dual-Mode Index Modulation for Secure Bandlimited VLC Systems

Yungui Nie [1,†], Wei Zhang [2,†], Yanbing Yang [3], Xiong Deng [4], Min Liu [1] and Chen Chen [1,*]

[1] School of Microelectronics and Communication Engineering, Chongqing University, Chongqing 400044, China; 202012021020t@cqu.edu.cn (Y.N.); liumin@cqu.edu.cn (M.L.)
[2] Institute of Cyberspace Security, China Electronic Technology Cyber Security Co., Ltd., Chengdu 610041, China; zhangw9761@cetcsc.com
[3] College of Computer Science, Sichuan University, Chengdu 610065, China; yangyanbing@scu.edu.cn
[4] Center for Information Photonics and Communications, School of Information Science and Technology, Southwest Jiaotong University, Chengdu 611756, China; xiongdeng@swjtu.edu.cn
* Correspondence: c.chen@cqu.edu.cn
† These authors contributed equally to this work.

Abstract: In this paper, for the first time, we propose and experimentally demonstrate a novel pairwise coding (PWC)-based multiband carrierless amplitude and phase (*m*CAP) modulation with chaotic dual-mode index modulation (DM) for secure bandlimited visible light communication (VLC) systems. The combination of *m*CAP and DM can sustain a higher spectral efficiency (SE) compared with *m*CAP with conventional index modulation (IM), while PWC can be employed to efficiently mitigate the signal-to-noise ratio (SNR) imbalance caused by the low-pass frequency response of light emitting diodes (LEDs). Moreover, the DM is enhanced by a two-dimensional (2D) chaotic encryption scheme to guarantee the security of the useful information in VLC systems. Simulation and experimental results successfully verify the superiority of the proposed PWC-based *m*CAP-DM scheme with two-level chaotic encryption over other benchmark schemes.

Keywords: visible light communication (VLC); carrierless amplitude and phase (CAP) modulation; pairwise coding (PWC); dual-mode index modulation (DM); chaotic encryption

Citation: Nie, Y.; Zhang, W.; Yang, Y.; Deng, X.; Liu, M.; Chen, C. Pairwise Coded *m*CAP with Chaotic Dual-Mode Index Modulation for Secure Bandlimited VLC Systems. *Photonics* **2022**, *9*, 141. https://doi.org/10.3390/photonics9030141

Received: 12 February 2022
Accepted: 25 February 2022
Published: 27 February 2022

1. Introduction

Visible light communication (VLC) utilizing commercially available light emitting diodes (LEDs) has become more appealing to the sixth generation (6G) and Internet of Things (IoT) systems, due to its enormous merits of abundant and unregulated spectrum resources, no electromagnetic radiation and high security [1,2]. However, the achievable capacity of practical VLC systems is greatly restrained by the small modulation bandwidth and the severe nonlinearity of off-the-shelf LEDs [3].

Many techniques have been proposed to expand the available bandwidth for VLC systems. For one thing, analog or digital equalization techniques can be applied to break the limited bandwidth of LEDs [4,5]. For another thing, various spectrally efficient modulation and multiple access techniques, including carrierless amplitude and phase (CAP) modulation, orthogonal frequency division multiplexing (OFDM) with high-order modulation formats, multiple-input multiple-output (MIMO) and nonorthogonal multiple access (NOMA), can also be considered to enhance the achievable data rate of VLC systems [6–9]. Moreover, CAP has the benefits of a lower implementation cost and lower peak-to-average power ratio (PAPR) in comparison to OFDM. To increase the link performance, multiband CAP (*m*CAP) has been further proposed, which splits the available signal bandwidth into m subbands so as to improve the tolerance against the low-pass effect of LEDs [10].

Recently, a novel *m*CAP with index modulation (*m*CAP-IM) technique has been proposed to obtain better bit error rate (BER) performance than classical *m*CAP. In *m*CAP-IM, the subbands of classical mCAP are divided into two parts: the active and inactive subbands. The active subbands can be modulated by constellation data symbols while the inactive subbands are nulled [11]. To further enhance the spectral efficiency (SE) of mCAP-IM, mCAP with dual-mode index modulation (mCAP-DM) has been proposed in [12], where all subbands are modulated to transmit data symbols. Moreover, the detection of signals plays an important role in optical wireless communication systems. In [13], a novel high-dimensional (HD) noncoherent detection scheme was proposed for multi-intensity-modulated ultraviolet communication (UVC) systems to address the intersymbol interference (ISI) issue. In [14], a sparse signal detection scheme exploiting the sparse reconstruction algorithms in compressed sensing (CS) was proposed for indoor VLC systems using generalized space shift keying (GSSK). For VLC systems employing mCAP-DM, the low-complexity near-optimal log-likelihood rate (LLR) detector is generally adopted [11,12]. For practical mCAP-DM VLC systems, the signal-to-noise ratio (SNR) of high-frequency subbands imposes inevitable degradation compared with that of low-frequency subbands. Hence, all subbands are in the condition of unbalanced SNR, which results in the degradation of the overall BER performance. In order to resolve the issue of unbalanced SNR, pairwise coding (PWC) can be introduced to improve the overall system performance, which requires no overhead and exhibits low computation complexity [15]. Although VLC has inherent security against eavesdroppers outside its coverage, the confidential information might still be eavesdropped by unintended or unauthorized users when they are located within the system coverage [16].

In this paper, we propose and investigate a novel PWC-based mCAP-DM scheme with two-dimensional (2D) chaotic encryption for practical VLC systems. The superiority of the proposed mCAP-DM scheme was successfully verified by both numerical simulations and hardware experiments.

2. PWC-Based mCAP-DM with 2D Chaotic Encryption

2.1. Principle

Figure 1a,b illustrate the block diagrams of the transmitter and receiver of the proposed PWC-based mCAP-DM with 2D chaotic encryption, respectively. In the transmitter, the input bits are firstly divided into G groups through a bit splitter and every group of b bits is used to generate an mCAP subblock of length N. Subsequently, every b bits is further partitioned into the index bits b_i and the constellation bits b_c, i.e., $b = b_i + b_c$. More specifically, the b_i bits are used to select the indices of k subbands out of N subbands in total via a chaotic index selector, while the b_c bits are employed to generate the corresponding constellation symbols to perform DM through a chaotic constellation mapper. Assuming the selected k subbands adopt constellation mode 1 while the remaining $N-k$ subbands adopt constellation mode 2, the two constellation sets corresponding to constellation modes 1 and 2 can be denoted by $\mathcal{M}_1 = [S_1^1, S_2^1, \cdots, S_{M_1}^1]^T$ with size M_1 and $\mathcal{M}_2 = [S_1^2, S_2^2, \cdots, S_{M_2}^2]^T$ with size M_2, respectively, where $(\cdot)^T$ denotes the transpose operation. Since $\mathcal{M}_1$ and $\mathcal{M}_2$ are two distinguishable constellation sets, we have $\mathcal{M}_1 \cap \mathcal{M}_2 = \varnothing$. The detailed principle of DM with 2D chaotic encryption is introduced in the following subsection. Subsequently, the mCAP block can be created by concatenating G subblocks through a block creator. Before executing upsampling, PWC encoding is performed to mitigate the SNR imbalance. After passing through the in-phase (I) and quadrature (Q) filters, the transmitted PWC-based mCAP-DM signal with 2D chaotic encryption is obtained. Specifically, the impulse responses of a pair of orthogonal I and Q filters for the n-th ($n = 1, \cdots, m$) subband can be, respectively, expressed as

$$f_n^I(t) = g(t)cos(2\pi f_{c,n}t), \tag{1}$$

$$f_n^Q(t) = g(t)sin(2\pi f_{c,n}t), \tag{2}$$

where $g(t)$ is the impulse response of the baseband shaping filter and $f_{c,n}$ denotes the center frequency of nth subband. The root raised cosine filter (RRCF) was adopted as the baseband shaping filter in this work [10].

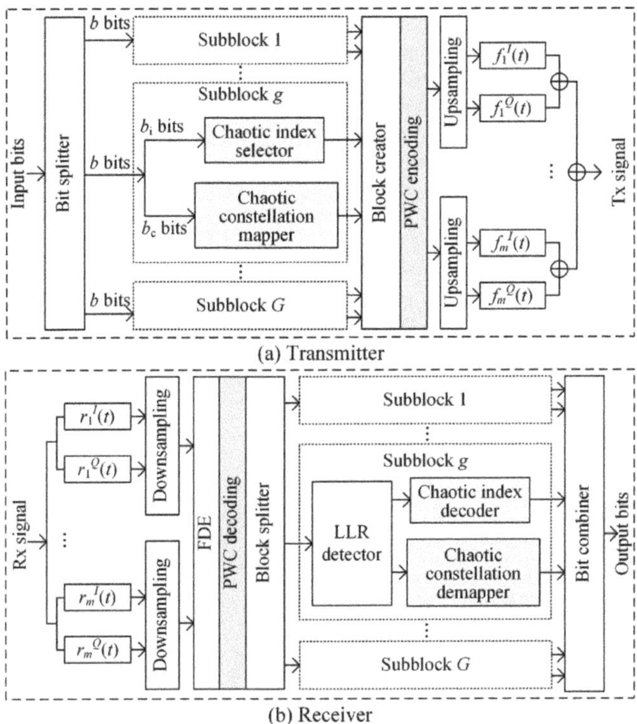

Figure 1. Block diagrams of the proposed PWC-based mCAP-DM with 2D chaotic encryption: (**a**) transmitter and (**b**) receiver.

In the receiver, as shown in Figure 1b, the received signal first passes through m pairs of matched filters and the impulse responses of the matched filters for the nth subband are given by

$$r_n^I(t) = f_n^I(-t), \tag{3}$$

$$r_n^Q(t) = f_n^Q(-t). \tag{4}$$

After downsampling, frequency-domain equalization (FDE) and PWC decoding are performed to recover the data symbols on each subband. Subsequently, the mCAP block is divided into G subblocks by a block splitter. The signal of each subblock can be detected by a low-complexity LLR detector. Letting y_g^η $(g = 1, \cdots, G; \eta = 1, \cdots, N)$ be the input signal, the corresponding LLR value for the gth subblock is given by

$$\lambda_g^\eta = C + \ln\left(\sum_{i=1}^{M_1} \exp\left(-\frac{\left|y_g^\eta - S_i^1\right|^2}{N_0}\right)\right) - \ln\left(\sum_{j=1}^{M_2} \exp\left(-\frac{\left|y_g^\eta - S_j^2\right|^2}{N_0}\right)\right), \tag{5}$$

where $C = \ln(k) - \ln(N-k)$ and N_0 denotes the noise power. After subblock detection in each subblock, the index bits and the constellation bits can be recovered via a chaotic index decoder and a chaotic constellation demapper, respectively. Finally, the obtained index bits and constellation bits are combined by a bit combiner to yield the output bits.

As a result, the SE of the mCAP-DM signal is calculated by

$$SE_{m\text{CAP-DM}} = \frac{b_i + b_c}{N} = \frac{\lfloor \log_2(C(N,k)) \rfloor + k\log_2(M_1) + (N-k)\log_2(M_2)}{N}, \tag{6}$$

where $\lfloor \cdot \rfloor$ represents the floor operator and $C(\cdot,\cdot)$ denotes the binomial coefficient. It should be noted that the SE of the mCAP-DM signal is not affected by the PWC coding and 2D chaotic encryption. In contrast, since only the selected subbands are used to modulate constellation symbols while the remaining subbands are left unmodulated in mCAP-IM, the SE of mCAP-IM is expressed by

$$SE_{m\text{CAP-IM}} = \frac{\lfloor \log_2(C(N,k)) \rfloor + k\log_2(M)}{N}, \tag{7}$$

where M is the size of the constellation modulated on the selected subbands in mCAP-IM.

2.2. DM with 2D Chaotic Encryption

The principle of DM with 2D chaotic encryption is described as follows. As can be seen from Figure 1a, the proposed 2D chaotic encryption scheme is performed with respect to each mCAP subblock, which mainly consists of the following two chaotic scrambling processes: one is the chaotic constellation scrambling (CCS) and the other is the chaotic mode scrambling (CMS). In each subblock, CCS and CMS are performed via the chaotic constellation mapper and the chaotic index selector, respectively.

For the mCAP-DM system with two distinguishable constellation sets, i.e., $\mathcal{M}_1$ and $\mathcal{M}_2$, the CCS process is conducted with respect to each constellation set. Letting $\mathbf{p}_g^1 = [p_{g,1}^1, p_{g,2}^1, \cdots, p_{g,M_1}^1]^T$ and $\mathbf{p}_g^2 = [p_{g,1}^2, p_{g,2}^2, \cdots, p_{g,M_2}^2]^T$ denote, respectively, the corresponding permutation vectors for $\mathcal{M}_1$ and $\mathcal{M}_2$ in the gth subblock with $g = 1, \cdots, G$, the resultant constellation sets corresponding to $\mathcal{M}_1$ and $\mathcal{M}_2$ after performing CCS are expressed as follows:

$$\mathcal{M}_1^{\text{CCS}} = \text{src}\{\mathcal{M}_1, \mathbf{p}_g^1\} = [S^1_{p^1_{g,1}}, S^1_{p^1_{g,2}}, \cdots, S^1_{p^1_{g,M_1}}]^T, \tag{8}$$

$$\mathcal{M}_2^{\text{CCS}} = \text{src}\{\mathcal{M}_2, \mathbf{p}_g^2\} = [S^2_{p^2_{g,1}}, S^2_{p^2_{g,2}}, \cdots, S^2_{p^2_{g,M_2}}]^T, \tag{9}$$

where $\text{src}\{\cdot,\cdot\}$ denotes the scrambling function that scrambles $\mathcal{M}_1$ and $\mathcal{M}_2$ according to their respective permutation vectors $\mathbf{p}_g^1$ and $\mathbf{p}_g^2$, where the permutation vector contains the new positions of the constellation points in the corresponding constellation set.

Moreover, the CMS process is performed with respect to the two constellation modes during the DM within each subblock. In the DM without applying CMS, each constellation mode corresponds to a fixed constellation set. However, in the CMS-encrypted DM, the corresponding relationship between two constellation modes, i.e., mode 1 and mode 2, and two encrypted constellation sets $\mathcal{M}_1^{\text{CCS}}$ and $\mathcal{M}_2^{\text{CCS}}$ is scrambled according to a binary scrambling vector. Letting $\mathbf{s} = [s_1, s_2, \cdots, s_G]^T$ denote the binary scrambling vector for the mCAP block with a total of G subblocks, the corresponding relationship between two constellation modes and two encrypted constellation sets in the gth ($g = 1, \cdots, G$) subblock is given in Table 1. As we can see, when $s_g = 0$, the two encrypted constellation sets corresponding to mode 1 and mode 2 are $\mathcal{M}_1^{\text{CCS}}$ and $\mathcal{M}_2^{\text{CCS}}$, respectively. However, when $s_g = 1$, the corresponding relationship becomes reversed, i.e., the two encrypted constellation sets corresponding to mode 1 and mode 2 become $\mathcal{M}_2^{\text{CCS}}$ and $\mathcal{M}_1^{\text{CCS}}$, respectively.

Table 1. Corresponding relationship of CMS in the gth subblock.

	$s_g = 0$	$s_g = 1$
Mode 1	$\mathcal{M}_1^{\text{CCS}}$	$\mathcal{M}_2^{\text{CCS}}$
Mode 2	$\mathcal{M}_2^{\text{CCS}}$	$\mathcal{M}_1^{\text{CCS}}$

By performing DM with 2D chaotic encryption, including CCS and CMS, the physical-layer security of mCAP-DM can be greatly enhanced. In order to successfully perform CCS and CMS, the permutation vectors $\mathbf{p}_g^1$ and $\mathbf{p}_g^2$ with respect to the gth subblock and the binary scrambling vector $\mathbf{s}$ with respect to the mCAP block with a total of G subblocks should be obtained in advance. In this work, the Hitzl–Zele chaotic map is adopted to simultaneously generate the permutation vectors $\mathbf{p}_g^1$ and $\mathbf{p}_g^2$ and the binary scrambling vector **s**, which is expressed by [17]

$$\begin{cases} x_{q+1} = 1 + y_q - z_q x_q^2 \\ y_{q+1} = \alpha x_q \\ z_{q+1} = \beta x_q^2 + z_q - 0.5 \end{cases}, \tag{10}$$

where α and β are the bifurcation parameters. When $\alpha = 0.25$ and $\beta = 0.87$, the Hitzl–Zele map exhibits chaotic behavior. Based on the Hitzl–Zele chaotic map, the permutation vectors $\mathbf{p}_g^1$ and $\mathbf{p}_g^2$ can be generated by using the states x_q and y_q as follows:

$$\mathbf{p}_g^1 = \text{sort}\{[x_{K+1}, x_{K+2}, \cdots, x_{K+M_1}]^T\}, \tag{11}$$

$$\mathbf{p}_g^2 = \text{sort}\{[y_{K+1}, y_{K+2}, \cdots, y_{K+M_2}]^T\}, \tag{12}$$

where sort$\{\cdot\}$ denotes the sorting function which returns the index vector of the elements of the input vector by sorting these elements in a descending order, and K is an integer parameter. Moreover, the gth element of the binary scrambling vector **s** can also be generated by using the state z_q by

$$s_g = \text{abs}(\text{mod}(\text{int}(z_{K+g} \times 10^7), 2))\}, \tag{13}$$

where abs$(\cdot)$ denotes the operation to obtain the absolute value of the input, mod$(\cdot, 2)$ returns the remainder of an input divided by 2, and int$(\cdot)$ returns the integer part of the input which truncates the input at the decimal point. During the 2D chaotic encryption as described above, the initial values x_0, y_0, z_0, and K can be set as the shared security keys between the transceiver. With numbers of about 16 decimal digits precision, the key space is more than 2^{172}, which is huge enough against eavesdropping using exhaustive attack methods [17].

2.3. PWC-Based mCAP

The schematic diagrams of the PWC encoding and decoding are illustrated in Figure 2a,b, respectively. Firstly, the quadrature amplitude modulation (QAM) constellation symbols corresponding to two paired subbands in an *m*CAP signal, i.e., subband 1 and subband 2, can be expressed by

$$x_{\text{SB1}} = a_n + jb_n, \tag{14}$$

$$x_{\text{SB2}} = c_n + jd_n, \tag{15}$$

where a_n and b_n are the I and Q parts of subband 1, and c_n and d_n are the I and Q parts of subband 2, respectively. In the encoding process, an angle rotation is firstly performed to obtain the following rotated constellation symbols:

$$x_{\text{SB1},\theta} = x_{\text{SB1}} e^{j\theta} = (a_n \cos\theta - b_n \sin\theta) + j(a_n \sin\theta + b_n \cos\theta), \tag{16}$$

$$x_{\text{SB2},\theta} = x_{\text{SB2}} e^{j\theta} = (c_n \cos\theta - d_n \sin\theta) + j(c_n \sin\theta + d_n \cos\theta), \tag{17}$$

where the rotation angle θ is usually set to $\theta = 45°$ [15]. Then, the I and Q components of each rotated constellation symbols are interleaved. After interleaving, the transmitted signal of each subband can be obtained by

$$x_{\text{SB1,PWC}} = \text{Re}(x_{\text{SB1},\theta}) + j\text{Re}(x_{\text{SB2},\theta}) = (a_n \cos\theta - b_n \sin\theta) + j(c_n \cos\theta - d_n \sin\theta), \tag{18}$$

$$x_{\text{SB2,PWC}} = \text{Im}(x_{\text{SB1},\theta}) + j\text{Im}(x_{\text{SB2},\theta}) = (a_n \sin\theta + b_n \cos\theta) + j(c_n \sin\theta + d_n \cos\theta), \tag{19}$$

where Re(·) and Im(·) denote the operations to extract the I and Q parts of a complex-valued input, respectively.

In the decoding process, as shown in Figure 2b, the I and Q parts of the corresponding received signals $y_{SB1,PWC}$ and $y_{SB2,PWC}$ are firstly separated and then deinterleaving is executed correspondingly. Finally, an angle rotation is further performed to recover the transmitted QAM constellation symbols.

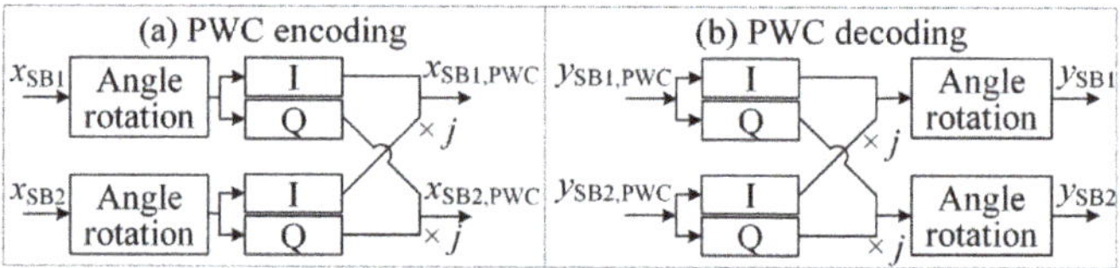

Figure 2. Schematic diagrams of (**a**) PWC encoding and (**b**) PWC decoding.

3. Results and Discussions

In this section, numerical simulations and hardware experiments are presented to investigate the performance of the proposed PWC-based mCAP-DM scheme with 2D chaotic encryption and compare it with other benchmark schemes. In our simulations and experiments, we considered the 4CAP case in which the overall bandwidth is divided into four subbands, i.e., $m = 4$. Moreover, each subblock was assumed to have two subbands, where one subband was selected to adopt constellation mode 1 while the remaining one adopted constellation mode 2 at each time slot, i.e., $N = 2$ and $k = 1$. The corresponding simulation parameters are shown in Table 2. To make a fair comparison, both 4CAP-DM and 4CAP-IM achieved the same SE of 2.5 bits/s/Hz. Specifically, the square 16-QAM constellation was used in 4CAP-IM, while the circular (7,1)-QAM constellation was adopted in 4CAP-DM. In our previous work [18], our simulation and experimental results showed that the circular (7,1)-QAM constellation achieved better BER performance than conventional square 8-QAM or 8-ary phase-shift keying (8-PSK) constellations in orthogonal frequency division multiplexing with dual-mode index modulation (OFDM-DM) based VLC systems. Therefore, the circular (7,1)-QAM constellation was adopted in 4CAP-DM in this work. Moreover, the interleaving-based constellation partitioning approach was considered to obtain two constellation modes from the circular (7,1)-QAM constellation. Please refer to our previous work [18] for more details. Since the basic 4CAP with general M-ary constellations cannot achieve a SE of 2.5 bits/s/Hz, it was not considered in our following simulations and experiments.

Table 2. Simulation Parameters.

Parameter	Value
Number of subbands (m)	4
Number of subbands in each subblock (N)	2
Number of activated subbands in each subblock (k)	1
Roll-off factor of RRCF	0.15
Upsampling factor	12
Spectral efficiency (SE)	2.5 bits/s/Hz
Simulation channel	AWGN

3.1. Simulation Results

Figure 3 shows the simulation BER versus SNR for 4CAP-IM and 4CAP-DM achieving the same SE of 2.5 bits/s/Hz over the additive white Gaussian noise (AWGN) channel. It can be clearly seen that 4CAP-DM requires a lower SNR to reach the 7% forward error correction (FEC) coding limit of BER = 3.8×10^{-3} in comparison to 4CAP-IM. More specifically, the required SNRs for 4CAP-IM and 4CAP-DM to reach the BER threshold are 10.8 and 9.2 dB, respectively, indicating that 4CAP-DM outperforms 4CAP-IM by an SNR gain of 1.6 dB.

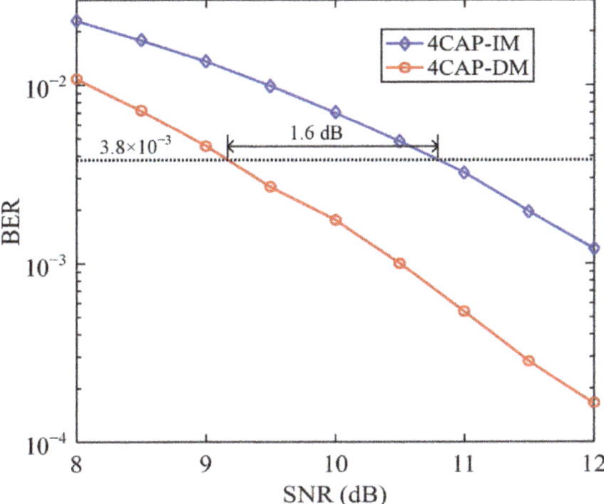

Figure 3. Simulation BER vs. SNR for 4CAP-IM and 4CAP-DM achieving the same SE of 2.5 bits/s/Hz over the AWGN channel.

3.2. Experimental Results

Based on the above simulation parameters, the corresponding hardware experiments were further performed. The experimental setup of a point-to-point VLC system using a blue mini-LED is illustrated in Figure 4, where the inset shows the photo of the overall experimental testbed. Firstly, the transmitted signal which was digitally generated offline by MATLAB was fed into an arbitrary waveform generator (AWG, Tektronix AFG31102) with a sampling rate of 250 MSa/s, where the signal bandwidth was set to 96 MHz. The AWG output signal was further combined with a 120 mA DC bias current via a bias-tee (MiniCircuits, ZFBT-6GW+) and the resultant signal was used to drive a blue mini-LED (HCCLS2021CHI03). Subsequently, a pair of biconvex lenses each with a diameter of 12.7 mm and a focal length of 20 mm were employed to ensure that most of the light emitted by the LED was focused on the active area of a photodetector (PD, Thorlabs PDA10A2). The PD had a bandwidth of 150 MHz and an active area of 0.8 mm^2. After that, the detected signal was recorded by a digital storage oscilloscope (DSO, LeCroy WaveSurfer432) with a sampling rate of 1 GSa/s, which was further processed offline by MATLAB. The detailed experimental parameters are given in Table 3.

Table 3. Experimental Parameters.

Parameter	Value
AWG sampling rate	250 MSa/s
Signal bandwidth	96 MHz
DC bias current	120 mA
Diameter of biconvex lenses	12.7 mm
Focal length of biconvex lenses	20 mm
Bandwidth of PD	150 MHz
Active area of PD	0.8 mm^2
DSO sampling rate	1 GSa/s

Figure 5a,b illustrate the measured nonlinear voltage–current curve and the low-pass frequency response of the system, respectively. It can be clearly observed that the system exhibits notable nonlinearity which is mainly caused by the blue mini-LED. We can also see that the system reflects a typical low-pass characteristic which can result in an unbalanced SNR between the high-frequency subbands and the low-frequency subbands in the 4CAP signal.

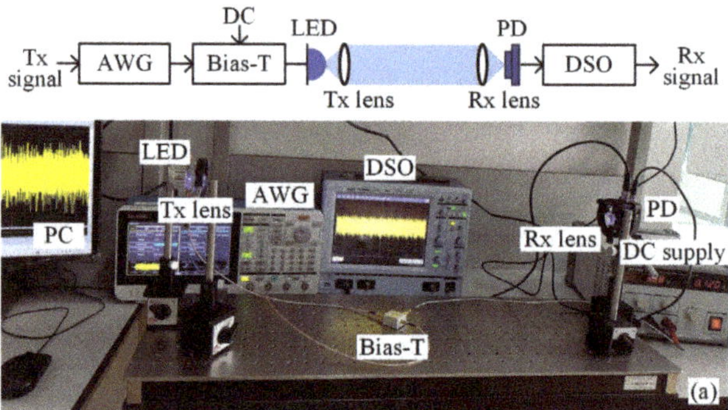

Figure 4. Experimental setup of a point-to-point VLC system using a blue mini-LED.

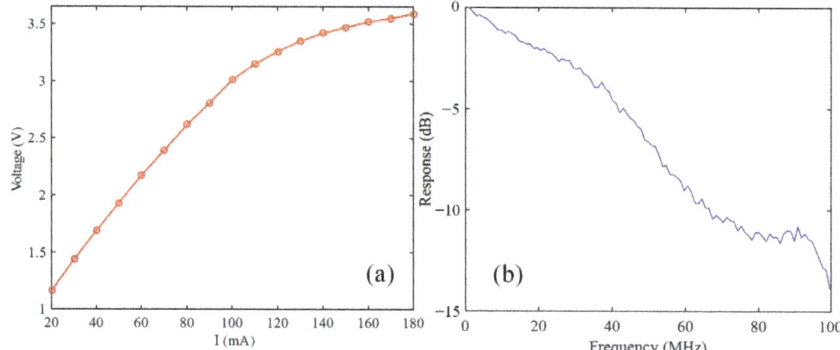

Figure 5. Measured (**a**) nonlinear voltage–current curve and (**b**) low-pass frequency response of the system.

Figure 6 shows the measured BER versus the peak-to-peak voltage (Vpp) of the AWG output signal for different cases, where the transmission distance is fixed at 80 cm. It can be clearly seen that 4CAP-DM greatly outperforms both 4CAP-IM and 4CAP-IM with PWC, achieving the lowest BER of 1.3×10^{-3} at Vpp = 2 V. Compared with 4CAP-DM, 4CAP-DM with PWC can achieve further significant BER improvement, obtaining the lowest BER of 4.45×10^{-4} at Vpp = 2 V. Moreover, nearly the same BER performance can be achieved for PWC-based 4CAP-DM with and without 2D chaotic encryption, suggesting that the application of 2D chaotic encryption does not affect the BER performance of PWC-based 4CAP-DM. It can also be observed that, for PWC-based 4CAP-DM with 2D chaotic encryption, the BER with no key and the BER with only the key for CMS are always about 0.38, and the BER with only the key for CCS is always about 0.21, which indicates that the application of 2D chaotic encryption can efficiently ensure the security of the transmitted 4CAP-DM signal. The insets (a)–(d) in Figure 6 depict the corresponding received constellation diagrams at Vpp = 2 V.

Figure 7 shows the measured BER versus the transmission distance for different cases with Vpp = 2 V. We can see that the BER of 4CAP-IM cannot reach the 7% FEC coding limit of BER = 3.8×10^{-3} when the distance is in the range from 80 to 110 cm. The maximum distances that can be transmitted by 4CAP-IM with PWC, 4CAP-DM, and 4CAP-DM with PWC at BER = 3.8×10^{-3} are 83.0, 87.0, and 98.8 cm, respectively. Hence, a 4.82% improvement of transmission distance can be achieved by 4CAP-DM in comparison to 4CAP-DM with PWC. Furthermore, distance extensions of 11.8 and 15.8 cm are obtained by 4CAP-DM with PWC compared with 4CAP-IM with PWC and 4CAP-DM, which correspond to remarkable 13.56% and 19.04% improvements of transmission distance, respectively. It can also be seen that the PWC-based 4CAP-DM signal with 2D chaotic encryption cannot be successfully eavesdropped if the eavesdropper does not have the keys for both CCS and CMS at the same time.

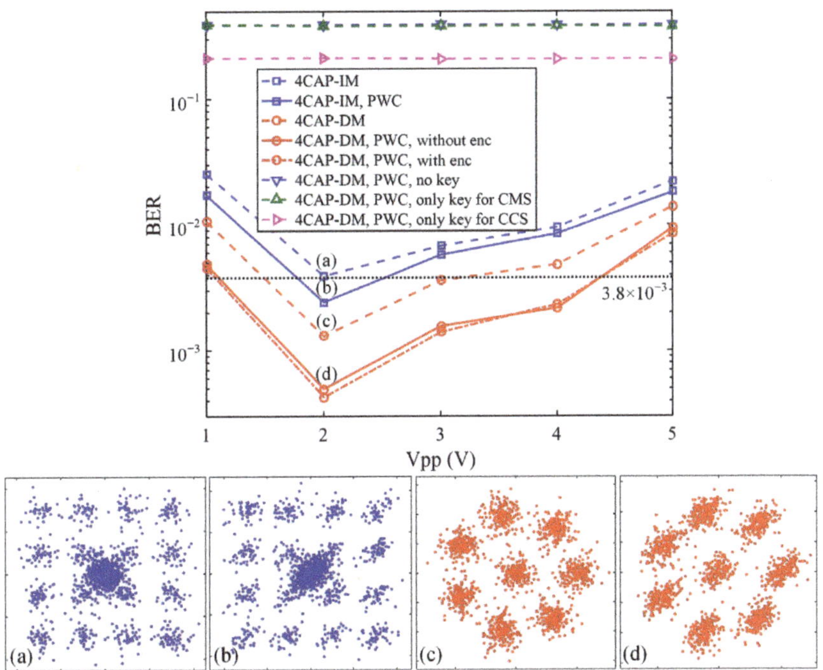

Figure 6. Measured BER vs. Vpp for different cases. Insets (**a–d**): the corresponding received constellation diagrams.

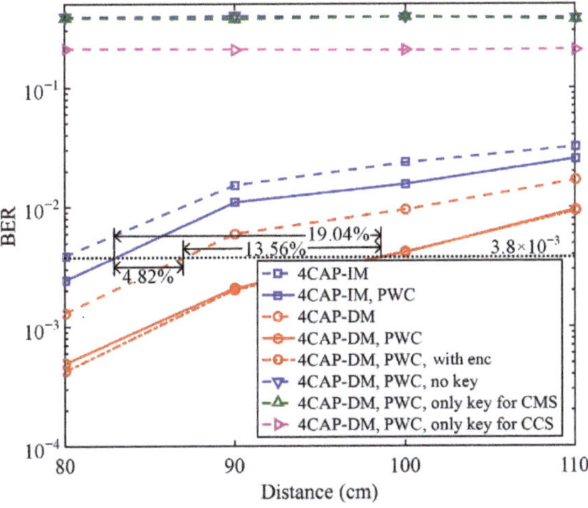

Figure 7. Measured BER vs. transmission distance for different cases.

4. Conclusions

In this paper, we proposed and evaluated a novel PWC-based mCAP-DM scheme with 2D chaotic encryption for secure, bandlimited VLC systems. The use of PWC coding can efficiently mitigate the SNR imbalance caused by the low-pass characteristic of LEDs, and the application of 2D chaotic encryption can successfully guarantee the security of the transmitted useful information in the VLC system. Simulation results validated the excellent BER performance of 4CAP-DM in comparison to 4CAP-IM, achieving the same SE of 2.5 bits/s/Hz over the AWGN channel. Furthermore, experimental results showed that PWC-based 4CAP-DM achieves much better BER performance than the benchmark schemes, while the physical-layer security of the system can be substantially enhanced by applying the 2D chaotic encryption approach. Therefore, the proposed PWC-based mCAP-DM scheme with 2D chaotic encryption can be a promising candidate for practical VLC systems.

Author Contributions: Conceptualization, C.C.; methodology. W.Z., Y.Y., X.D. and M.L.; validation, Y.Y.; formal analysis, W.Z. and X.D.; investigation, Y.N.; resources, Y.Y.; data curation, Y.N.; writing—original draft preparation, Y.N.; writing—review and editing, W.Z., X.D. and C.C.; supervision, M.L.; project administration, M.L.; funding acquisition, C.C. All authors have read and agreed to the published version of the manuscript.

Funding: This research was funded by the National Natural Science Foundation of China (61901065), the Natural Science Foundation of Chongqing (cstc2021jcyj-msxmX0480), and the Fundamental Research Funds for the Central Universities (2021CDJQY-013).

Institutional Review Board Statement: Not applicable.

Informed Consent Statement: Not applicable.

Data Availability Statement: Not applicable.

Acknowledgments: The authors would like to thank the anonymous reviewers for their valuable comments and suggestions.

Conflicts of Interest: The authors declare no conflict of interest.

References

1. Chi, N.; Zhou, Y.; Wei, Y.; Hu, F. Visible light communication in 6G: Advances, challenges, and prospects. *IEEE Veh. Technol. Mag.* **2020**, *15*, 93–102. [CrossRef]
2. Chen, C.; Fu, S.; Jian, X.; Liu, M.; Deng, X.; Ding, Z.G. NOMA for energy-efficient LiFi-enabled bidirectional IoT communication. *IEEE Trans. Commun.* **2021**, *69*, 1693–1706. [CrossRef]
3. Rajagopal, S.; Roberts, R.D.; Lim, S.K. Visible light communication: Modulation schemes and dimming support. *IEEE Commun. Mag.* **2012**, *3*, 72–82. [CrossRef]
4. Minh, H.L; Brien, D.O.; Faulkner, G.; Zeng, L.; Lee, K.; Jung, D.; Oh, Y. High-speed visible light communications using multiple-resonant equalization. *IEEE Photonics Technol. Lett.* **2008**, *20*, 1243–1245. [CrossRef]
5. Chen, C.; Nie, Y.G.; Liu, M.; Du, Y.F.; Liu, R.S.; Wei, Z.X.; Fu, H.Y. Digital pre-equalization for OFDM-based VLC systems: Centralized or distributed? *IEEE Photonics Technol. Lett.* **2021**, *33*, 1081–1084. [CrossRef]
6. Wu, F.M.; Lin, C.T.; Wei, C.C.; Chen, C.W.; Huang, H.-T.; Ho, C.-H. 1.1-Gb/s White-LED-based visible light communication employing carrier-less amplitude and phase modulation. *IEEE Photonics Technol. Lett.* **2012**, *24*, 1730–1732. [CrossRef]
7. Chen, C.; Tang, Y.R.; Cai, Y.P.; Liu, M. Fairness-aware hybrid NOMA/OFDMA for bandlimited multi-user VLC systems. *Opt. Express* **2021**, *29*, 42265–42275. [CrossRef]
8. Elgala, H.; Mesleh, R.; Haas, H. Indoor broadcasting via white LEDs and OFDM. *IEEE Trans. Consum. Electron.* **2009**, *55*, 1127–1134. [CrossRef]
9. Chen, C.; Zhong, W.D.; Yang, H.L.; Du, P.F. On the performance of MIMO-NOMA based visible light communication systems. *IEEE Photonics Technol. Lett.* **2018**, *30*, 307–310. [CrossRef]
10. Haigh, P.A.; Le, S.T.; Zvanovec, S.; Ghassemlooy, Z.; Luo, P.; Xu, T.; Chvojka, P.; Kanesan, T.; Giacoumidis, E.; Canyelles-Pericas, P.; et al. Multi-band carrier-less amplitude and phase modulation for bandlimited visible light communications systems. *IEEE Wirel. Commun.* **2015**, *22*, 46–53. [CrossRef]
11. Akande, K.O.; Popoola, W.O. Subband index carrierless amplitude and phase modulation for optical communications. *J. Lightwave Technol.* **2018**, *36*, 4190–4197. [CrossRef]

12. Wang, Z.; Zhang, J.W.; He, Z.X.; Zou, P.; Chi, N. Subcarrier index modulation super-Nyquist carrierless amplitude phase modulation for visible light communication systems. *J. Lightwave Technol.* **2021**, *39*, 6420–6433. [CrossRef]
13. Hu, W.X.; Zhang, M.; Li, Z.; Popov, S.; Leeson, M.S.; Xu, T. High-dimensional feature based non-coherent detection for multi-intensity modulated ultraviolet communications. *J. Lightwave Technol.* **2021**. [CrossRef]
14. Zuo, T.; Wang, F.S.; Zhang, J.K. Sparsity signal detection for indoor GSSK-VLC system. *IEEE Trans. Veh. Technol.* **2021**, *70*, 12975–12984. [CrossRef]
15. He, J.; He, J.; Shi, J. An enhanced adaptive scheme with pairwise coding for OFDM-VLC system. *IEEE Photonics Technol. Lett* **2018**, *30*, 1254–1257. [CrossRef]
16. Yang, Y.B.; Chen, C.; Zhang, W.; Deng, X.; Du, P.F.; Yang, H.L.; Zhong, W.D.; Chen, L.Y. Secure and private NOMA VLC using OFDM with two-level chaotic encryption. *Opt. Express* **2018**, *26*, 34031–34042. [CrossRef] [PubMed]
17. Kordov, K.; Stoyanov, B. Least significant bit steganography using Hitzl-Zele chaotic map. *Int. J. Electron. Telecommun.* **2017**, *63*, 417–422. [CrossRef]
18. Chen, C.; Nie, Y.G.; Ahmed, F.; Zeng, Z.H.; Liu, M. On the Constellation Design of DFT-S-OFDM with Dual-Mode Index Modulation in VLC. 2022. Available online: https://www.researchgate.net/publication/357747021_On_the_constellation_design_of_DFT-S-OFDM_with_dual-mode_index_modulation_in_VLC (accessed on 11 February 2022).

Article

Constant Transmission Efficiency Dimming Control Scheme for VLC Systems

Jia-Ning Guo, Jian Zhang *, Gang Xin and Lin Li

National Digital Switching System Engineering & Technological Research Center, Zhengzhou 450000, China; 14291003@bjtu.edu.cn (J.-N.G.); downxg@163.com (G.X.); wclilin@163.com (L.L.)
* Correspondence: Zhang_xinda@126.com

Abstract: As a novel mode of indoor wireless communication, visible light communication (VLC) should consider the illumination functions besides the primary communication function. Dimming control is one of the most crucial illumination functions for VLC systems. However, the transmission efficiency of most proposed dimming control schemes changes as the dimming factor changes. A block coding-based dimming control scheme has been proposed for constant transmission efficiency VLC systems, but there is still room for improvement in dimming range and error performance. In this paper, we propose a dimming control scheme based on extensional constant weight codeword sets to achieve constant transmission efficiency. Meanwhile, we also provide a low implementation complexity decoding algorithm for the scheme. Finally, comparisons show that the proposed scheme can provide a wider dimming range and better error performance.

Keywords: visible light communication (VLC); dimming control; constant transmission efficiency; error performance; light-emitting diode (LED)

Citation: Guo, J.-N.; Zhang, J.; Xin, G.; Li, L. Constant Transmission Efficiency Dimming Control Scheme for VLC Systems. *Photonics* **2021**, *8*, 7. https://doi.org/10.3390/photonics8010007

Received: 23 November 2020
Accepted: 25 December 2020
Published: 31 December 2020

1. Introduction

With the increasing shortage of radio frequency spectrum and the development of light-emitting diodes (LEDs), visible light communication (VLC) has attracted extensive attention from many scholars [1,2]. Compared to conventional wireless communications, VLC has higher rates, lower power consumption, and less electromagnetic interference [3–5]. As a combination of communication and illumination, VLC achieves reliable communication and high-quality illumination through the fast response characteristic of LEDs. For convenience, VLC systems realize high-speed communication utilizing intensity modulation and direct detection (IM/DD). There is no doubt that VLC is a novel wireless communication mode with great potential. At the same time, illumination function plays an essential role in the indoor VLC system. Dimming control is one of the most crucial illumination functions that can adjust the brightness according to the users' requirements under the condition of ensuring normal communication functions [6]. However, the dimming control function may affect communication performance to some extent. Therefore, how to balance the dimming control function and the communication performance is a challenge for VLC systems. In order to meet this challenge, lots of researchers have proposed many dimming control schemes.

Variable on-off keying (VOOK) scheme and variable pulse position modulation (VPPM) scheme are two basic dimming control schemes proposed by the IEEE 802.15.7 task group [7]. The frames of the VOOK scheme can be divided into data frames and free frames. Data frames are utilized to realize communication, while free frames are responsible for dimming control. VPPM scheme is the combination of 2-pulse position modulation (2-PPM) and pulse width modulation (PWM), in which 2-PPM is used to realize data transmission, and PWM realizes dimming control. Paper [8] proposed multiple pulse position modulation (MPPM) scheme, which can further improve the spectral efficiency and error performance for dimmable indoor VLC systems. When n approaches infinity, the MPPM scheme can reach the theoretical limit of transmission efficiency. Those modulation based schemes realize dimming control and data transmission through different modulation mode. A rate-compatible punctured code (RCPC) scheme has been proposed in [9]. The RCPC scheme can provide a wide range of brightness and simple coding structure for all different rates. Paper [10] provides a modified Reed-Muller (RM) code based scheme which utilizes a coset constructed from a bent function and RM codes for providing dimming control function. Paper [11] proposed a dimming control scheme based on weight threshold check code (WTCC), which can achieve dimming control and improve spectral efficiency further. The scheme is proposed for binary data transmission but can not provide constant transmission efficiency. The transmission efficiency will change when the dimming factor varies. Those coding-based schemes realize dimming control and data transmission by different encoding/decoding constructions. A dimming control scheme based on multilevel parity check codes (ML-PCC) proposed in paper [12], and paper [13] introduced a dimming control scheme based on multilevel incremental constant weight codes (ML-ICWC). Both of the two schemes are proposed for multilevel transmission and can not provide constant transmission efficiency. Significantly, the ML-PCC scheme can provide better error performance than the ML-ICWC scheme, but the dimming factor of ML-PCC can not cover the whole range.

However, all the dimming control schemes we mentioned above have the same problem: The transmission efficiency is not constant when the dimming factors are different. To solve this problem, paper [14] presents a block coding-based dimming control scheme that can provide constant transmission efficiency. Nevertheless, the dimming range and the error performance of the block coding-based scheme still has room for improvement. In this paper, we propose a dimming control scheme based on extensional constant weight codeword sets (ECWCS) to realize dimming control with constant transmission efficiency. The simulation results show that when the transmission efficiency and dimming factor are both the same, the error performance of the proposed scheme is better than that of the block coding scheme. We can also conclude that the proposed scheme has a wider dimming range than the block coding scheme. The ECWCS scheme can provide constant transmission efficiency, a wider dimming range, and a lower division value of the dimming factor. Therefore, the ECWCS scheme is a better choice for the indoor scene with variable brightness, such as office, conference room, classroom, and movie theater.

The rest of this paper can be settled as follows: Section 2 proposes the system model of the proposed ECWCS scheme. In Section 3, the motivation, codeword construction, and the encoding/decoding procedure of the proposed scheme are given. Section 4 shows the simulation results of the proposed scheme and the contrast scheme. The conclusion of the whole paper is shown in Section 6.

2. System Model

The system model of the proposed ECWCS scheme will be presented in this section. In this paper, the scheme is proposed for the single-input single-output VLC system, which contains a single LED and a single photodetector (PD). The system model is shown as Figure 1. First, the codeword set is constructed according to the dimming factor. After generated by the massage generator, the binary data is encoded by the dimming encoder. Then the coded binary data are modulated by the OOK modulator to the LED. After passing through the optical channel with additive white Gaussian noise (AWGN), the received signals are detected by the PD and then decoded by the dimming decoder. There is a point that should be noted: This scheme can not only be applied to OOK modulation but also pulse amplitude modulation (PAM) and multi-carrier modulation (MCM). Figure 1 is just an example of the ECWCS scheme applied to OOK modulation for the convenience of understanding. When the algorithm is applied to PAM, we should also adjust the average code weight of the codeword set to realize dimming control. Because PAM has a higher modulation order, the codeword set will be further expanded, and the spectral efficiency will be further increased. However, the corresponding error performance will have a certain amount of loss. When the algorithm is applied to MCM, we should ensure the average code weight of the codeword set for every subcarrier is constant and adjust it to realize dimming control. When MCM is utilized in the proposed ECWCS scheme, the throughput will be increased, and the resistance to interference will be enhanced. However, the implementation complexity and cost will also be significantly increased. Modern Discrete Multi-Tone (DMT) modulation is a kind of MCM and can be utilized in the proposed scheme. Paper [15] reported the first visible light link based on WDM and DMT modulation of a single RGB-type white LED, and paper [16] proposed 1-Gb/s VLC systems using a white LED and the DMT signal. Both of the two schemes improved the throughput to a large extent. However, the implementation complexity and the cost of the electronic devices are weaknesses for the DMT modulation.

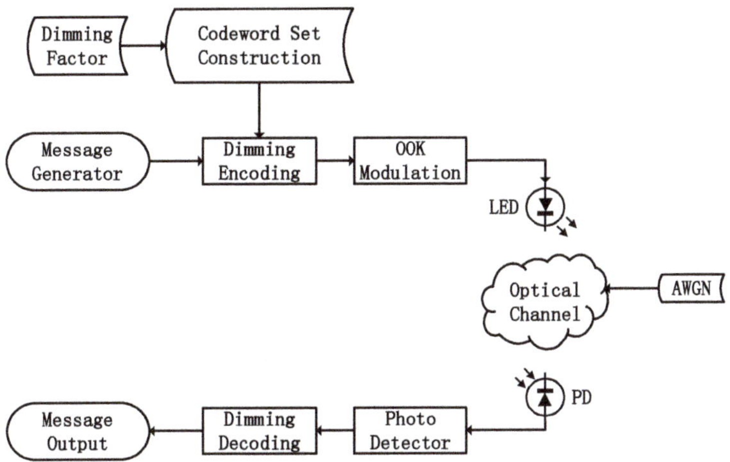

Figure 1. System model of the proposed dimming control scheme applied to on-off keying (OOK) modulation.

At last, the message will be output to the users. In this system model, there are a few assumptions that should be noted:

a. The channel state information (CSI) is available both at the receiver and the transmitter.
b. Compared with the direct light, the reflected light is much weaker in the indoor VLC systems [1,17]. For the convenience of computer analysis, we only consider the LOS path. Therefore, the multipath influence may not be considered in the proposed indoor dimmable VLC system.
c. A static link should be supposed in the proposed scheme because the channel is slow time-varying.

From Figure 1 and the assumptions we mentioned before, we can conclude that the received signal can be expressed as

$$y = \mu h x + n, \tag{1}$$

where y represents the received signal, x denotes the transmitted signal, μ is the photoelectric conversion factor which can be normalized as $\mu = 1$, h is the channel gain we will introduce in this section, and n is the additive white Gaussian noise of which the mean is 0 and the variance is σ^2.

From the contents in [1], the channel can be modeled as the Lambertian model, and the channel can be calculated by:

$$h = \begin{cases} \frac{A(u+1)}{2\pi D^2} \cos^u(\phi) T_S(\psi) g(\psi) \cos\psi, & 0 \leq \psi \leq \Psi_C \\ 0, & \psi > \Psi_C \end{cases} \tag{2}$$

where h is the channel gain between the receiver and the transmitter, D is the distance from the LED to the PD, u is the order of the Lambertian model, and A represents the physical area. The other physical quantities are about illumination. ψ is the incidence angle, and ϕ denotes the irradiance angle. $T_S(\psi)$ represents the optical filter gain of the LED, $g(\psi)$ is the concentrator gain of the LED, and the field of view (FOV) of the PD can be represented by Ψ_C [1].

For the dimmable VLC systems, there are two crucial constraints. The first constraint is the peak power constraint. The LEDs utilized for the VLC system have the optimal operating range due to the characteristics of nonlinear. We should also take the eye safety of the users into account [18]. Therefore, the peak power constraint is necessary, and it can be expressed as $0 \leqslant x \leqslant P_{peak}$, where P_{peak} is the peak power. In this paper, we can normalize the peak power without loss of generality, which can be expressed as $P_{peak} = 1$. So the normalized power limitation is

$$0 \leqslant x \leqslant 1. \tag{3}$$

The other constraint is the dimming control constraint. In essence, the dimming control function of the peak power limited VLC systems is to adjust the average power when the peak power is fixed. The dimming factor is the ratio of the average power to peak power. We adjust the average power to meet the requirements of the users by communication technologies [9]. Therefore, the dimming control constraint can be expressed as

$$\bar{P} = E(\mathbf{x}) = \gamma P_{peak}, \tag{4}$$

where γ denotes the dimming factor of the dimmable VLC systems with the range $\gamma \in (0, 1)$.

3. The Proposed Dimming Control Scheme

This paper introduces the motivation, the codeword set construction, and the encoding/decoding procedure of the proposed ECWCS scheme.

3.1. Motivation of the ECWCS Scheme

In order to realize dimming control for VLC systems with constant transmission efficiency, paper [14] proposed a scheme based on block coding via the bitwise AND operation. The error performance and spectral efficiency still have room for improvement. Motivated by the works in [12,13], we propose a dimming control scheme for VLC systems with constant transmission efficiency. In the proposed scheme, the code weight of the codewords in a set are all odd or all even. Therefore, the minimum Hamming distance of a codeword set is 2, which can improve the error performance to a certain extent. Meanwhile, the extensional constant weight codeword set scheme can achieve arbitrary dimming control within a wide range.

3.2. Construction of the Codeword Set

In this subsection, we provide the construction of the codeword set. Firstly, some notations should be defined. In the proposed ECWCS scheme, the transmitted binary data should be divided into several binary data sequences with the same length. We define the length of the original binary sequence is k, thus the original binary sequence can be expressed as $\mathbf{b} = [b_1, b_2, \cdots, b_k]$. The length of the dimming coded binary sequence is defined as n, thus the coded binary sequence is represented by $\mathbf{c} = [c_1, c_2, \cdots, c_n]$. From the definitions provided above, it is not difficult to know that the most important point of the proposed scheme is to find the proper value of n and construct the codeword set when k is known.

We know that when i is a positive integer and $i \leqslant n/2$, the combinatorial number have the relationship as follows: $C_n^1 < C_n^2 < \cdots < C_n^i$, where C_n^m represents the combinatorial number of selecting m elements from n elements. The essence of the proposed scheme's codeword set construction is to increase the number of codewords available while keeping the transmission efficiency constant and dimming control function. In order to ensure the dimming control function, the average code weight of a codeword set should be fixed. In order to obtain a constant transmission efficiency, the ratio of k to n should be a constant when the dimming factor γ varies. Based on the above two points, when the length of the original binary sequence k is fixed, the length of the coded binary sequence need to satisfy the condition:

$$n \geqslant 2^k. \tag{5}$$

We define $\mathcal{S}$ is the codeword set constructed by the proposed scheme, and m represents the total number of bit '1' of all elements in set $\mathcal{S}$. Thus the dimming factor can be expressed as

$$\gamma = \frac{m}{n \times 2^k}. \tag{6}$$

From Equation (6) and the analyses in this subsection, we know that the dimming range of the proposed dimming control scheme is $[\frac{1}{n}, \frac{n-1}{n}]$ and the division value of the dimming factor is $\frac{1}{n \times 2^{k-1}}$. The reason for the division value is $\frac{1}{n \times 2^{k-1}}$ rather than $\frac{1}{n \times 2^k}$ is as follows: In order to improve the error performance, we make all the code weight of the codewords in the codeword set odd or even to ensure the Hamming distance of the codewords in the codeword set is 2. For example, if 111000 is a codeword in the codeword set $\mathcal{S}$, the code weight of the rest codewords in $\mathcal{S}$ are all odd. Meanwhile, 2^k is an even, and m is the total number of bit '1' of all elements in set $\mathcal{S}$. In other words, it represents the total code weight of codewords in $\mathcal{S}$. The summation of 2^k odd numbers (or even numbers) is an even number, thus the division value of the dimming factor is $\frac{2}{n \times 2^k} = \frac{1}{n \times 2^{k-1}}$. For the sake of calculation, in the rest of this paper, the dimming factor $\gamma = a \times \frac{1}{2^k}$, where a is an even integer with $0 < a \leqslant 2^k \times \frac{n-1}{n}$. According to the value of the dimming factor, the encoding procedure can be divided into two cases:

For the convenience of expression, we define i is a positive integer and $0 < i < n$.

(1) When the dimming factor $\gamma = \frac{i}{n}$, the code weight of every codeword in the codeword set $\mathcal{S}$ is i. Therefore, we can not only guarantee the dimming factor $\gamma = \frac{i}{n}$ but also ensure the Hamming distance of the codewords in the codeword set $\mathcal{S}$ is 2 to improve the error performance.

(2) When the dimming factor $\gamma \neq \frac{i}{n}$ and $\frac{1}{n} < \gamma < \frac{n-1}{n}$, the construction of the codeword set is more complicated. First we find the range $\frac{i}{n} < \gamma < \frac{i+1}{n}$. Then we encode the original binary codes like the last case which makes all the code weight of every codeword in set $\mathcal{S}$ is i. At last, we calculate $m = n \times 2^k \times \gamma$, $q = (m - 2^k i)/2$, select q codewords in $\mathcal{S}$ and replace two bit '0' with two bit '1' of the q codewords respectively.

To make it easier to understand, we fix $k = 3$ provide two examples for the two cases. When $k = 3$, we can calculate $n \geqslant 8$ from Equation (5), thus we fix $n = 8$. The examples are as follows:

Example 1. *When the dimming factor $\gamma = 1/8$, $i = 1$. Therefore, the mapping between the original binary sequence and the dimming coded binary sequence is shown in Table 1.*

Table 1. The mapping between the original binary sequence and the dimming coded binary sequence with $k = 3$, $n = 8$, and $\gamma = 1/8$.

Original Binary Sequence	Dimming Coded Binary Sequence	Original Binary Sequence	Dimming Coded Binary Sequence
000	00000001	100	00010000
001	00000010	101	00100000
010	00000100	110	01000000
011	00001000	111	10000000

Example 2. *When the dimming factor $\gamma = 5/16$, we firstly find the range $2/8 < 5/16 < 3/8$ and conclude that $i = 2$. Then we construct a codeword set $\mathcal{S}$ with $i = 2$ like the last example and calculate that $m = n \times 2^k \times \gamma = 8 \times 2^3 \times 5/16 = 20$, $q = (m - 2^k i)/2 = (20 - 2^3 \times 2)/2 = 2$. Thus we select 2 codewords in $\mathcal{S}$ and replace two bit '0' with two bit '1' of the 2 codewords respectively. Therefore, the mapping between the original binary sequence and the dimming coded binary sequence is shown in Table 2.*

Table 2. The mapping between the original binary sequence and the dimming coded binary sequence with $k = 3$, $n = 8$, and $\gamma = 5/16$.

Original Binary Sequence	Dimming Coded Binary Sequence	Original Binary Sequence	Dimming Coded Binary Sequence
000	00000011	100	00110000
001	00000110	101	01100000
010	00001100	110	11000011
011	00011000	111	10000111

The construction of the codeword set can be summarized in Algorithm 1.

Algorithm 1: Construction of the Codeword Set.

Input:The dimming factor γ, the length of original binary sequence k and the original binary signal $\mathbf{s}$
Fix proper the length of the coded binary sequence n by the constraint: $n \geqslant 2^k$.
If $\gamma = \frac{i}{n}$, where i is a positive integer and $0 < i < n$
Construct the codeword set $\mathcal{S}$ in which the code weight of every codeword is i.
Else If $\gamma \neq \frac{i}{n}$ and $\frac{1}{n} < \gamma < \frac{n-1}{n}$
Find the range $\frac{i}{n} < \gamma < \frac{i+1}{n}$
Construct the codeword set $\mathcal{S}$ in which the code weight of every codeword is i.
Calculate $m = n \times 2^k \times \gamma$ and $q = (m - 2^k i)/2$
Select q codewords in $\mathcal{S}$ and replace two bit '0' with two bit '1' of the q codewords respectively.
End
End
Output: The codeword set $\mathcal{S}$

3.3. *Encoding/Decoding Procedure the ECWCS Scheme*

In the last subsection, the codeword set $\mathcal{S}$ is constructed by Algorithm 1. By the construction of the codeword set, we can obtain the mapping relation between the original binary sequence and the dimming coded binary sequence from the table like Tables 1 and 2. In this subsection, the encoding/decoding procedure will be introduced.

For the encoding procedure, we first divide the original binary signal $\mathbf{s}$ into several length-k binary sequences. Then we map the binary sequence $\mathbf{b}$ to the length-n dimming coded binary sequence $\mathbf{c}$. In the end, we can obtain the transmitted binary signal x.

For the decoding procedure, paper [19] provide a fast decoding algorithm. Motivated by the contents in paper [19], we provide a decoding algorithm to decrease complexity. We know that the traditional Maximum Likelihood (ML) decoding algorithm is to compare the probability density function of the received sequence $\mathbf{r}$ conditioned on the coded binary sequence $\mathbf{c}$ when the system is a single-input single-output system, which is expressed as [20]:

$$p(\mathbf{r}|\mathbf{c}) = \frac{1}{(\sqrt{2\pi\sigma^2})^N}\exp(-\frac{\|\mathbf{r}-\mathbf{c}\|^2}{2\sigma^2}). \tag{7}$$

The decoder can be simplified as:

$$\hat{\mathbf{r}} = \arg\min \|\mathbf{r}-\mathbf{c}\|^2. \tag{8}$$

In order to decrease the complexity of the decoding procedure, the decoder can also be given as:

$$\|\mathbf{r}-\mathbf{c}\|^2 = \|\mathbf{r}\|^2 + \|\mathbf{c}\|^2 - 2\mathbf{r}^T\mathbf{c}. \tag{9}$$

When the received signal is the same one and the code weight is a constant, the decoder is represented by:

$$\hat{\mathbf{r}} = \arg\max \mathbf{r}^T\mathbf{c}. \tag{10}$$

Due to the coded sequence $\mathbf{c}$ is a binary sequence, we should only find the position of bit '1' in $\mathbf{c}$ and sum the elements at the same position in $\mathbf{r}$. For example, when $\mathbf{c} = [1,1,0,0]$ and $\mathbf{r} = [1.05, 0.83, 0.35, 0.21]$, the summation is $1.05 + 0.83 = 1.88$. After we get the estimated sequence $\hat{\mathbf{r}}$ of $\mathbf{r}$, we can recover the original binary data by looking up from the table like Tables 1 and 2.

The proposed decoding algorithm is for the constant weight codes. Therefore, according to the two cases in the last subsection, there are two cases for the decoding algorithm. The two cases are as follows:

(1) When the dimming factor $\gamma = \frac{i}{n}$, the code weight of every codeword in the codeword set $\mathcal{S}$ is i. Thus we can utilize the proposed decoding algorithm directly.

(2) When the dimming factor $\gamma \neq \frac{i}{n}$ and $\frac{1}{n} < \gamma < \frac{n-1}{n}$, the code weight of every codeword in set $\mathcal{S}$ is not a constant. First we find the range $\frac{i}{n} < \gamma < \frac{i+1}{n}$ and calculate $m = n \times 2^k \times \gamma$, $q = (m - 2^k i)/2$. Then we find the q codewords the code weight of which is not i in $\mathcal{S}$. At last, replace two bit '1' with two bit '0' of the q codewords respectively. That is the inverse process of the codeword set construction and requires the decoder to know the details of the construction process. In this way, we can utilize the proposed decoding algorithm.

The decoding algorithm can be summarized in Alogrithm 2.

Algorithm 2: Decoding Procedure.

Input:The dimming factor γ, the length of original binary sequence k,
the length of received binary sequence n, and the received sequence $\mathbf{r}$
If $\gamma \neq \frac{i}{n}$ and $\frac{1}{n} < \gamma < \frac{n-1}{n}$
Find the range $\frac{i}{n} < \gamma < \frac{i+1}{n}$
Calculate $m = n \times 2^k \times \gamma$ and calculate $q = (m - 2^k i)/2$
Replace two bit '1' with two bit '0' of the q codewords respectively.
Else If $\gamma = \frac{i}{n}$, where i is a positive integer and $0 < i < n$
End
End
Find the position of bit '1' in $\mathbf{c}$, sum the elements at the same position in $\mathbf{r}$, and obtain the estimated sequence $\hat{\mathbf{r}}$ of $\mathbf{r}$.
Recover the original binary data by looking up from the table like Tables 1 and 2.
Output: The estimated signal $\hat{\mathbf{s}}$ of the original binary signal $\mathbf{s}$.

4. Simulation Results

In this subsection, we will provide the simulation results, including dimming range, error performance, and spectral efficiency. The analysis will be provided at last.

4.1. Dimming Range

Dimming range is the range that can be achieved by a dimming control scheme, which can be expressed as:

$$\gamma_r = \gamma_{max} - \gamma_{min}, \tag{11}$$

where γ_{min} is the minimum dimming factor, γ_{max} represents the maximum dimming factor, and γ_r is the dimming range.

For both the two dimming control schemes with constant transmission efficiency, including the block coding scheme and the proposed ECWCS scheme, the dimming range depends on the value of n and the value of n is affected by the value of k. Thus we provide the relationship between k and the dimming range γ_r. What needs illustration is that the length of the coded binary sequence n of the proposed ECWCS scheme is not unique when k is fixed. For convenience, we utilize $n = 2^k$. In paper [14], the dimming range of the block coding dimming control scheme can be summarized as Table 3, and the dimming factor's division value of the block coding scheme is $\frac{1}{k}$.

Table 3. The dimming range of the block coding scheme under different k.

k	γ_{min}	γ_{max}	γ_r
2	$\frac{1}{4}$	$\frac{3}{4}$	$\frac{1}{2}$
3	$\frac{1}{6}$	$\frac{5}{6}$	$\frac{2}{3}$
4	$\frac{1}{8}$	$\frac{7}{8}$	$\frac{3}{4}$
k	$\frac{1}{2k}$	$\frac{2k-1}{2k}$	$\frac{k-1}{k}$

The dimming range of the proposed ECWCS dimming control scheme is shown in Table 4.

Table 4. The dimming range of the proposed ECWCS scheme under different k.

k	γ_{min}	γ_{max}	γ_r
2	$\frac{1}{4}$	$\frac{3}{4}$	$\frac{1}{2}$
3	$\frac{1}{8}$	$\frac{7}{8}$	$\frac{3}{4}$
4	$\frac{1}{16}$	$\frac{15}{16}$	$\frac{7}{8}$
k	$\frac{1}{2^k}$	$\frac{2^k-1}{2^k}$	$\frac{2^{k-1}-1}{2^{k-1}}$

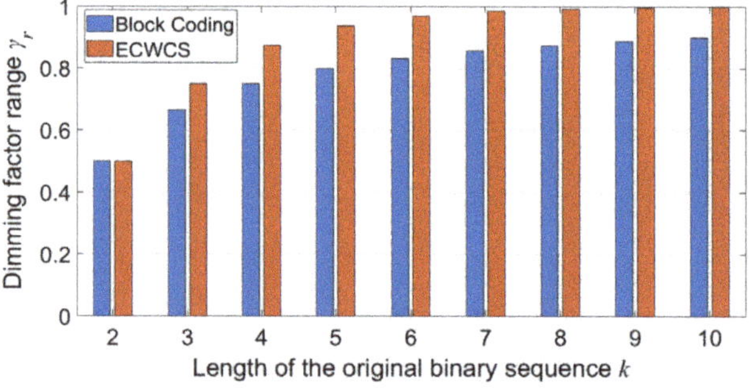

Figure 2. Dimming range of the block coding scheme and the extensional constant weight codeword sets (ECWCS) scheme under different k.

Figure 2 shows the dimming range under different k. From Tables 3 and 4 and Figure 2, we can know that the proposed ECWCS scheme has a wider dimming range and a lower division value compared with the block coding scheme. We can also see that under the condition of $k \geqslant 7$, the dimming range is nearly all covered when the ECWCS scheme is utilized.

4.2. Error Performance

Error performance is another essential index for VLC systems. It is represented by curves of the bit error rate (BER). The curves describe the relationship between the signal-to-noise ratio (SNR) and BER. The SNR can be expressed as [21]:

$$\text{SNR} = 10 \lg \frac{1}{R_C \sigma^2}, \tag{12}$$

where R_C is the code rate.

In this paper, we compare the block coding dimming control scheme and the proposed scheme. From the discussion before and the contents in the last subsection, we know that the length of the coded binary sequence n of the proposed ECWCS scheme is not unique when k is fixed. As shown in Equation (5), for the proposed scheme $n \geqslant 2^k$. While for the block coding scheme, $n = k^2$. Therefore, we fix the value of n, k, and γ to compare the error performance of the two schemes for fairness.

From Figure 3, we can conclude that when the value of n, k, and γ are the same, the error performance of the ECWCS scheme is better than that of the block coding scheme. That is because the minimum Hamming distance between the codewords of the ECWCS scheme is larger than that of the block coding scheme. We can also see that for the proposed ECWCS scheme, the error performance with $k = 2, n = 4$ outperformed $k = 3, n = 9$ under the same dimming factor.

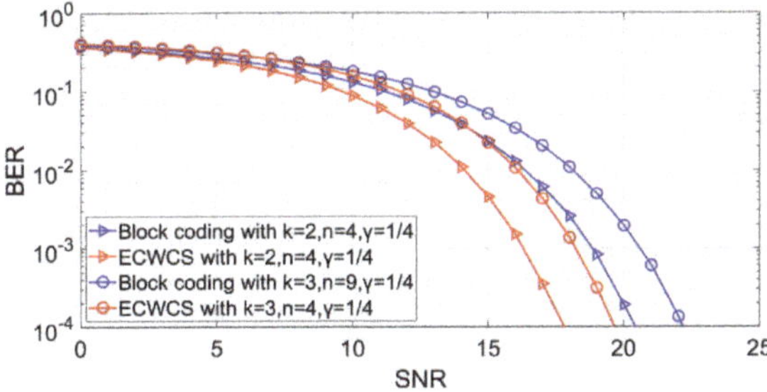

Figure 3. Error performance of the block coding scheme and the ECWCS scheme.

For clarity, we demonstrate the adaptability of this system in terms of provided illuminance. First, we introduce the relationship between the dimming factor and illuminance. We know that the dimming factor is the ratio of the average power to peak power, and the range of dimming factor is $0 < \gamma < 1$. The value of the dimming factor reflects the illuminance of the LED. With the increase of the dimming factor, the illuminance of the LED increases gradually. Then we demonstrate the relationship between BER and illuminance. We can assume a set $\mathcal{D}$ consists of the Euclidean distance between any two distinct received signals. The minimum Euclidean distance (MED) of the received signals can be defined as the minimum of the elements in set $\mathcal{D}$. Therefore, the MED of the received signals can be expressed as:

$$d_R = \min \sqrt{\|y^a - y^b\|^2}, \tag{13}$$

where y^a and y^b are two distinct received signals.

Similarly, the MED of the transmitted signals is:

$$d_T = \min \sqrt{\|x^a - x^b\|^2}, \tag{14}$$

where x^a and x^b are two distinct transmitted signals.

We know that the received signal can be expressed as Equation (1) in Section 2. We substitute Equation (1) into Equation (13) and the MED of the received signals can be also expressed as

$$d_R = \min \sqrt{\|(\mu h x^a + n) - (\mu h x^b + n)\|^2} = \min \sqrt{\mu h \|x^a - x^b\|^2} = d_T \sqrt{\mu h}. \tag{15}$$

There is a point that should be noted: In terms of sequences, the received signal is the received sequence **r** and the transmitted signal is the coded binary sequence **c**. The distance between the signals means the distance between the sequences, and the numbers in the sequence represent the intensity level of the optical signal. From the contents in paper [22], we know that the BER curve can represent the error performance, and the error performance is determined by the MED of the received signals when the SNR is fixed. Therefore, in the proposed system model of this paper, the error performance is determined by the MED of the transmitted signals when the SNR is fixed. The way we explain the BER metric according to minimum Euclidean distance is not only for OOK modulation but also for pulse amplitude modulation. In this paper, we provide OOK modulation as an example to introduce the ECWCS scheme. For the proposed ECWCS scheme, the MED of the transmitted signals is constant when the dimming factor varies. Therefore, the BER curves are the same with different dimming factors. However, when the dimming factor varies, the average power of the transmitted signals changes. The SNR will be affected by the change of the average power of the transmitted signals. With the increase of the average power, SNR increases gradually. Therefore, the error performance will be affected by the dimming factor. From the above reasoning and the BER curves provided in Figure 3, we can conclude that with the increase of the dimming factor, the error performance gets better gradually.

4.3. Spectral Efficiency

The dimming control scheme based on extensional constant weight codeword sets is proposed for indoor VLC systems with constant transmission efficiency in this paper. The other dimming control schemes widely utilized can not realize constant transmission efficiency when the dimming factor varies. Spectral efficiency indicates the effective bit rate R_b can be realized when the bandwidth B is fixed and can be expressed as:

$$\nu = \frac{R_b}{B}. \tag{16}$$

When the bandwidth B is fixed, the spectral efficiency is only related to the effective bit rate R_b. Therefore, the spectral efficiency can reflect the transmission efficiency and the effective throughput. The proposed ECWCS scheme guarantee that the spectral efficiency is not affected by the variation of the dimming factor through the unique way of encoding. Therefore, we can say that the proposed ECWCS scheme realizes constant spectral efficiency.

Figure 4 shows the spectral efficiency comparison between the proposed ECWCS scheme and other dimming control schemes without constant transmission efficiency when the length of the coded binary sequence $n = 8$.

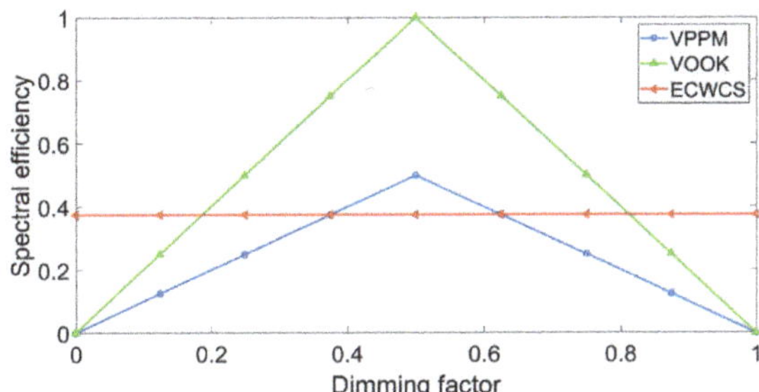

Figure 4. Spectral efficiency of variable on-off keying (VOOK), variable pulse position modulation (VPPM) and ECWCS.

We can see from Figure 4 that the spectral efficiency of the VOOK scheme and the VPPM scheme is greater than that of the ECWCS scheme within a specific dimming range. However, when the dimming factor is close to 0 or 1, the spectral efficiency of the ECWCS scheme is greater than that of the VOOK scheme and the VPPM scheme. It can be construed that the constant transmission efficiency is guaranteed by reducing the spectral efficiency within a specific dimming range.

We can assume a scenario that the users need to change the dimming factor of the indoor VLC system from 0.1 to 0.5. When we utilize the VOOK scheme or the VPPM scheme, the transmission efficiency varies according to the value of the dimming factor, and the communication quality would be influenced. On the other hand, for example, when the length of the coded binary sequence $n = 8$, the dimming factor's division value of the VOOK scheme and the VPPM scheme is $1/8$. From the contents in Section 3, we know that the dimming factor's division value of the proposed ECWCS scheme is $\frac{1}{n \times 2^{k-1}} = 1/32$. Therefore, based on the above two points, the proposed ECWCS scheme is a better choice for the indoor scene with variable brightness.

4.4. Analysis

In this section, we have provided a comparison to the currently existing variable-weight coding scheme, which is called block coding dimming control scheme [14]. The code weight of the block coding dimming control scheme is variable when the dimming factor is fixed. Thus we can classify it as the variable-weight coding scheme. We can know that compared with the block coding dimming control scheme, the ECWCS scheme has a wider dimming range, a smaller division value, and better error performance.

The currently existing constant-weight schemes can not provide a constant transmission. All of those schemes have the same code weight when the dimming factor is fixed. However, when the dimming factor varies, the code weight and transmission efficiency will change. Thus it is unfair to compare those schemes with the proposed ECWCS scheme. When we utilize constant-weight codes to realize dimming control with constant transmission efficiency, it will add redundancy compared with the ECWCS scheme. For example, we fix the dimming factor is $\gamma = 3/8$ and the length of the original binary sequence is $k = 2$. The length of the coded binary sequence is $n = 4$ when the ECWCS scheme is utilized, and the length of the coded binary sequence is $n = 8$ when utilizing constant-weight codes.

5. Discussion

We have done much research on the dimming control function of VLC systems in different scenes. Paper [12] proposed a dimming control scheme based on multilevel parity check codes (ML-PCC) for multilevel transmission. A dimming control scheme based on multi-LED phase-shifted space-time codes (MP-STC) were proposed for multi-LED VLC systems in paper [23]. Paper [24] provided a dimming control scheme based on constant weight space-time codes (CWSTC) for Multiple Input Multiple Output (MIMO) VLC systems. In this paper, we propose the ECWCS scheme for the dimmable VLC systems with constant transmission efficiency and a low complexity decoding algorithm for the ECWCS scheme. In addition to this, it can also provide a wider dimming range, a smaller division value, and better error performance compared with the previous constant transmission efficiency dimming control scheme [14]. We think that the ECWCS scheme is a better choice for dimmable VLC systems with constant transmission efficiency.

6. Conclusions

In this paper, a dimming control scheme based on extensional constant weight codeword sets has been proposed for the constant transmission efficiency VLC systems. The proposed ECWCS scheme can realize reliable transmission and maintain constant transmission efficiency when the dimming factor varies by optimizing the codeword set. From the simulation results, the proposed ECWCS scheme has a wider dimming range, a smaller division value, and better error performance compared with the block coding dimming control scheme. Meanwhile, a low complexity decoding algorithm has been proposed for the ECWCS scheme to enhance practicality. Therefore, the ECWCS scheme will be considered as a potential choice for the constant transmission efficiency dimmable VLC systems.

Author Contributions: J.-N.G. has contributed to the scientific part of this work. J.-N.G. and J.Z. contributed to the conception and design of the study. G.X. and L.L. have critically reviewed the paper. All authors have read and agreed to the published version of the manuscript.

Funding: This work is supported by Grant 161100210200 from Major Scientific and Technological of Henan Province, China, Grant 2016B010111001 Major Scientific and Technological of Guangdong Province, China, the National Key Research and Development Project (2018YFB1801903) and the National Natural Science Foundation of China (NSFC) under Grant (62071489,61671477,61901524).

Acknowledgments: The authors wish to thank the anonymous reviewers for their valuable suggestions.

Conflicts of Interest: The authors declare no conflict of interest.

References

1. Komine, T.; Nakagawa, M. Fundamental analysis for visible-light communication system using LED lights. *IEEE Trans. Consum. Electron.* **2004**, *50*, 100–107. [CrossRef]
2. Andrews, J.G.; Buzzi, S.; Choi, W.; Hanly, S.V.; Lozano, A.; Soong, A.C.K.; Zhang, J.C. What Will 5G Be? *IEEE J. Sel. Areas Commun.* **2014**, *32*, 1065–1082. [CrossRef]
3. O'Brien, D.; Minh, H.L.; Zeng, L.; Faulkner, G.; Lee, K.; Jung, D.; Oh, Y.; Won, E.T. Indoor visible light communications: Challenges and prospects. In *Free-Space Laser Communications VIII*; Majumdar, A.K., Davis, C.C., Eds.; International Society for Optics and Photonics, SPIE: Bellingham, WA, USA, 2008; Volume 7091, pp. 60–68. [CrossRef]
4. Elgala, H.; Mesleh, R.; Haas, H. Indoor optical wireless communication: Potential and state-of-the-art. *IEEE Commun. Mag.* **2011**, *49*, 56–62. [CrossRef]
5. Wang, Z.Y.; Yu, H.Y.; Wang, D.M. Energy Efficient Transceiver Design for NOMA VLC Downlinks with Finite-Alphabet Inputs. *Appl. Sci.* **2018**, *8*, 1823.
6. Karunatilaka, D.; Zafar, F.; Kalavally, V.; Parthiban, R. LED Based Indoor Visible Light Communications: State of the Art. *IEEE Commun. Surv. Tutor.* **2015**, *17*, 1649–1678. [CrossRef]
7. Roberts, R.D.; Rajagopal, S.; Lim, S. IEEE 802.15.7 physical layer summary. In Proceedings of the 2011 IEEE GLOBECOM Workshops (GC Wkshps), Houston, TX, USA, 5–9 December 2011; pp. 772–776.
8. Lee, K.; Park, H. Modulations for Visible Light Communications With Dimming Control. *IEEE Photonics Technol. Lett.* **2011**, *23*, 1136–1138. [CrossRef]
9. Kim, J.; Park, H. A Coding Scheme for Visible Light Communication With Wide Dimming Range. *IEEE Photonics Technol. Lett.* **2014**, *26*, 465–468. [CrossRef]
10. Kim, S.; Jung, S. Modified Reed–Muller Coding Scheme Made From the Bent Function for Dimmable Visible Light Communications. *IEEE Photonics Technol. Lett.* **2013**, *25*, 11–13. [CrossRef]
11. Zuo, Y.; Zhang, J.; Zhang, Y.; Chen, R. Weight Threshold Check Coding for Dimmable Indoor Visible Light Communication Systems. *IEEE Photonics J.* **2018**, *10*, 1–11. [CrossRef]
12. Guo, J.N.; Zhang, J.; Zhang, Y.Y.; Li, L.; Zuo, Y.; Chen, R.H. Multilevel transmission scheme based on parity check codes for VLC with dimming control. *Opt. Commun.* **2020**, *467*, 125733. [CrossRef]
13. Zuo, Y.; Zhang, J.; Zhang, Y.Y. A spectral-efficient dimming control scheme with multi-level incremental constant weight codes in visible light communication systems. *Opt. Commun.* **2018**, *426*, 531–534. [CrossRef]

14. Zuo, Y.; Zhang, J. A Novel Coding Based Dimming Scheme with Constant Transmission Efficiency in VLC Systems. *Appl. Sci.* **2019**, *9*, 803. [CrossRef]
15. Vučić, J.; Kottke, C.; Habel, K.; Langer, K. 803 Mbit/s visible light WDM link based on DMT modulation of a single RGB LED luminary. In Proceedings of the 2011 Optical Fiber Communication Conference and Exposition and the National Fiber Optic Engineers Conference, Los Angeles, CA, USA, 6–10 March 2011; pp. 1–3.
16. Khalid, A.M.; Cossu, G.; Corsini, R.; Choudhury, P.; Ciaramella, E. 1-Gb/s Transmission Over a Phosphorescent White LED by Using Rate-Adaptive Discrete Multitone Modulation. *IEEE Photonics J.* **2012**, *4*, 1465–1473. [CrossRef]
17. Lee, K.; Park, H.; Barry, J.R. Indoor Channel Characteristics for Visible Light Communications. *IEEE Commun. Lett.* **2011**, *15*, 217–219. [CrossRef]
18. Wang, J.; Wang, J.; Chen, M.; Song, X. Dimming scheme analysis for pulse amplitude modulated visible light communications. In Proceedings of the 2013 International Conference on Wireless Communications and Signal Processing, Ilmenau, Germany, 27–30 August 2013; pp. 1–6. [CrossRef]
19. Zuo, Y.; Zhang, J.; Zhang, Y. The fast detection algorithm suitable for fast dimmable indoor visible light communication system. In Proceedings of the 2017 9th International Conference on Advanced Infocomm Technology (ICAIT), Chengdu, China, 22–24 November 2017; pp. 210–213. [CrossRef]
20. Proakis, J.; Salehi, M. *Digital Communications*; McGraw-Hill: New York, NY, USA, 2008.
21. Kim, S.; Jung, S. Novel FEC Coding Scheme for Dimmable Visible Light Communication Based on the Modified Reed–Muller Codes. *IEEE Photonics Technol. Lett.* **2011**, *23*, 1514–1516. [CrossRef]
22. Zhang, Y.; Yu, H.; Zhang, J.; Zhu, Y. Signal-Cooperative Multilayer-Modulated VLC Systems for Automotive Applications. *IEEE Photonics J.* **2016**, *8*, 1–9. [CrossRef]
23. Guo, J.; Zhang, J.; Zhang, Y.; Xin, G. Efficient multi-LED dimming control scheme with space–time codes for VLC systems. *Appl. Opt.* **2020**, *59*, 8553–8559. [CrossRef]
24. Guo, J.N.; Zhang, J.; Zhang, Y.Y.; Xin, G.; Li, L. Constant Weight Space-Time Codes for Dimmable MIMO-VLC Systems. *IEEE Photonics J.* **2020**, *12*, 1–15. [CrossRef]

Article

Deep Learning-Assisted Index Estimator for Generalized LED Index Modulation OFDM in Visible Light Communication

Manh Le-Tran † and Sunghwan Kim *,†

Department of Electrical, Electronic and Computer Engineering, University of Ulsan, Ulsan 44610, Korea; tranmanh@mail.ulsan.ac.kr
* Correspondence: sungkim@ulsan.ac.kr
† These authors contributed equally to this work.

Abstract: In this letter, we present the first attempt of active light-emitting diode (LED) indexes estimating for the generalized LED index modulation optical orthogonal frequency-division multiplexing (GLIM-OFDM) in visible light communication (VLC) system by using deep learning (DL). Instead of directly estimating the transmitted binary bit sequence with DL, the active LEDs at the transmitter are estimated to maintain acceptable complexity and improve the performance gain compared with those of previously proposed receivers. Particularly, a novel DL-based estimator termed index estimator-based deep neural network (IE-DNN) is proposed, which can employ three different DNN structures with fully connected layers (FCL) or convolution layers (CL) to recover the indexes of active LEDs in a GLIM-OFDM system. By using the received signal dataset generated in simulations, the IE-DNN is first trained offline to minimize the index error rate (IER); subsequently, the trained model is deployed for the active LED index estimation and signal demodulation of the GLIM-OFDM system. The simulation results show that the IE-DNN significantly improves the IER and bit error rate (BER) compared with those of conventional detectors with acceptable run time.

Keywords: visible light communications; deep learning; bit error rate; orthogonal frequency division multiplexing; index modulation

Citation: Le-Tran, M.; Kim, S. Deep Learning-Assisted Index Estimator for Generalized LED Index Modulation OFDM in Visible Light Communication. *Photonics* **2021**, *8*, 168. https://doi.org/10.3390/photonics8050168

Received: 1 April 2021
Accepted: 17 May 2021
Published: 19 May 2021

1. Introduction

Visible light communication (VLC) is a promising alternative for complementing the overcrowded radio frequency bandwidth of wireless communication owing to its license-free operations, relative low cost, high spatial diversity, high bandwidth efficiency, and electromagnetic interference-free transmission [1–7]. Orthogonal frequency division multiplexing (OFDM) has been extensively used in VLC to maintain a better transmission rate due to its high spectral efficiency and low-complexity implementation to achieve better transmission rates [8,9]. Recently, the combination of multiple-input multiple-output (MIMO) and index modulation [10–13] to convey additional bits has gained a lot of attention in VLC literature. There are several published papers that investigated different aspects of the index modulation technique, such as non-coherent detection [14], low complexity detection [15], or precoding [16]. In addition, index modulation was proposed in optical communication [17] and later in [18] to implement generalized 4×4 light-emitting diode (LED) index modulation (GLIM-OFDM); in which the four transmitter side LEDs were divided into groups to transmit real-imaginary and positive-negative signals separately. As a result, depending on every OFDM symbol, just one LED in each group was chosen to convey the OFDM symbol parts. In [19], an LED selection algorithm for GLIM-OFDM was introduced to determine the best active LED combination to improve the system performance.

In general, a maximum likelihood (ML) detector is cost-ineffective and unpractical for an OFDM-based communication system. According to [18,19], maximum a posteriori (MAP) detectors are better than zero-forcing (ZF) and minimum mean square error (MMSE)

detectors because of their better bit error rate (BER) performance and reasonable demodulation complexity. However, because all previously mentioned detectors only consider the instantaneous signal received at the receiver, the underlying relationship between the generated OFDM symbols is not fully exploited in the demodulation process.

Deep learning (DL) has had an enormous impact on various research fields over the last couple of years; it has been applied in wireless communication since then [20–23]. Many researchers have already suggested DL solutions to wireless communication problems, e.g., modulation detection and signal detection [24]. VLC channel models, similar to radio frequency channel models, have been developed according to probability and signal processing theories. Furthermore, DL techniques have recently made a number of advancements in the VLC system. Since the VLC transmission must be able to support a variety of constraints, such as dimming and lighting, according to user preferences in practical applications, the implementation of DL becomes more challenging. In [25], a data-driven approach based on DL was proposed to predict the outage events in a line-of-sight link. In [23], DL-based constellation design scheme for MIMO-VLC systems was introduced, which can significantly reduce complexity compared to existing schemes while maintaining a near-optimal performance. Despite a wide range of promising applications, the performance of VLC systems has been impaired by the high SNR required to achieve a low error rate for a variety of scenarios. Therefore, DL approach can be applied to VLC systems to improve performances in practical high-speed and low error-rate indoor transmission application, such as hospital facility. To the best of the authors' knowledge, no researcher has used DL to improve the BER performance of the GLIM-OFDM system.

In this paper, a deep neural network (DNN)-aided active LED index estimator (IE) for improving the GLIM-OFDM detector performance in the VLC system is proposed. Unlike traditional DNN-based detectors, which directly estimate the transmitted bits at the transmitter side, the proposed method extracts the feature of the active LED index set with the time-domain signal of the receiver. Estimating the GLIM system's active LED indexes is crucial for detecting the transmitted signal; this step has usually been performed by the ZF, MMSE, and MAP detectors in previous studies [18,19]. Although an ML-based detector is performance-optimal but impractical, a DNN-based solution is complexity-efficient and still useful for performance enhancement. Briefly, the key contributions of the proposed method are as follows:

- An IE-based DNN (IE-DNN) is introduced into the GLIM (multiple-input multiple-output) MIMO-VLC demodulator;
- Only a single IE-DNN module without the need to change the transmitter and the signal demodulation part in the receiver structure;
- Three different structures of the IE-DNN are proposed and compared to demonstrate that the CNN-based estimator delivers the best performance;
- In comparison with conventional detectors, a remarkable active LED index estimation accuracy significantly improves the BER performance at acceptable complexity costs.

This paper is organized as the following. In Section 2, we describe the system model of the GLIM with the mention of conventional detection techniques. In Section 3, we present the proposed DL-based index estimator include different network structures, the sample generation, and training specification stages. In Section 4, we provide the simulation results to demonstrate the advantage of the proposed index estimator in comparison with the previous ones. Finally, we provide the concluding remark in Section 5.

2. GLIM System Model

This section presents the GLIM system model in Figure 1, which focuses on the MIMO transmission of $n_T \times n_R$; n_T are n_R are the numbers of LEDs and PDs, respectively. Without loss of generality, it can be assumed that $n_T = n_R = n$, where n is an even number. In the GLIM, the OFDM modulator directly processes the complex frequency-domain OFDM frame through inverse fast Fourier transform (IFFT) operation. After the parallel-to-serial (P/S) operation, the time-domain OFDM signals x_l $(l = 1, \cdots, N)$ are separated into their real and imaginary parts ($x_l = x_{l,R} + jx_{l,I}$) and converted into the time-domain signal vector $\mathbf{t} = [x_{1,R}, x_{1,I}, x_{2,R}, x_{2,I}, \ldots, x_{N,R}, x_{N,I}]$ of length $2N$. Therefore, the GLIM technique transmission can help to convey complex and bipolar signal to the real and unipolar signal to be transmitted through the VLC channel. Consequently, the vector $\mathbf{z} = [z_1 \cdots z_n]^T$ is the signal transmitted by the LEDs. In the receiver side, after estimating the active LEDs and recovering the transmitted signal, the serial-to-parallel (S/P) and fast Fourier transform (FFT) operations convert the received signal to the estimated symbol. More specifically, N is the number of subcarriers and $4N$ is divisible by n; consequently, the vector $\mathbf{t}$ is transformed into a matrix $\mathbf{T}$ of size $4N/n \times n/2$. At each time instant i, each column of $\mathbf{T}$ (a signal vector of length $n/2$)) is transmitted by all n LEDs as $\mathbf{z}_i$. More specifically, let

$$sgn(\mathrm{a}) = \begin{cases} 1, & \text{if} \quad a \geq 0 \\ -1, & \text{if} \quad a < 0' \end{cases} \tag{1}$$

the j-th signal of the i-th column of $\mathbf{T}$ can be transmitted with all n LEDs as

$$\begin{aligned} z_i(2(j-1)+1) &= \frac{sgn(\mathbf{T}(i,j))+1}{2}\mathbf{T}(i,j), \\ z_i(2j) &= \frac{sgn(\mathbf{T}(i,j))-1}{2}\mathbf{T}(i,j). \end{aligned} \tag{2}$$

Consequently, by activating a half number of LEDs while the other half are inactive, the signal $\mathbf{z}_i$ is transmitted over the $n \times n$ optical MIMO channel as

$$\mathbf{y}_i = \mathbf{H}\mathbf{z}_i + \mathbf{n}_i, \tag{3}$$

where $\mathbf{y}_i = [y_i(1), \cdots, y_i(n)]$ is the received signal, $\mathbf{H} \in \mathbb{R}^{n \times n}$ is the optical communication channel matrix, and $\mathbf{n}_i$ is real-valued additive white Gaussian noise with elements distributed as $\mathcal{N}(0, \sigma_w^2)$. In addition, a flat fading channel is considered in this letter [18].

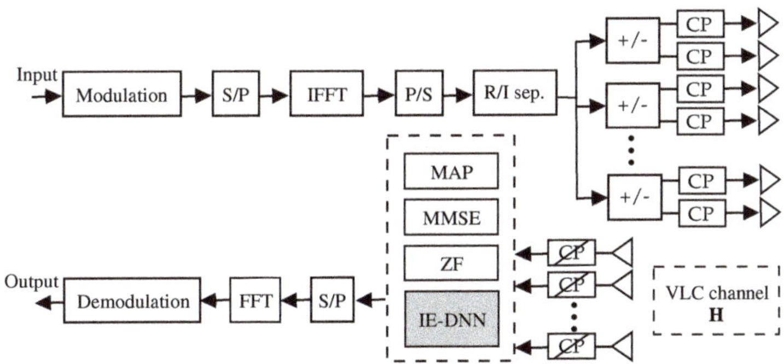

Figure 1. System model with proposed estimator.

Among the three detection techniques, e.g., ZF, MMSE, and MAP, MAP is regarded as the preferred receiver as it can achieve the best BER performance [18]. Moreover, the number of transmitted signals ($4N$) is generally high owing to a large number of subcarriers (N) used for transmission. For example, in the OFDM modulation of eight subcarriers, the transmitted signal vector at the LEDs has a size of 32 and this number is 128 with 32 subcarriers. Meanwhile, a small number of LEDs in MIMO-VLC systems, usually four LEDs [18], or eight LEDs [19], are used in transmission, leads to only four or eight signals that are simultaneously transmitted by LEDs and detected by the MAP detectors at the same time. To improve the demodulation performance, all the received signals (e.g., the vector $\mathbf{y} = [\mathbf{y}_1 \ldots, \mathbf{y}_i, \ldots \mathbf{y}_{4N/n}]$) instead of only the n-length received signals $\mathbf{y}_i$ should be simultaneously considered in the demodulation process. In the next section, three model-driven DNN architectures as alternatives to conventional receivers (e.g., ZF, MMSE, and MAP detectors) are proposed; the methods combine DNN with expert knowledge for wireless communication.

3. Structure and Operation of DL-Based Index Estimator

3.1. Proposed DL-Based Index Estimator

To consider the BER and computational complexity of different network structures and parameters for the active LED IE, three IE-DNN structures are considered. The three IE-DNN structures are constructed with an LED-based fully connected DNN (LFC-DNN), a subcarrier-based fully connected DNN (SFC-DNN), and a convolutional neural network (CNN), respectively. The DNNs contain several sub-blocks that are sequentially connected (the so-called hidden layers). The three IE-DNNs are presented in Figure 2.

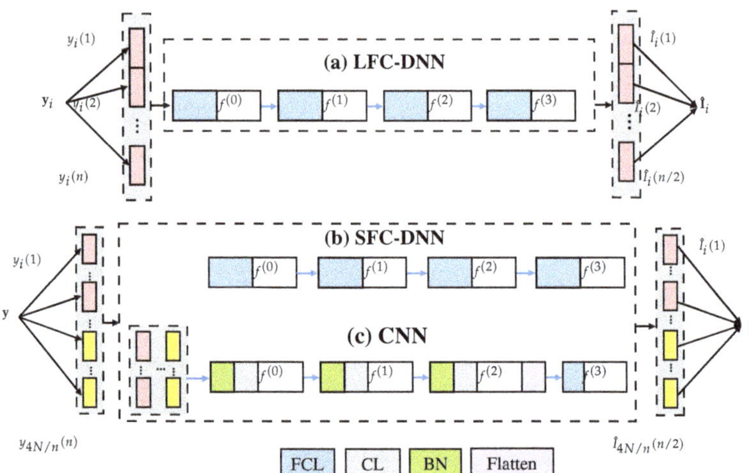

Figure 2. IE-DNN-assisted active LED IE structures for GLIM-MIMO system: (**a**) LFC-DNN, (**b**) SFC-DNN, (**c**) CNN.

More specifically, for a 4×4 system and OFDM symbols with 16 subcarriers, the size of the time-domain signal vector $\mathbf{t}$ is 32 because the real and imaginary parts are transmitted separately. Thus, for the LFC-DNN in Figure 2, the input is a vector of length four, and the output is a vector of length two. For the SFC-DNN and CNN, the outputs are vectors of length 16; the inputs are vectors of length 64 and matrices of size 16×4, respectively.

3.1.1. LFC-DNN

The LFC-DNN takes the instantaneous received signal at the PDs $(\mathbf{y}_i)$ as the input. The predicted instantaneous active LED index vector for the corresponding input is obtained as a vector of length $n/2$ at the output. For a 4×4 system, the received signal vector of length four $(\mathbf{y}_i)$ is taken as the DNN network input. Subsequently, the LFC-DNN outputs a vector of length two as the estimated LED index value, for example, $\tilde{\mathbf{I}}_i = \left[\tilde{I}_i(1)\ \tilde{I}_i(2)\right]^T$ with $\tilde{I}_1(i)$ as the estimated active index for the $\mathbf{T}(i,1)$ signal. The estimated active index $\hat{\mathbf{I}}_i$ can be derived with hard decision decoding as

$$\hat{I}_i(j) = \begin{cases} 1, & \tilde{I}_i(j) \geq 0.5 \\ 0, & \tilde{I}_i(j) < 0.5, \end{cases} \tag{4}$$

If $\hat{I}_i(1) = 1$, the signal $\mathbf{T}(i,1)$ has a positive value and was transmitted by the first LED. If $\hat{I}_i(1) = 0$, the signal $\mathbf{T}(i,1)$ has a negative value and was transmitted by the second LED. For $i = 1, \ldots, 4N/n$, the estimated active index vectors $\hat{\mathbf{I}}_i$ are concatenated to determine the estimated vector $\hat{\mathbf{I}} = \left[\hat{\mathbf{I}}_1 \ldots, \hat{\mathbf{I}}_i, \ldots \hat{\mathbf{I}}_{4N/n}\right]$. Finally, with the estimated active LED indexes, the demodulation process is generally conducted with FFT and QAM demodulation steps to obtain the original binary information sequence, similar to the previous demodulator [18]. All four sub-blocks are composed of fully connected layers (FCLs) and activation functions $f(0), .., f(3)$. The numbers of neurons in the FCLs are 32, 16, 16, and 2, respectively. Moreover, the activation functions for the first three sub-blocks $f(0), f(1), f(2)$ are "ReLU" activation functions, while the activation function for the last sub-block $f(3)$ is a "Sigmoid" activation function. The "ReLU" function is as following form

$$\mathrm{ReLU}(x) = \begin{cases} x, & x \geq 0 \\ 0, & x < 0' \end{cases} \tag{5}$$

while the "Sigmoid" function has the following form

$$\mathrm{S}(x) = \frac{1}{1 + e^{-x}}. \tag{6}$$

3.1.2. SFC-DNN

Because the time-domain signal $\mathbf{t}$ after the IFFT step is a vector that the elements contain underlying information about the relationship between the QAM symbols mapped onto the OFDM subcarriers, the input for the network should be the entire received time-domain signal vector $\mathbf{y} = [\mathbf{y}_1 \cdots \mathbf{y}_{4N/n}]$ to exploit the underlying features of the time-domain generated signals. Intuitively, a structure that uses the received signal vector $\mathbf{y}$ at once is more complex and requires higher computational complexity cost. Nevertheless, it will be demonstrated in the simulation section that this cost will be accommodated by the performance gains, which is of most importance since the computational capacity of recent mobile devices has been enormously enlarged. The SFC-DNN is identical to the LFC-DNN, except that the input of the network is the vector $\mathbf{y}$, which is composed of all time-domain received signal at all LEDs (it is a vector of length $4N$). To facilitate the increase in the number of inputs, the numbers of neurons in the FCLs are 48, 32, 32, and 16, respectively. The remaining parameters (e.g., the activation functions) are identical to those of the proposed LFC-DNN.

3.1.3. CNN

The CNN structure contains four sub-blocks, such as the previously presented DNN structures. However, instead of using the FCL in all sub-blocks, a batch normalization (BN), convolution layer (CL), and an activation function are employed in the first three sub-blocks while the last sub-block remains identical to those of the previously proposed structures. The CL contains 16 kernels with sizes of 1; the number of neurons in the last FCL is 16. Such as the SFC-DNN structure, the CNN structure receives the received time-domain signal $[\mathbf{y}_1, \cdots, \mathbf{y}_{4N/n}]^T$ as the network input in the form of a matrix to exploit the underlying features of the received signals. It is well known that the CL can perform neighborhood filtering on the inputs to capture the input features [26]. Moreover, to ensure that the inputs in the consecutive layer follow the same distribution, to improve the speed of CNN training significantly, and to alleviate the initial parameter dependency of the learning process, BN is considered before using the CL in each sub-block.

3.2. Sample Generation

Training samples are obtained by randomly generating information bits and processing according to the system model as QAM mapping, IFFT. Then, the cyclic prefix (CP) is added to IFFT symbols to effectively mitigate multipath which induces intersymbol interference. Subsequently, the time-domain OFDM symbols are mapped onto the corresponding active and inactive LEDs; for each signal, a positive value indicates that the index value is one and vice versa. More specifically, one training sample can be denoted as $\mathbf{u} = \{\mathbf{y}, \mathbf{I}\}$. It should be mentioned that due to the difference between the proposed networks, the sizes of input vectors are different. However, a similar process of training and testing sample set generation can be applied. Moreover, to speed up the training process and improve the prediction performance, we usually notice that a small number of layers will outperform the more complicated network when the i-th element of l-th normalization training sample $\mathbf{d}_i^{(l)}$ [23] is expressed as

$$\mathbf{d}_i^{(l)} = \frac{\mathbf{u}_i^{(l)} - \mathbb{E}\{u_i\}}{\max(u_i) - \min(u_i)}. \tag{7}$$

3.3. Training Specification

The proposed IE-DNN estimators need to be trained offline prior to the online deployment stage. The parameters of the IE-DNN network are optimized during offline training. The binary cross-entropy loss function [26] is used to quantify the difference between the predicted and actual training indexes as

$$\mathcal{L}(\mathbf{I}, \hat{\mathbf{I}}) = \sum_{\tau=1}^{\mathcal{N}} \frac{I(\tau)\log \hat{I}(\tau) + (1 - I(\tau))\log(1 - \hat{I}(\tau))}{-\mathcal{N}}, \tag{8}$$

where $\mathcal{N} = n/2$ for the LFC-DNN network and $\mathcal{N} = 4N/n$ for the SFC-DNN and CNN networks, respectively. In the next step, the network parameters can be optimized with the adaptive moment estimation optimizer (ADAM), with early termination and ReduceLROnPlateau callbacks in Keras. More specifically, ADAM [27] is an algorithm based on adaptive estimates of lower-order moments to first-order gradient-based optimize stochastic objective function. After offline training, the optimal parameters for the IE-DNN estimator can be used to estimate the active LED index at the receiver side. If the received time-domain OFDM symbols are input into the trained IE-DNN estimator, the active LED index can be estimated in accordance with the framework defined in Section 2. With the estimated active LED indexes, the demodulation is similar to that of the conventional demodulators [18].

3.4. Complexity Analysis

The computational complexity is generally measured by the number of operations based on the active LED index detector and the OFDM algorithm [18]. The required number of FLOPs operations of the algebraic expressions [28] can be summarized in Table 1. Moreover, the real operations for GLIM with MAP detector consist of $76N + 4N\log_2(N)$ real multiplications and $32N + 4\log_2(N)$ real additions, which can be approximated as $\mathcal{O}(N\log_2 N)$ [18]. On the other hand, the computational complexity of the proposed active LED index detectors is dominated by the structure of the networks, such as the dimension of each input, output of layer, and the number of hidden layers. More specifically, the operations for the GLIM with SFC-DNN also consist of $98N + 4N\log_2(N)$ real multiplications and $64N + 4\log_2(N)$ real additions, while the operations for the GLIM with CNN can be composed of $80N + 2N^2 + 4N\log_2(N)$ real multiplications and $48N + N^2 + 4\log_2(N)$ real additions. However, most of the network operations are the feed-forward computations with sparse vector-matrix and matrix-matrix multiplications. Therefore, the asymptotic complexity for the proposed SFC-DNN and CNN detectors can be similarly expressed as $\mathcal{O}(N\log_2 N)$ and $\mathcal{O}(N^2)$, respectively.

Table 1. Complexity cost of matrix-vector operations [28].

Expression	Description	Multiplications	Summations	Total Flops
$\alpha\mathbf{a}$	Vector Scaling	N		N
$\alpha\mathbf{A}$	Matrix Scaling	MN		MN
Ab	Matrix-Vector Prod.	MN	$M(N-1)$	$2MN-M$
AB	Matrix-Matrix Prod.	MNL	$ML(N-1)$	$2MNL-ML$
AD	Matrix-Diagonal Prod.	MN		MN
$\mathbf{a}^H\mathbf{b}$	Inner Prod.	N	$N-1$	$2N-1$
$\mathbf{a}\mathbf{c}^H$	Outer Prod.	MN		MN
$\mathbf{A}^H\mathbf{A}$	Gram	$\frac{MN(N+1)}{2}$	$\frac{N(M-1)(N+1)}{2}$	$MN^2+N\left(M-\frac{N}{2}\right)-\frac{N}{2}$
$\|\|\mathbf{A}\|\|^2$	Euclidean norm	MN	$MN-1$	$2MN-1$
$\mathbf{Q}^{-1}$	Inverse of Pos. Definite	$\frac{N^3}{2}+\frac{3N^2}{2}$	$\frac{N^3}{2}-\frac{N^2}{2}$	N^3+N^2+N Including N roots

4. Results and Discussion

In this section, the estimation accuracy of the active LED indexes is evaluated with the index error rate (IER), and the BER characteristics of the proposed methods (LFC-DNN, SFC-DNN, and CNN) are compared with those of the conventional index estimation algorithms (the best method is MAP detection). Besides, the complexity costs are also compared in terms of the run time measurement. More specifically, a MIMO VLC system's performance composed of 4 LEDs and 4 PDs are shown. For convenience, the system configurations of the simulation are similar to those in [18]: a $5\,\text{m} \times 5\,\text{m} \times 3\,\text{m}$ room, positions of LEDs, PDs, and the channel gain matrix are similar to the physical channel C in [18] is used for the simulation. At the transmitter, OFDM is employed with two cases. Case 1 and 2 have $N=8$ and $N=32$ subcarriers, respectively. In both cases, 4-QAM symbols are mapped onto the subcarrier before the IFFT. A CP length of six is chosen for the computer simulation to ensure that it covers the maximal delay spread of the channel.

Subsequently, the IE-DNNs are trained over 200 iterations with a batch size of 400 samples. Moreover, the networks are trained with 200 epochs during each iteration. When the validation loss value stops improving, the iteration can be terminated early. When the validation loss does not improve after 50 epochs, the learning rate is reduced by $1/3$ until it reaches the minimum (0.0002).

Figure 3 compares the index error rates (IER) of the different algorithms. Nevertheless, the MAP estimator only achieves an IER below 0.1 while the proposed SFC-DNN and CNN are better in estimating the active LED index. More specifically, an accuracy of 0.001 can be obtained in the high SNR regime. In addition, the higher the number of subcarriers is, the better the index accuracy is. Because the LFC-DNN and MAP detection have similar results, the simultaneous extraction of all received signals at the PDs and the additional signal features remarkably improve the index estimation accuracy.

Figure 4 presents the BERs of Case 1. As the IER values of the MAP detection and LFC-DNN are similar, their BERs are the worst among those of all compared estimators. The CNN-based estimator exhibits significantly better BERs than the others. More specifically, at the SNR value of 10 dB, the proposed CNN, FCN, and the conventional MAP detector achieve a BER value of 10^{-4}, 10^{-3}, and 10^{-2}, respectively. Nevertheless, the SNR gain in the BER is only 4 dB when a CNN instead of MAP detection is used. This is because the total number of correctly estimated active LED indexes is not the sole dominant factor that affects the BERs of the demodulator. Because the estimator only estimates the active LED index, after the CNN-based and other estimators, the signal magnitude of the active LEDs is also an essential factor that impacts the BER. With its lower IER, MAP detection can estimate a good number of active LEDs with large signal magnitudes, which leads to an acceptable BER. Moreover, owing to the remarkably high accuracy of SFC-DNN and CNN regarding the IER, the additional correct LED index estimation usually lower in signal magnitude and not as important as the higher ones. Therefore, only an SNR gain of 4 dB can be achieved with the CNN-based estimator. On the other hand, with lower complexity cost, the proposed FCN can only provide a SNR gain of 3 dB at the BER value of 10^{-6}.

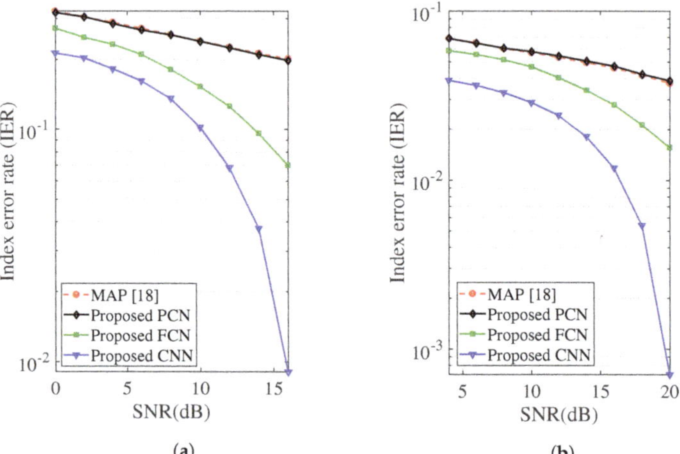

Figure 3. Comparison of IER: (**a**) Case 1 and (**b**) Case 2.

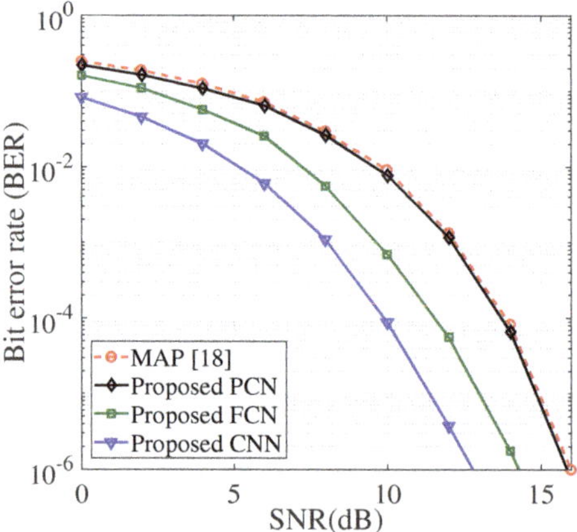

Figure 4. Comparison of BERs for Case 1.

The same observation is made in Case 2. In Figure 5, the CNN-based estimator achieves an SNR gain of approximately 3 dB, and SFC-DNN achieves an SNR gain of only 2 dB. More specifically, the proposed PCN provides a same performance with the conventional MAP detector in almost all SNR regions, except the SNR value between 0 and 5 dB. Additionally, in comparison with Case 1, due to a large number of employed subcarriers, an additional SNR increment of around 3 dB is required to maintain same value of 10^{-6} BER. Therefore, the performance observed in all figures demonstrates the advantage of the proposed detector over the conventional MAP detector in terms of error rate performance, which is most important to the reliability of VLC systems.

In Table 2, the complexity costs of the compared estimators are analyzed. To evaluate the run time of both IE-DNN estimators and the MAP estimator, we use MATLAB to convert the trained model from Keras and estimate the run time for each sample. It should be noted that only the active LED index estimation results are evaluated in this simulation because all other operations (e.g., the costly IFFT computation) are the whole part and identical for all estimators. For all cases, the run time increases with the number of subcarriers hence the number of trainable parameters. The MAP estimator with its simple algorithm and poorer performance has the shortest estimation time; by contrast, the SFC-DNN with the highest number of trainable parameters has the longest estimation time. Interestingly, the CNN exhibits the best performance. At first glance, it seems that the SFC-DNN and CNN estimators are computational costly for GLIM-VLC systems. However, compared with other operations, such as IFFT and QAM demodulation (which significantly increase complexity), the active LED index estimation increase does not significantly impact the total complexity cost. Moreover, with the ever-growing computational power of mobile devices and required quality of services, the BER improvement is much more important than complexity costs.

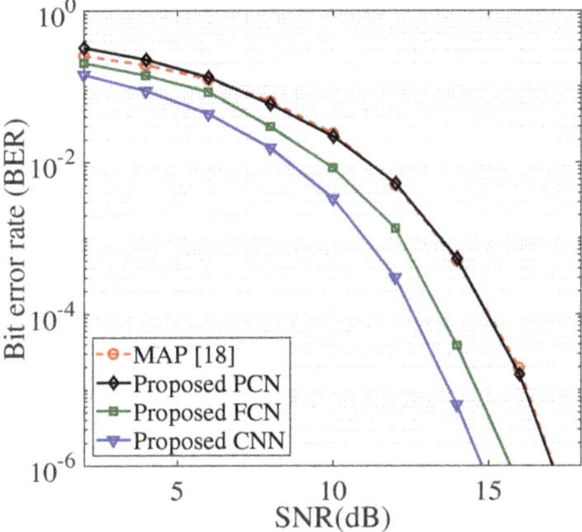

Figure 5. Comparison of BERs for Case 2.

Table 2. Comparison of run time of active LED IEs.

Detector	Numbers of Trainable Parameters (Case 1/Case 2)	Estimation Time (Case 1/Case 2)
LFC-DNN	994/994	13 ms/44 ms
SFC-DNN	4736/9544	28 ms/78 ms
CNN	3096/6328	22 ms/67 ms
MAP	-/-	6 ms/23 ms

5. Conclusions

This paper presents an IE-DNN for active LED index estimation in GLIM-MIMO-VLC demodulation. The proposed network structure promotes signal demodulation by estimating the correct active LED indexes. Moreover, the simulation results of the proposed LFC-DNN, SFC-DNN, and CNN are compared. According to the results, the relationship between the time-domain OFDM signals can be exploited to a higher extent with the IE-DNN structures. Especially, with an acceptable complexity cost, the network with CNN one is the one that can provide the best IER and BER performance improvement.

Author Contributions: All authors discussed the contents of the manuscript and contributed to its presentation. M.L.-T. designed and implemented the proposed scheme, analyzed the simulation results and wrote the paper under the supervision of S.K. All authors have read and agreed to the published version of the manuscript.

Funding: This work was supported by the 2021 Research Fund of University of Ulsan.

Institutional Review Board Statement: Not applicable.

Informed Consent Statement: Not applicable.

Data Availability Statement: Not applicable.

Conflicts of Interest: The authors declare no conflict of interest.

Abbreviations

The following abbreviations are used in this manuscript:

AWGN	Additive white Gaussian noise
CL	Convolution layers
CNN	Convolutional neural network
DNN	Deep neural network
FCL	Fully connected layers
MIMO	Multiple-input multiple-output
LED	Light emitting diode
GLIM	Generalized LED index modulation optical orthogonal frequency-division multiplexing
PD	Photo detector
MAP	Maximum a posteriori
ZF	Zero-forcing
VLC	Visible light communication

References

1. Komine, T.; Nakagawa, M. Fundamental Analysis for Visible-Light Communication System Using LED Lights. *IEEE Trans. Consum. Electron.* **2004**, *50*, 100–107. [CrossRef]
2. Matheus, L.E.M.; Vieira, A.B.; Vieira, L.F.M.; Vieira, M.A.M.; Gnawali, O. Visible Light Communication: Concepts, Applications and Challenges. *IEEE Commun. Surv. Tutor.* **2019**, *21*, 3204–3237. [CrossRef]
3. Gong, C. Visible Light Communication and Positioning: Present and Future. *Electronics* **2019**, *8*, 788. [CrossRef]
4. Le Tran, M.; Kim, S. Effective Receiver Design for MIMO Visible Light Communication with Quadrichromatic LEDs. *Electronics* **2019**, *8*, 1383. [CrossRef]
5. Guo, J.N.; Zhang, J.; Xin, G.; Li, L. Constant Transmission Efficiency Dimming Control Scheme for VLC Systems. *Photonics* **2021**, *8*, 7. [CrossRef]
6. Wang, Q.; Giustiniano, D.; Zuniga, M. In Light and in Darkness, in Motion and in Stillness: A Reliable and Adaptive Receiver for the Internet of Lights. *IEEE J. Sel. Areas Commun.* **2018**, *36*, 149–161. [CrossRef]
7. Li, D.C.; Chen, C.C.; Liaw, S.K.; Afifah, S.; Sung, J.Y.; Yeh, C.H. Performance Evaluation of Underwater Wireless Optical Communication System by Varying the Environmental Parameters. *Photonics* **2021**, *8*, 74. [CrossRef]
8. Armstrong, J. OFDM for Optical Communications. *J. Light. Technol.* **2009**, *27*, 189–204. [CrossRef]
9. Pathak, P.H.; Feng, X.; Hu, P.; Mohapatra, P. Visible Light Communication, Networking, and Sensing: A Survey, Potential and Challenges. *IEEE Commun. Surv. Tutor.* **2015**, *17*, 2047–2077. [CrossRef]
10. Basar, E.; Aygolu, U.; Panayirci, E.; Poor, H.V. Orthogonal Frequency Division Multiplexing With Index Modulation. *IEEE Trans. Signal Process.* **2013**, *61*, 5536–5549. [CrossRef]
11. Basar, E. Reconfigurable Intelligent Surface-Based Index Modulation: A New Beyond MIMO Paradigm for 6G. *IEEE Trans. Commun.* **2020**, *68*, 3187–3196. [CrossRef]
12. Mao, T.; Wang, Q.; Wang, Z.; Chen, S. Novel Index Modulation Techniques: A Survey. *IEEE Commun. Surv. Tutor.* **2018**, *21*, 1. [CrossRef]
13. Ishikawa, N.; Sugiura, S.; Hanzo, L. 50 Years of Permutation, Spatial and Index Modulation: From Classic RF to Visible Light Communications and Data Storage. *IEEE Commun. Surv. Tutor.* **2018**, *20*, 1905–1938. [CrossRef]
14. Fazeli, A.; Nguyen, H.H.; Hanif, M. Generalized OFDM-IM with Noncoherent Detection. *IEEE Trans. Wirel. Commun.* **2020**, *4*, 1323–1337. [CrossRef]
15. Zheng, J.; Liu, Q. Low-Complexity Soft-Decision Detection of Coded OFDM with Index Modulation. *IEEE Trans. Veh. Technol.* **2018**, *67*, 7759–7763. [CrossRef]
16. Belaoura, W.; Althunibat, S.; Qaraqe, K.A.; Ghanem, K. Precoded Index Modulation Based Multiple Access Scheme. *IEEE Trans. Veh. Technol.* **2020**, *69*, 12912–12920. [CrossRef]
17. Mesleh, R.; Elgala, H.; Haas, H. Optical Spatial Modulation. *J. Opt. Commun. Netw.* **2011**, *3*, 234. [CrossRef]
18. Yesilkaya, A.; Basar, E.; Miramirkhani, F.; Panayirci, E.; Uysal, M.; Haas, H. Optical MIMO-OFDM with Generalized LED Index Modulation. *IEEE Trans. Commun.* **2017**, *65*, 3429–3441. [CrossRef]
19. Le Tran, M.; Kim, S.; Ketseoglou, T.; Ayanoglu, E. LED Selection and MAP Detection for Generalized LED Index Modulation. *IEEE Photon. Technol. Lett.* **2018**, *30*, 1695–1698. [CrossRef]
20. Chen, M.; Challita, U.; Saad, W.; Yin, C.; Debbah, M. Artificial Neural Networks-Based Machine Learning for Wireless Networks: A Tutorial. *IEEE Commun. Surv. Tutor.* **2019**, *21*, 3039–3071. [CrossRef]

21. Luong, T.V.; Ko, Y.; Vien, N.A.; Nguyen, D.H.N.; Matthaiou, M. Deep Learning-Based Detector for OFDM-IM. *IEEE Wireless Commun. Lett.* **2019**, *8*, 1159–1162. [CrossRef]
22. Wang, J.; Jiang, C.; Zhang, H.; Ren, Y.; Chen, K.C.; Hanzo, L. Thirty Years of Machine Learning: The Road to Pareto-Optimal Wireless Networks. *IEEE Commun. Surv. Tutor.* **2020**, *22*, 1472–1514. [CrossRef]
23. Le-Tran, M.; Kim, S. Deep Learning-Based Collaborative Constellation Design for Visible Light Communication. *IEEE Commun. Lett.* **2020**, *24*, 2522–2526. [CrossRef]
24. Sun, L.; Wang, Y. CTBRNN: A Novel Deep-Learning Based Signal Sequence Detector for Communications Systems. *IEEE Signal Process. Lett.* **2019**, *27*, 21–25. [CrossRef]
25. Wu, Z.Y.; Ismail, M.; Serpedin, E.; Wang, J. Efficient Prediction of Link Outage in Mobile Optical Wireless Communications. *IEEE Trans. Wirel. Commun.* **2020**, *20*, 882–896. [CrossRef]
26. Goodfellow, I.; Bengio, Y.; Courville, A. *Deep Learning Adaptive Computation and Machine Learning*; The MIT Press: Cambridge, MA, USA, 2016.
27. Kingma, D.P.; Ba, J. Adam: A Method for Stochastic Optimization. *arXiv* **2017**, arXiv:cs/1412.6980.
28. Golub, G.H.; Loan, C.F.V. *Matrix Computations*, 4th ed.; Johns Hopkins Studies in the Mathematical Sciences; Johns Hopkins University Press: Baltimore, MD, USA, 2012.

MDPI

Article

Experimental Demonstration and Simulation of Bandwidth-Limited Underwater Wireless Optical Communication with MLSE

Jialiang Zhang [1], Guanjun Gao [1,*], Jingwen Li [2], Ziqi Ma [1] and Yonggang Guo [3]

[1] State Key Laboratory of Information Photonics and Optical Communications, Beijing University of Posts and Telecommunications, Beijing 100876, China; jialiangzhang@bupt.edu.cn (J.Z.); mzq2021@bupt.edu.cn (Z.M.)
[2] China Mobile Group Zhejiang Co., Ltd., Hangzhou 310012, China; lijingwen7@chinamobile.zj.com
[3] State Key Laboratory of Acoustic, The Institute of Acoustics of the Chinese Academy of Sciences, Beijing 100190, China; guoyg@mail.ioa.ac.cn
* Correspondence: ggj@bupt.edu.cn

Abstract: Underwater wireless optical communication (UWOC) is able to provide large bandwidth, low latency, and high security. However, there still exist bandwidth limitations in UWOC systems, with a lack of effective compensation methods. In this paper, we systematically study the bandwidth limitation due to the transceiver and underwater channel through experiments and simulations, respectively. Experimental results show that by using the 7-tap maximum likelihood sequence estimation (MLSE) detection, the maximum bitrate of the simple rectangular shape on–off-keying (OOK) signaling is increased from 2.4 Gb/s to 4 Gb/s over 1 GHz transceiver bandwidth, compared to the conventional symbol-by-symbol detection. For the bandwidth limitation caused by the underwater channel, we simulate the temporal dispersion in the UWOC by adopting a Monte Carlo method with a Fournier–Forand phase function. With MLSE adopted at the receiver, the maximum available bitrate is improved from 0.4 to 0.8 Gb/s in 12 m of harbor water at the threshold of hard-decision forward-error-correction (HD-FEC, 3.8×10^{-3}). Moreover, when the bitrate for 0.4 Gb/s 12 m and 0.8 Gb/s 10 m OOK transmission remains unchanged, the power budget can be reduced from 33.8 dBm to 30 dBm and from 27.8 dBm to 23.6 dBm, respectively. The results of both experiments and simulations indicate that MLSE has great potential for improving the performance of bandwidth-limited communication systems.

Keywords: underwater wireless optical communication; temporal dispersion; bandwidth limitation; Monte Carlo method; maximum likelihood sequence estimation

Citation: Zhang, J.; Gao, G.; Li, J.; Ma, Z.; Guo, Y. Experimental Demonstration and Simulation of Bandwidth-Limited Underwater Wireless Optical Communication with MLSE. *Photonics* **2022**, *9*, 182. https://doi.org/10.3390/photonics9030182

Received: 8 February 2022
Accepted: 10 March 2022
Published: 12 March 2022

Publisher's Note: MDPI stays neutral with regard to jurisdictional claims in published maps and institutional affiliations.

1. Introduction

High-bandwidth underwater wireless communication has gained increasing interest recently and is considered to be applicable in many marine engineering fields, such as underwater high-definition video transmission and deep-sea observation. At present, the mature and available technology is based on acoustic communication. However, the capacity of underwater acoustic communication is limited to Kbps because of the low bandwidth, which cannot meet increasing requirements such as those for underwater video communication. Underwater wireless optical communication (UWOC) using blue/green wavelength (400–600 nm) lasers has become a better choice and is capable of offering larger bandwidth, lower latency, and higher security [1–3]. Many previous underwater optical transmission experiments have proposed several solutions to increase the channel bandwidth from Megahertz to Gigahertz (GHz). In [4–11], several experiments have demonstrated from hundreds Mbps to more than ten Gbps for meters' UWOC transmission with on-off-keying (OOK) or high order modulation. X. Liu et al. have experimentally demonstrated a 2.70 Gbps data rate over a 34.5 m underwater transmission distance by using non-return-to-zero on-off keying (NRZ-OOK) modulation scheme [4]. C. Lu et al. have demonstrated a 2.5 Gbps UWOC system over 60 m underwater transmission distance at the BER level of 3.5×10^{-3} by using NRZ-OOK modulation and digital nonlinear equalization technology [5]. C. Li et al. have achieved a 25 Gbps UWOC system over a 5 m highly turbid harbor water link [6]. Moreover, spatial multiplexing based on the 2 or 4 parallel orbit momentum multiplexing (OAM) mode is also used to increase the bitrate and have demonstrated a GHz class capacity for UWOC [12,13].

Although there exist many underwater optical transmission experiments which achieve significant results on communication bitrate and distance. Most existing experiments are mainly focusing on the efficient coding, high-order modulation format, and high-bandwidth transceivers to improve the performance [4–11] and there is a lack of studies on bandwidth limitation and methods to solve it for UWOC systems, especially for simple OOK modulations.

In this paper, we systematically analyze bandwidth limitation of UWOC system caused by transceiver and channel temporal dispersion, and its mitigation through experiments and simulation, respectively. Experimental results show that by using the 7-tap maximum likelihood sequence estimation (MLSE) detection, the maximum bitrate of the simple rectangular shape OOK signaling is increased from 2.4 Gb/s to 4 Gb/s over 1 GHz transceiver bandwidth, compared to the conventional symbol-by-symbol detection. For the bandwidth limitation caused by the underwater channel temporal dispersion, Monte Carlo simulation with Fournier–Forand phase function is carried out for turbid harbor water. Simulation results show that the maximum available bitrate is improved from 0.4 to 0.8 Gb/s over 12 m harbor water by using MLSE detection, where the available bandwidth is limited to about 0.12 GHz.

2. Principle and Models

It is noteworthy that the actual UWOC system is a time-variant system, because the parameters of the channel are affected by the temperature, velocity fields, and turbidity of the seawater. To simplify the analysis of the problem, we focus on the impact of important communication parameters such as the seawater optical attenuation coefficient, communication distance on the UWOC system. Here, we consider that the UWOC system has a precisely aligned line-of-sight (LOS) link and the photodetector is perpendicular to the beam [14]. A schematic diagram for the UWOC system model is given in Figure 1.

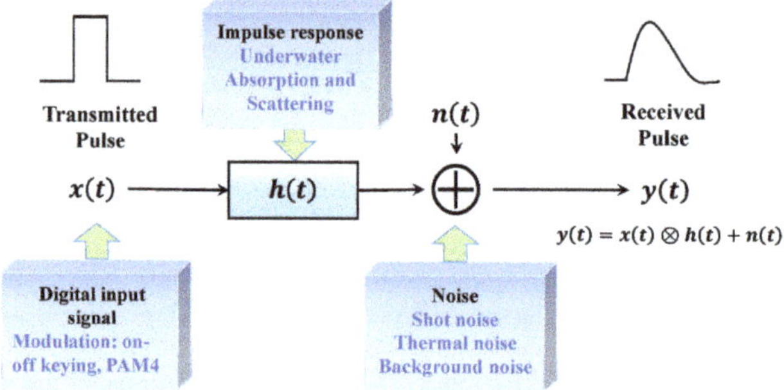

Figure 1. Model of UWOC channel.

Many plankton and suspended particles are present in the water, such as mineral components, organic matter, water molecules, and dissolved salts, which affect the propagating photon in two ways, absorption and scattering [15–19]. These effects on the beam are reflected in the underwater channel impulse response. After being sent into the underwater channel, the beam is deteriorated by absorption and scattering in underwater environments before arriving at the receiver. The noise in the UWOC system depends on the type of receiver. The device of the receiver itself produces background radiation noise and dark current noise. The formula for a simplified model of a UWOC system is shown in Figure 1, where $x(t)$ is the transmitted signal, $h(t)$ is the channel response function, $n(t)$ is the noise at the receiver, and $\otimes$ indicates the convolution operation [15]. The specific calculation of $h(t)$ and $n(t)$ will be described in the next part of the paper.

2.1. Channel Model

In this part, we model the underwater channel characteristics of the UWOC system by adopting the Monto Carlo method [17–21] together with a Fournier–Forand (FF) phase function [21].

We use the Monte Carlo method to simulate the process of photon propagation in water to obtain the channel response function $h(t)$ mentioned above. The Monte Carlo method is proposed as a kind of numerical calculation method guided by probability theory, commonly solving many computational problems using random or pseudo-random numbers. It is of great significance for UWOC to study the multipath transmission of light and temporal characteristics of the underwater channel in seawater. The Monte Carlo method is widely used in the research of scattering characteristics of a seawater media [22–26]. The propagation of light in water is mainly affected by absorption and scattering [16–19]. The absorption can be interpreted as a process in which a photon collides with other particles, which reduces the energy of the photons especially in a turbid medium, leading to the attenuation of the amplitude at the receiver [23]. The scattering can be interpreted as an interaction of the photon with other suspended particles, which changes the direction of the optical beams, resulting in temporal dispersion and reduction in the available bandwidth [17].

Monte Carlo simulates the interaction of each photon with plankton and suspended particles in the water and records the state of each photon including the photon position in Cartesian coordinates (x, y, z), direction of transmission described by the zenith angle θ and azimuth angle φ, propagation time t, and weight w [18]. Each photon is initialized with weight (w = 1), location (0, 0, 0), and propagation time (t = 0). After several interactions, the photon is detected by the photoelectrical detector (PD) or absorbed completely by the suspended particles. Figure 2 shows the simulation flow chart for each photon motion in the Monte Carlo method [18].

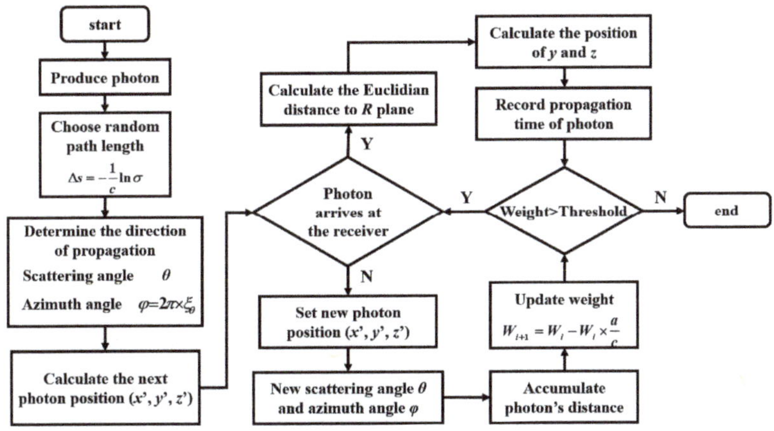

Figure 2. Flow chart for Monte Carlo simulation.

2.2. Volume Scattering Phase Function

The core of the Monte Carlo simulation of seawater scattering is the interaction between a single photon and scattering particles, with the photon scattering depending on the volume scattering phase function of the scattering medium. Therefore, the analysis and simulation of the scattering phase function of seawater media is a key issue to study the characteristics of seawater scattering.

At present, the scattering phase function (SPF) commonly used in research is the Henyey Greenstein function (HG function) derived by Henyey and Greenstein [27]. The HG function has the advantage of simulating the scattering of photons in water with a random nature of collisions. However, the HG function has an inevitable deficiency in that it cannot be well fitted for both forward and back scattering peaks [17].

Therefore, we adopted a Fournier–Forand (FF) phase function [21]. Compared with HG function, FF function can better fit the scattering situation of underwater particles, so the scattering Angle is determined based on FF phase function in the subsequent Monte Carlo simulation of seawater channel characteristics. The detailed contents about formulas and derivation please refer to [18].

2.3. MLSE Algorithm

The MLSE is an equalization algorithm and considers all possible convolutions of the transmitted sequence with the channel impulse response, and it finds the sequence with the minimum European distance to the received signal. At the receiver, the MLSE algorithm can mitigate the ISI introduced by channel temporal dispersion according to a sequence of bits [28,29].

We define the input sequence as [28,30]:

$$a(D) = a_0 + a_1 D + a_2 D^2 + \ldots, \tag{1}$$

where D represents the delay operator in a symbol period.

Similarly, the received sequence can be expressed as:

$$y(D) = y_0 + y_1 D + y_2 D^2 + \ldots, \tag{2}$$

The principle of MLSE is to choose the right $a(D)$ to maximize the probability $p[y(D) \mid a(D)]$. The symbol $s(D)$ is the state sequence and has a one-by-one correspondence with $a(D)$; therefore, it is equivalent to choosing the optimum state sequence $s(D)$ to maximize the probability $p[y(D) \mid s(D)]$.

Considering that the noise is independently distributed, we calculate the log-likelihood value $\ln p[y(D) \mid a(D)]$ by decomposing it into the accumulation of independent increments:

$$\ln p[y(D)|s(D)] = \sum_k \ln f[y_k - x(s_{k-1}, s_k)] \tag{3}$$

where $f(\cdot)$ represents the probability density function of the noise sequence and y_k is the received sequence and state at time k. $x(s_{k-1}, s_k)$ is the input sequence from state s_{k-1} to state s_k [31,32].

The symbol m indicates the level of the signal, and v indicates the tap number for MLSE. During the entire iteration, mv paths are obtained from the previous moment from time 0 to time k at each iteration and m^{v+1} paths are generated at the current time when all the state sequences are traversed. The most possible path is chosen according to the stored weight value and the Euclidean distance between $y_{(k+1)}$ and $a_{(k+1)}$ at a time of $k+1$. Then, the symbols before time $k+1$ can be determined for accumulation the next time by comparing the path storage weights in each state $s_{(k+1,i)}$. Finally, we obtain the optimum $s(D)$ to maximize the probability $p[y(D) \mid s(D)]$.

3. Experimental Setup

The schematic of the experimental setup is illustrated in Figure 3a. At the transmitter, binary electrical OOK signal is generated from an arbitrary waveform generator (Agilent M8190A, 12 Sa/s) using a pseudo-random bit sequence (PRBS) with length of 2^{15}. After a broadband power amplifier (SHF 100 BP, 3 dB bandwidth of 22 GHz, gain of 19 dB), the amplified electrical NRZ-OOK signal is filtered by a 1.5 GHz electrical low pass filter to exclude the excess noise. Through a bias-tee, the electrical NRZ-OOK signal, together with the direct bias current, is combined to drive the compact 405 nm blue laser diode (Thorlabs L405P20) for generating the free-space blue light signal. Using a micro lens with focal distance of 4.6 mm, the free-space optical signal is transmitted through a common 1-meter length glass tank filled with tap water. The experiment is carried out under common illumination condition, without special treatment for preventing the illumination light.

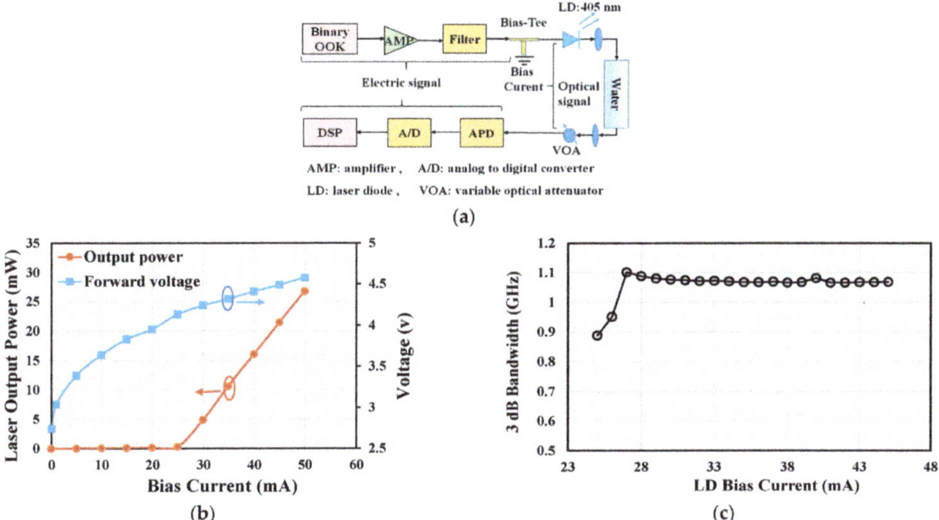

Figure 3. (**a**) Schematic diagram of the experimental setup. (**b**) Laser diode output power and voltage vs. bias current. (**c**) LD bias current vs. 3 dB bandwidth of the system.

At the receiver side, a neutral density filter (NDF) is used as variable optical attenuator to adjust the attenuation of the received optical power. A Si amplified fixed detector (Thorlabs APD210, 400–1000 nm, 3 dB bandwidth of 1 GHz) is used to detect the received optical signal. After optical-electrical conversion, the waveform of the received electrical signal is captured by a digital oscilloscope (Agilent DSO 9320A, 80 GSa/s) working with sampling frequency of 10 GSa/s, and sharing the same 10 MHz reference clock with the AWG to avoid the sampling frequency drift.

Before the signal transmission, we first characterize the specifications of the LD, and bandwidth of the system, as shown in Figure 3b,c. The relations between laser output power, forward voltage, and bias current for driving the LD is characterized to make sure the LD is driven in its linear region. As shown in Figure 3b, the threshold current of the 405 nm LD is 25 mA, and the output power reaches 27 mW at bias current of 50 mA. We, then, measure the overall bandwidth of the experimental system by using a vector network analyzer (Agilent 8722ES), as shown in Figure 3c. For measuring the system bandwidth, both the broadband power amplifier and the 1.5 GHz low pass filter are removed. As shown in Figure 3c, by adjusting the bias current of the LD from 25 to 45 mA, the maximum 3 dB bandwidth of the system is 1.1 GHz for bias current of 27 mA and 39 mA. For bias current within 27–45 mA, the 3-dB bandwidth of the overall system is about 1.08 GHz. In the experiment, the received data are processed offline by using either traditional symbol-by-symbol (SBS) detection or the MLSE detection, which is used for mitigation the ISI caused by the bandwidth limitations when transmitting high bitrate signal. The bit-error-ratio (BER) is measured from more than half million symbols.

4. Experimental Results and Discussion

To obtain the optimal system working conditions, we first investigate the influence of LD bias current and amplitude of the modulation bandwidth on the system performance. The influence of LD bias current on the BER for 2 Gb/s, 2.5 Gb/s, and 3 Gb/s OOK signals is shown in Figure 4, by using either SBS detection and 7-tap MLSE detection. During the measurement, the received optical power is fixed at 0.6 mW by adjusting the attenuation of the neutral density filter, and the amplitude of the electrical OOK signal sent from the AWG is fixed at Vpp = 0.4 V.

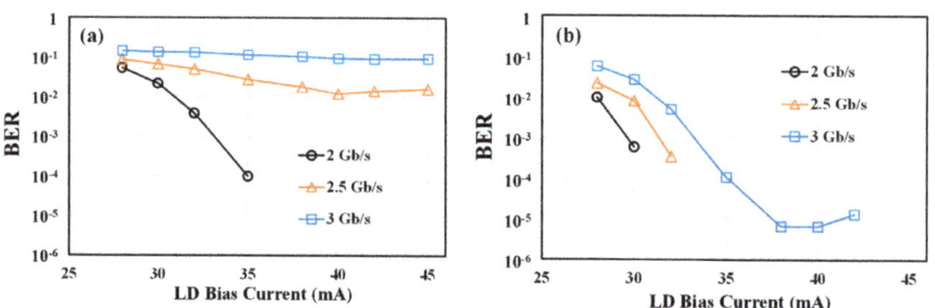

Figure 4. BER vs. LD bias current for various bitrate using (**a**) symbol by symbol detection and (**b**) 7-tap MLSE detection.

As shown in Figure 4a, by using SBS detection, the BER reduces from 2.1×10^{-2} to 1×10^{-4} when the LD bias current increase from 30 mA to 35 mA. For larger bias current, no bit error is detected for 2 Gb/s signal. However, for higher bitrate, the minimum BER increases to 1.2×10^{-2} and 9.4×10^{-2} for 2.5 Gb/s and 3 Gb/s signal, respectively, due to the large inter-symbol interference introduced by the limited system bandwidth. Figure 4b shows the BER performance by using the MLSE detection, which are used to combat with large ISI. As shown in Figure 4b, the BER reduce significantly by using the MLSE detection. For instance, the BER of 3 Gb/s OOK signal with MLSE detection reduces from 9.4×10^{-2} to 6.6×10^{-6}, compared to the traditional SBS detection. Moreover, for 2 Gb/s and 2.5 Gb/s OOK signal, no bit error is observed when the LD bias current is above 30 and 32 mA, respectively.

By keeping the laser bias current of 35 mA, we further investigate the influence of signal amplitude on the BER performance for 2.5 and 3 Gb/s OOK signal, still using either SBS detection or 7-tap MLSE detection, as shown in Figure 5. As shown in Figure 5a, the minimum BER of 2.5 and 3 Gb/s OOK transmission by using SBS detection is 1.2×10^{-2} and 9.2×10^{-2}, respectively, obtained at optimal RF voltage of 350 mV. By using MLSE detection, the optimal RF voltage is still about 350 mV for 3 Gb/s OOK signal, as shown in Figure 5b and the BER of 3 Gb/s OOK signal is reduced from 1×10^{-2} to 6.6×10^{-6} at RF voltage of 450 mV, compared to the traditional SBS detection. For 2.5 Gb/s signal, no bit error is observed when RF voltage is below 550 mV. From the results as shown in Figures 4 and 5, it can be observed either small LD bias current and large overdriving voltage can cause noticeable system degradation and, thus, the bias current and RF voltage need to be optimized for achieving appropriate system performance.

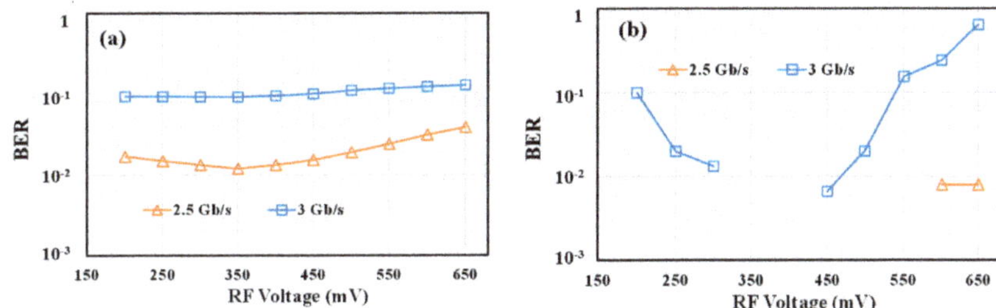

Figure 5. BER vs. amplitude voltage of RF OOK signal by using (**a**) symbol by symbol (SBS) detection and (**b**) MLSE detection, the LD bias current is 35 mA and received optical power is 0.6 mW.

By fixing the laser bias current of 35 mA and OOK amplitude voltage of 350 mV, we measure the BER performance for various received optical power by adjusting the neutral density filter, as shown in Figure 6. It is shown that with SBS detection, the received optical power only slightly influences the BER of OOK signal. Moreover, no BER difference is observed when the received optical power is larger than 0.6 mW (−2.2 dBm).

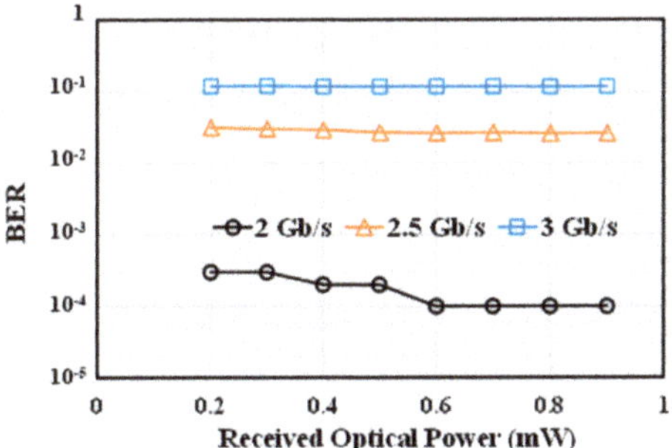

Figure 6. BER vs. received optical power by using SBS detection, the LD bias current is 35 mA and RF voltage is 350 mV.

After thoroughly system optimization, we fix the LD bias current of 40 mA, RF voltage of 350 mV and received optical power of 0.6 mW for further investigation. By changing the transmitted OOK bitrate from 2 Gb/s to 5 Gb/s, the BER performance for both SBS and MLSE detection are obtained, as shown in Figure 7. Assuming BER threshold of 3.8×10^{-3} for FEC with 7% overhead, it can be observed the maximum signal bitrate that can be transmitted by using SBS detection is roughly 2.4 Gb/s. As shown in Figure 7, by using the MLSE detection, the maximum tolerable OOK bitrate is 4 Gb/s, reaching spectral efficiency of almost 4 b/s/Hz by using OOK signaling.

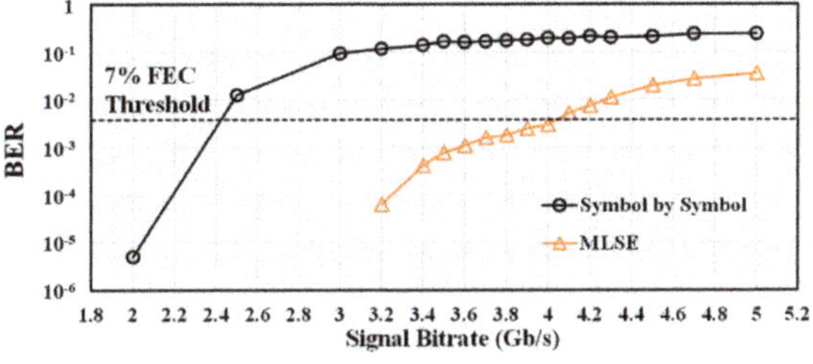

Figure 7. BER vs. signal bitrate using symbol by symbol detection and 7-tap MLSE detection.

The current state of the art bandwidth-limited OOK transmissions in high-speed fiber optical transmissions use MLSE algorithm with tap number of 7–10 [33]. Larger tap number would require unavoidable processing complexity and costs. Thus, it is necessary to find the minimum required tap number of MLSE algorithm for bandwidth limited underwater optical wireless OOK transmission with various bitrates. The relation between MLSE tap number and BER performances for 3, 3.5, and 4 Gb/s OOK signals are shown in Figure 8.

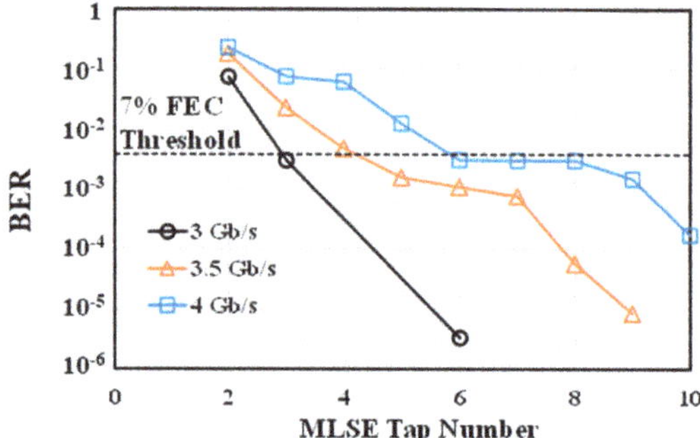

Figure 8. BER vs. signal bitrate using symbol by symbol detection and 7-tap MLSE detection.

With the increase in the bitrate, the required number of taps of MLSE detection for achieving the same BER performance also increases. For instance, as shown in Figure 8, the minimum required tap number of for MLSE is 3, 5, and 6 for signal bitrate of 3, 3.5, and 4 Gb/s, respectively. By further increasing the MLSE tap number, lower BER can still be obtained with the increase in processing complexity. By using 9 tap MLSE, the BER that can be obtained for 3.5 Gb/s and 4 Gb/s OOK is 8.5×10^{-6} and 1.5×10^{-3}, respectively. Thus, under the serious bandwidth limitation of the UWOC communication, the OOK signaling with MLSE detection can realize high transmission bitrate and spectral efficiency, with the advantage for its simplicity, and cost efficiency, exhibiting great potential for high-speed UWOC applications.

5. Simulation Results and Discussion

5.1. Channel Characteristics

Based on the Monte Carlo method together with the volume scattering phase function mentioned above, we simulate the beam propagation of various link ranges in harbor water and obtain the impulse response curve for the channel. In [32], we have obtained some simulation results of this work. The parameter settings for Monte Carlo simulation and the noise model are listed in (a) of the Table 1 [14], and (b) of the Table 1 [34], respectively.

To observe the impact of the transmission distance on the channel impulse response, we normalized the simulation results with link range distances of 8 m, 10 m, 12 m, 14 m, and 16 m, respectively, as shown in Figure 9a. After comparing the curves for the impulse response for different transmission distances in harbor water shown in Figure 9a, we believe that as the transmission distance increases, the temporal dispersion becomes increasingly large, which leads to the power of the pulse spreading into adjacent symbols. An obvious reason for this is that the longer transmission distance gives photons more of a chance to collide, thus leading to more photons being absorbed and scattered by the planktons and suspended particles. It is obvious that the temporal dispersion becomes very large when the transmission distance is 16 m, the full width at half maximum (FWHM) of the response exceeds 4ns, as shown in Figure 9a.

Table 1. (**a**) Parameter settings for Monte Carlo simulation [14]. (**b**) Parameter settings for the noise model [34].

(a)	
Coefficient	**Value**
absorption coefficient a	0.366 m^{-1}
scattering coefficient b	1.824 m^{-1}
refractive index n	1.33
PMT cumulative integration time t_d	0.1 ns
radius of receive aperture R	50 cm
asymmetry factor g	0.9
photon number simulation	10^8
various fields of view (FOV)	30°
(b)	
Coefficient	**Value**
electronic bandwidth B	20 GHz
quantum efficiency of the detector η	0.8
wavelength of the source λ	532 nm
the electron charge q	1.6×10^{-19} coulombs
dark current I_{dc}	1.226 nA
equivalent temperature T_e	290 K
noise figure F	4
load resistance R_L	100 Ω

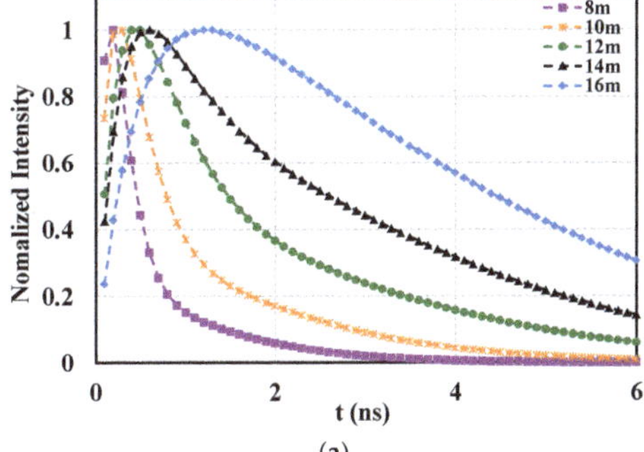

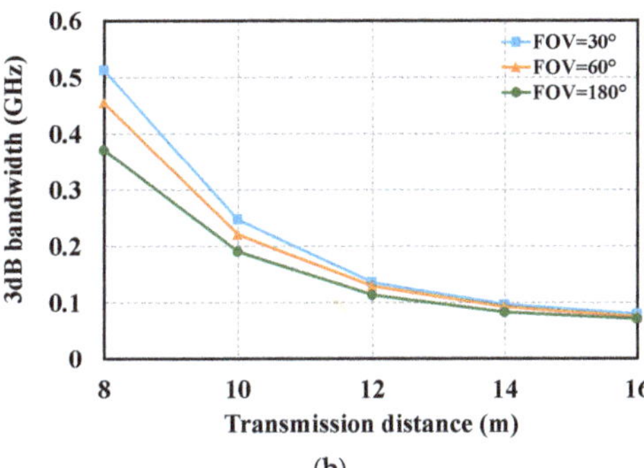

Figure 9. (**a**) The normalized channel impulse response with different link range. (**b**) 3 dB channel bandwidth versus transmission distance under FOV of 30°, 60°, and 180°.

To obtain the channel performance in harbor water in the frequency domain, we take a Fourier transform of the channel time domain response and calculate the 3-dB bandwidth for different transmission distances and different FOVs, as shown in Figure 9b. The results are obtained under the condition of harbor water, which has a rigorous limitation on channel bandwidth. According to Figure 9b, with the transmission distance increasing, the channel bandwidth decreases dramatically. When the FOV is 30°, the channel bandwidth is reduced from 0.5 to 0.1 GHz for distances from 8 to 14 m.

5.2. *Performance Evaluation*

In this section, we present the performance evaluation for a UWOC system based on the channel impulse response. First, we predicate a reasonable tap number for MLSE, and, then, we evaluate the effect of ISI caused by temporal dispersion, as well as BER performance. Finally, we discuss the power budget.

At the receiver, the MLSE algorithm is adopted to compensate for the ISI induced by the limited bandwidth underwater channel, resulting in improved system performance.

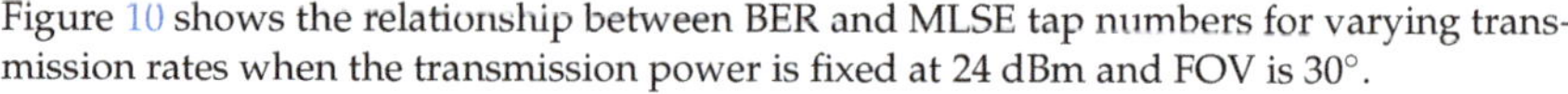

Figure 10 shows the relationship between BER and MLSE tap numbers for varying transmission rates when the transmission power is fixed at 24 dBm and FOV is 30°.

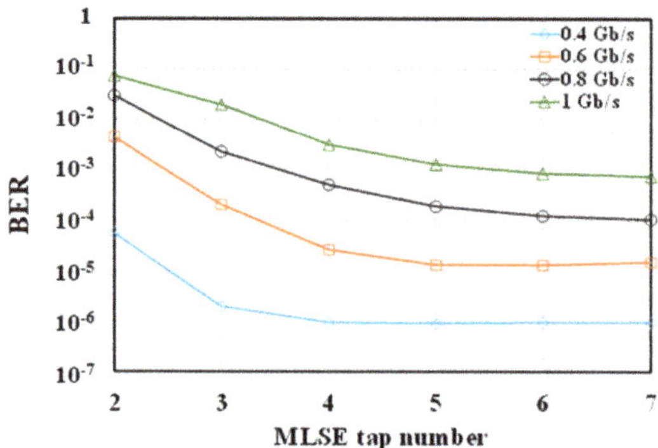

Figure 10. BER vs. MLSE tap number by transmitting 0.4, 0.6, 0.8, and 1 Gb/s Signal bitrate in harbor water of 10 m with transmit power of 24 dBm and 30° FOV.

As shown in Figure 10, the BER performance is considerably improved as the MLSE tap number increases. The BER of the 0.6 Gb/s OOK signal decreases from 4.5×10^{-3} to 1.34×10^{-5} for increasing MLSE tap number from 2 to 5. However, the system performance improvement is not obvious when the MLSE tap number increases from 5 to 7. Considering that an increase in the tap number will improve the system complexity, we adopt a 5-taps MLSE as a trade-off between compensation effect and complexity.

Due to the absorption and scattering of photons caused by particles in the water, the channel response shows temporal dispersion, i.e., temporal spread of the beam pulse, thus, leading to ISI. The OOK and pulse amplitude modulation (PAM) signals are investigated, respectively, to evaluate the effect of ISI. We evaluate the system performance by sending a pseudo random binary sequence (PRBS) with a length of 220 through the harbor water underwater channel. The BER performance without MLSE is also investigated and compared. Figure 11 shows the BER performance versus transmission power for various transmission distances and signal bitrates when the FOV is 30°. The transmission power is defined as the average transmission power of the pulse slots after modulation [14]. To investigate the system performance influenced by various modulation formats and MLSE, we conduct an UWOC system simulation changing from OOK modulation to PAM4 modulation. A simulation for OOK and PAM4 under different transmission distances is conducted to assess the underwater channel quality. At the receiver, for OOK transmission and PAM4 transmission, MLSE is adopted to equalize the temporal dispersion influence.

From Figure 11a, ISI leads to BER become lower than 7% FEC threshold for high data rates, but the system performance shows an obvious improvement after using MLSE. ISI mitigation with MLSE is evident in Figure 11b. From Figure 11b, when the transmission distance is 14 m, the BER performance of the UWOC cannot meet the threshold of 7% FEC without using MLSE. However, after using 5-tap MLSE, a transmission rate of 0.6 Gb/s can be achieved. Additionally, by adopting 5-tap MLSE, the maximum available bitrate with HD-FEC is improved from 0.4 to 0.8 Gb/s for 12 m transmission distance.

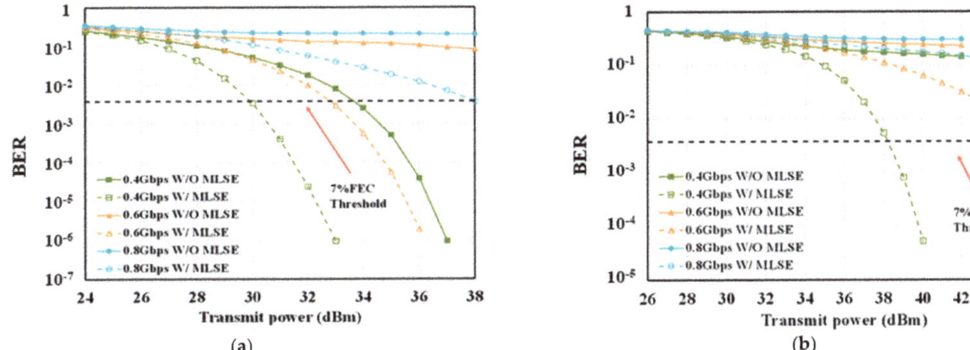

Figure 11. (**a**) 12 m transmission distance's BER performance versus transmit power with OOK modulation in harbor water with 30° FOV. (**b**) 14 m transmission distance's BER performance versus transmit power with OOK modulation in harbor water with 30° FOV.

As shown in Figure 12, PAM4 modulation could transmit at a high bitrate, but it only achieves a 0.8 Gb/s transmission for 8 m transmission distance without MLSE at 23 dBm transmit power. After adopting MLSE, the transmission capacity comes to 2 Gb/s for 10 m transmission distance at 25 dBm transmit power.

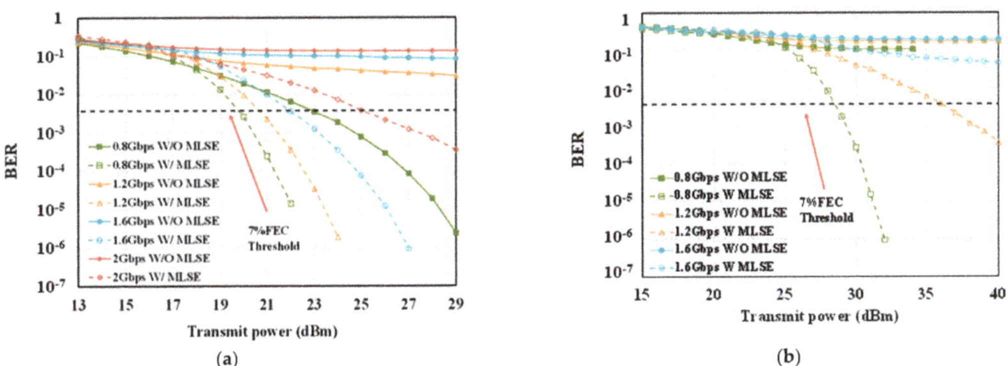

Figure 12. (**a**) 8 m transmission distance's BER performance versus transmit power with PAM4 modulation in harbor water with 30° FOV. (**b**) 10 m transmission distance's BER performance versus transmit power with PAM4 modulation in harbor water with 30° FOV.

To take a full consideration of the required transmission power of the laser in the experiments, we investigated the relationship between the required transmit power and the transmit distance, as well as the transmission rate using the OOK modulation. The power budget is defined as the minimum transmit power required that enables the BER performance to meet the threshold of 7% FEC. The dotted line shows the results for the power budget versus the transmission distance after adopting MLSE, and the solid line shows the results obtained without MLSE, as shown in Figure 13.

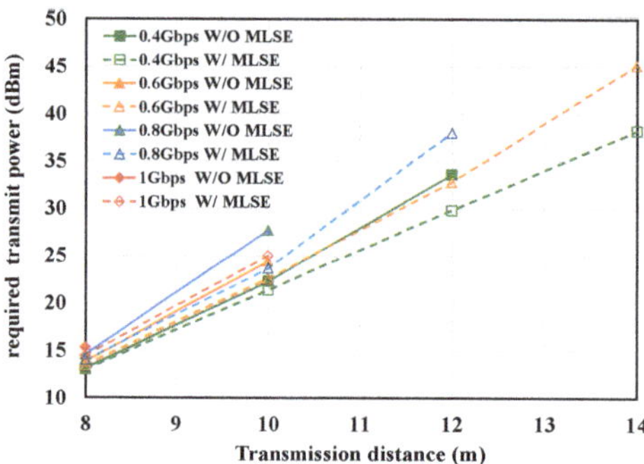

Figure 13. Required transmit power versus transmission distance with OOK modulation in harbor water with 30° FOV.

From Figure 13, it is observed that when the transmission distance is short, the MLSE can effectively reduce the power budget. For example, the power budget is reduced by 4 dB after using MLSE for a transmission rate and distance of 0.8 Gb/s and 10 m, respectively. When the signal could not be transmitted due to temporal dispersion, we are able to achieve a 14-m transmission by utilizing MLSE. Therefore, MLSE is a feasible algorithm for realizing transmission with a large bandwidth and to effectively reduce the power budget in turbid harbor water.

6. Conclusions

For UWOC systems, there still exist bandwidth limitations, with a lack of studies to investigate its influence on system's performance and effective compensation methods for its mitigation. In this paper, we investigate the bandwidth limitation for UWOC system caused by transceiver and underwater temporal dispersion, as well as how to mitigate it through experiments and simulations.

Experimental results show that, compared with conventional symbol-by-symbol detection, the maximum bitrate of the simple rectangular shape OOK signaling is raised from 2.4 Gb/s to 4 Gb/s by adopting the 7-tap MLSE detection. In the simulation, we use Monte Carlo method with Fournier–Forand phase function to simulate the temporal dispersion and bring in the bandwidth limitation in underwater channel. With MLSE adopted at the receiver, the maximum available bitrate is improved from 0.4 to 0.8 Gb/s in 12 m of harbor water. Moreover, the power budget can be reduced from 27.8 dBm to 23.6 dBm for 0.8 Gb/s 10 m OOK transmission with MLSE equalization. Both the experimental and simulation results show that MLSE method has the potential for enhancing the performance for UWOC systems faced with bandwidth limitation.

Author Contributions: J.Z. performed the experiment and realized simulation models about MLSE; G.G. performed the investigation and wrote the original draft. Z.M. performed analytical calculations and plotting. J.L. performed the Monte Carlo simulation. Y.G. discussed the experimental results and revised the manuscript. All authors have read and agreed to the published version of the manuscript.

Funding: This work was supported in part by the National Natural Science Foundation of China under Grant U1831110, in part by the Fundamental Research Funds for the Central Universities under Grant 2019XD-A15-2, and in part by the State Key Laboratory of Information Photonics and Optical Communications Funds under Grant IPOC2020ZZ02.

Data Availability Statement: Not applicable.

Conflicts of Interest: The authors declare no conflict of interest.

References

1. Zeng, Z.; Fu, S.; Zhang, H.; Dong, Y.; Cheng, J. A survey of underwater optical wireless communications. *IEEE Commun. Surv. Tutor.* **2016**, *19*, 204–238. [CrossRef]
2. Kaushal, H.; Kaddoum, G. Underwater optical wireless communication. *IEEE Access* **2016**, *4*, 1518–1547. [CrossRef]
3. Saeed, N.; Celik, A.; Al-Naffouri, T.Y.; Alouini, M.S. Underwater optical wireless communications, networking, and localization: A survey. *Ad Hoc Netw.* **2019**, *94*, 101935. [CrossRef]
4. Liu, X.; Yi, S.; Zhou, X.; Fang, Z.; Qiu, Z.J.; Hu, L.; Cong, C.; Zheng, L.; Liu, R.; Tian, P. 34.5 m underwater optical wireless communication with 2.70 Gbps data rate based on a green laser diode with NRZ-OOK modulation. *Opt. Express* **2017**, *25*, 27937–27947. [CrossRef] [PubMed]
5. Lu, C.; Wang, J.; Li, S.; Xu, Z. 60 m/2.5 Gbps Underwater Optical Wireless Communication with NRZ-OOK Modulation and Digital Nonlinear Equalization. In Proceedings of the 2019 Conference on Lasers and Electro-Optics (CLEO), San Jose, CA, USA, 5–10 May 2019; pp. 1–2. [CrossRef]
6. Li, C.-Y.; Lu, H.-H.; Tsai, W.-S.; Wang, Z.-H.; Hung, C.-W.; Su, C.-W.; Lu, Y.-F. A 5 m/25 Gbps Underwater Wireless Optical Communication System. *IEEE Photonics J.* **2018**, *10*, 1–9. [CrossRef]
7. Fei, C.; Zhang, J.; Zhang, G.; Wu, Y.; Hong, X.; He, S. Demonstration of 15-M 7.33-Gb/s 450-nm Underwater Wireless Optical Discrete Multitone Transmission Using Post Nonlinear Equalization. *J. Lightwave Technol.* **2018**, *36*, 728–734. [CrossRef]
8. Du, J.; Wang, Y.; Fei, C.; Chen, R.; Zhang, G.; Hong, X.; He, S. Experimental demonstration of 50-m/5-Gbps underwater optical wireless communication with low-complexity chaotic encryption. *Opt. Express* **2021**, *29*, 783–796. [CrossRef] [PubMed]
9. Wang, J.; Lu, C.; Li, S.; Xu, Z. 100 m/500 Mbps underwater optical wireless communication using an NRZ-OOK modulated 520 nm laser diode. *Opt. Express* **2019**, *27*, 12171–12181. [CrossRef] [PubMed]
10. Chen, X.; Yang, X.; Tong, Z.; Dai, Y.; Li, X.; Zhao, M.; Zhang, Z.; Zhao, J.; Xu, J. 150 m/500 Mbps Underwater Wireless Optical Communication Enabled by Sensitive Detection and the Combination of Receiver-Side Partial Response Shaping and TCM Technology. *J. Lightwave Technol.* **2021**, *39*, 4614–4621. [CrossRef]
11. Zhang, L.; Tang, X.; Sun, C.; Chen, Z.; Li, Z.; Wang, H.; Jiang, R.; Shi, W.; Zhang, A. Over 10 attenuation length gigabits per second underwater wireless optical communication using a silicon photomultiplier (SiPM) based receiver. *Opt. Express* **2020**, *28*, 24968–24980. [CrossRef] [PubMed]
12. Baghdady, J.; Miller, K.; Morgan, K.; Byrd, M.; Osler, S.; Ragusa, R.; Li, W.; Cochenour, B.M.; Johnson, E.G. Multi Gigabit/s underwater optical communication link using orbital angular momentum multiplexing. *Opt. Express* **2016**, *24*, 9794–9805. [CrossRef]
13. Ren, Y.; Li, L.; Wang, Z.; Kamali, S.M.; Arbabi, E.; Arbabi, A.; Zhao, Z.; Xie, G.; Cao, Y.; Ahmed, N.; et al. Orbital Angular Momentum-based Space Division Multiplexing for High capacity Underwater Optical Communications. *Sci. Rep.* **2016**, *6*, 33306. [CrossRef] [PubMed]
14. Tang, S.; Dong, Y.; Zhang, X. Impulse response modeling for underwater wireless optical communication links. *IEEE Trans. Commun.* **2014**, *62*, 226–234. [CrossRef]
15. Jaruwatanadilok, S. Underwater wireless optical communication channel modeling and performance evaluation using vector radiative transfer theory. *IEEE J. Sel. Areas Commun.* **2008**, *26*, 1620–1627. [CrossRef]
16. Gabriel, C.; Khalighi, M.; Bourennane, S.; Leon, P.; Rigaud, V. Channel modeling for underwater optical communication. In Proceedings of the IEEE GC Wkshps, Houston, TX, USA, 5–9 December 2011; pp. 833–837.
17. Gabriel, C.; Khalighi, M.; Bourennane, S.; Leon, P.; Rigaud, V. Monte-Carlo-based channel characterization for underwater optical communication systems. *IEEE/OSA J. Opt. Commun. Net.* **2013**, *5*, 1–12. [CrossRef]
18. Cox, W.C. Simulation, Modeling and Design of Underwater Optical Communication Systems. In *Dissertations & Theses—Gradworks*; North Carolina State University: Raleigh, NC, USA, 2012; Volume 34, pp. 930–942.
19. Li, J.; Ma, Y.; Zhou, Q.; Zhou, B.; Wang, H. Monte Carlo study on pulse response of underwater optical channel. *Opt. Eng.* **2012**, *51*, 066001. [CrossRef]
20. Dalgleish, F.; Caimi, F.; Vuorenkoski, A. Efficient laser pulse dispersion codes for turbid undersea imaging and communications applications. *Ocean Sens. Monit. II* **2010**, *7678*, 1–12. [CrossRef]
21. Mobley, C. *Light and Water: Radiative Transfer in Natural Waters*; Academic Press (Elsevier Science): Amsterdam, The Netherlands, 1994.
22. Lillycrop, W.J.; Parson, L.E.; Irish, J.L. Development and Operation of the SHOALS Airborne Lidar Hydrographic System. In *CIS Selected Papers: Laser Remote Sensing of Natural Waters: From Theory to Practice*; SPIE: Bellingham, DC, USA, 1996; Volume 2964, pp. 26–37. [CrossRef]
23. Cochenour, B.M.; Mullen, L.J. Free-space optical communications underwater. In *Advanced Optical Wireless Communon System*; Cambridge University Press: Cambridge, UK, 2012; pp. 201–239.
24. Hickman, G.D.; Hogg, J.E. Application of an airborne pulsed laser for near shore bathymetric measurements. *Remote Sens. Environ.* **1969**, *1*, 47–58. [CrossRef]
25. Lillycrop, W.J.; Parson, L.E.; Estep, L.L. Field testing of the U S Army Cops of engineers airborne lidar: Hydrography survey system. In Proceedings of the US Hydrographic Conference, Norfolk, VA, USA, 18–23 April 1994; pp. 144–151.
26. Wang, L.; Jacques, S.L.; Zheng, L. MCML, Monte Carlo modeling of light transport in multi-layered tissues. *Comput. Meth. Prog. Biomed.* **1995**, *47*, 131–146. [CrossRef]
27. Henyey, L.C.; Greenstein, J.L. Diffuse radiation in the Galaxy. *Astrophys. J.* **1941**, *93*, 70–83. [CrossRef]

28. Manor, H.; Arnon, S. Performance of an optical wireless communication system as a function of wavelength. *Appl. Opt.* **2003**, *42*, 4285–4294. [CrossRef] [PubMed]
29. Chen, S.; Xie, C.; Zhang, J. Comparison of advanced detection techniques for QPSK signals in super-Nyquist WDM systems. *IEEE Photonics Technol. Lett.* **2014**, *27*, 105–108. [CrossRef]
30. Gao, G.; Li, J.; Zhao, L.; Guo, Y.; Zhang, F. 1 Gb/s underwater optical wireless On-off-keying communication with 167 MHz receiver bandwidth. In Proceedings of the 2018 OCEANS—MTS/IEEE Kobe Techno-Oceans (OTO), Kobe, Japan, 28–31 May 2018; pp. 1–4.
31. Xu, C.; Gao, G.; Chen, S.; Zhang, J.; Luo, M.; Hu, R.; Yang, Q. Sub-symbol-rate sampling for PDM-QPSK signals in super-Nyquist WDM systems using quadrature poly-binary shaping. *Opt. Express* **2016**, *24*, 26678–26686. [CrossRef] [PubMed]
32. Li, J.; Gao, G.; Xu, C.; Bai, J.; Guo, Y. Influence of Temporal Dispersion on the Undersea Wireless Optical Communication and Its Mitigation using MLSE. In Proceedings of the OCEANS 2018 MTS/IEEE Charleston, Charleston, SC, USA, 22–25 October 2018; pp. 1–5. [CrossRef]
33. Karinou, F.; Stojanović, N.; Yu, Z.; Zhao, Y. Toward cost-efficient 100G metro networks using IM/DD 10 GHz components and MLSE receiver. *J. Lightwave Technol.* **2015**, *33*, 4109–4117. [CrossRef]
34. Hamza, T.; Khalighi, M.-A.; Bourennane, S.; Léon, P.; Opderbecke, J. Investigation of solar noise impact on the performance of underwater wireless optical communication links. *Opt. Express* **2016**, *24*, 25832–25845. [CrossRef] [PubMed]

MDPI

Article

Experimental Demonstration of High-Sensitivity Underwater Optical Wireless Communication Based on Photocounting Receiver

Chao Li [1,*], Zichen Liu [2], Daomin Chen [3], Xiong Deng [4], Fulong Yan [5], Siqi Li [1] and Zhijia Hu [1]

1 Information Materials and Intelligent Sensing Laboratory of Anhui Province, Anhui University, Hefei 230601, China; sqli@ahu.edu.cn (S.L.); zhijiahu@ahu.edu.cn (Z.H.)
2 Wuhan National Laboratory for Optoelectronics, Huazhong University of Science and Technology, Wuhan 430074, China; lzc8888@hust.edu.cn
3 Shanghai Huawei Technology Co., Ltd., Shanghai 200001, China; chendaomin@huawei.com
4 Center for Information Photonics and Communications, School of Information Science and Technology, Southwest Jiaotong University, Chengdu 611756, China; xiongdeng@swjtu.edu.cn
5 Alibaba Cloud, Alibaba Group, Beijing 100102, China; yanfulong.yfl@alibabainc.com
* Correspondence: chao.li@ahu.edu.cn

Abstract: In this paper, we propose a high-sensitivity long-reach underwater optical wireless communication (UOWC) system with an Mbps-scale data rate. Using a commercial blue light-emitting diode (LED) source, a photon counting receiver, and return-to-zero on–off keying modulation, a receiver sensitivity of −70 dBm at 7% FEC limit is successfully achieved for a 5 Mbps intensity modulation direct detection UOWC system over 10 m underwater channel. For 1 Mbps and 2 Mbps data rates, the receiver sensitivity is enhanced to −76 dBm and −74 dBm, respectively. We further investigate the system performance under different water conditions: first type of seawater ($c = 0.056\ \mathrm{m}^{-1}$), second type ($c = 0.151\ \mathrm{m}^{-1}$), and third type ($c = 0.398\ \mathrm{m}^{-1}$). The maximum distance of the 2 Mbps signal can be extended up to 100 m in the first type of seawater.

Keywords: underwater optical wireless communication (UOWC); long-reach; photon counting

Citation: Li, C.; Liu, Z.; Chen, D.; Deng, X.; Yan, F.; Li, S.; Hu, Z. Experimental Demonstration of High-Sensitivity Underwater Optical Wireless Communication Based on Photocounting Receiver. *Photonics* **2021**, *8*, 467. https://doi.org/10.3390/photonics8110467

Received: 8 September 2021
Accepted: 21 October 2021
Published: 22 October 2021

1. Introduction

With the expanding area explored by human beings, the observation and utilization of the underwater world is growing increasingly important. Various underwater sensors, unmanned vehicles, and nodes are deployed underwater to transfer and collect information. To build an underwater transmission link, both cable- and wireless-based methods are utilized. Cables or fibers can offer a stable communication link with high transmission speed but limit the freedom of the communication terminal for a long-reach link.

The traditional method for underwater communication is to use acoustics, which is a medium of sound. The attenuation of the sound wave in water is acceptable, which is competent for ultralong-reach communication up to tens of kilometers. However, underwater acoustic communication is limited by the huge transmitted power, low data rate, and large latency [1]. Due to the skin effect, electromagnetic waves suffer from huge attenuation when propagating in water. Thus, it is hard to realize long-reach underwater communication using electromagnetic waves. Studies have shown that the visible spectrum from blue–green wavelengths suffers less attenuation caused by underwater absorption and scattering than electromagnetic waves [2]. Benefiting from the rich bandwidth resource of a laser diode (LD), a Gbps-scale underwater optical wireless communication (UOWC) system within tens of meters is feasible [3]. However, a strict tracking and alignment system is required after long-distance transmission due to the narrow beam and small divergence angle of the LD source. Moreover, most of the reported UOWC links are conducted in tap water with avalanche photodetectors (APDs) for optical signal detection, which may not be so attractive and available for some long-distance transmission scenarios requiring a large optical power budget and photon-scale detection, e.g., internal communications with Mbps data rates between autonomous underwater vehicles or underwater sensor nodes in underwater dynamic conditions [4]. Thus, a high-sensitivity detector combined with a large-coverage-area light source is indispensable to build a reliable communication link with respect to unpredictable channel obstructions and the various conditions of the sea. Photomultiplier tubes (PMTs), possessing the capability of single-photon detection, are the most widespread vacuum electronic devices in every field of experimental studies including optical communication, biology, space research, and chemistry. Compared with silicon photomultipliers, PMT needs high voltage to drive the device. However, the PMTs are not sensitive to temperature and have a lower noise level. Due to the sensitivity of PMT to background noise and magnetic fields, it is more suitable to build a PMT-based long-range UOWC link for deep-sea implementation.

Before building a long-range experimental UOWC system, the underwater channel conditions need to be investigated to establish the system parameters such as the optimal transmitted optical wavelength, modulation scheme, signal baud rate, and beam aperture. Because underwater data transmission using a light beam is not an easy mission in the presence of high water absorption and scattering, characterizing the underwater optical channel property to achieve appropriate system parameters is of crucial importance to enable a high-reliability and high-quality UOWC link.

In this paper, we consider a comprehensive underwater channel model to simulate the property of an underwater optical communication link by taking the practical system parameters into account. Under the guidance of the simulation results, we propose and experimentally demonstrate a long-reach Mbps-scale UOWC scheme with high receiver sensitivity based on a light-emitting diode (LED) transmitter and a PMT receiver. The proposed system can significantly relax the alignment requirement especially after long-distance transmission. Bit-error-ratio (BER) performance enhancements for 1 Mbps, 2 Mbps, and 5 Mbps after 10 m transmission are experimentally investigated under different water turbidities with an adaptive decision threshold (DT). The receiver adapts to the changing of signal level. With added attenuation, the maximum link loss at an attenuation coefficient of 1.33 m^{-1} is up to 99 dB at λ = 448 nm. The achievable maximum distances for a 2 Mbps data rate in the first type of seawater (c = 0.056 m^{-1}) are up to 100 m and 134 m at 1 W and 10 W transmitted electrical power, respectively.

2. Operation Principle

Compared with free-space atmospheric laser communication, the UOWC system faces some unique challenges.

(i) Spectrum for communication: blue or green wavelengths should be dedicated to the UOWC link due to the water absorption effect, rather than infrared wavelengths (C-band 1530–1565 nm and L-band 1565–1625 nm) for an atmospheric free-space link enabled by well-established fiber-optic technologies and optoelectronic devices and components.

(ii) Channel condition: affected by seawater, the underwater optical transmission channel is quite complicated. When the modulated light propagates through seawater, it suffers from absorption and scattering. Seawater absorption means that part of the photon energy launched into the seawater is converted into other forms of energy, such as thermal and chemical. Scattering refers to the interaction between light and seawater, which changes the optical transmission path. Both absorption and scattering cause the loss of optical signal energy at the receiver, resulting in a reduction in signal-to-noise ratio and communication distance. As illustrated in [5], the link loss for a realistic 10 m green-light UOWC system can vary from 6.6 dB to 95.5 dB due to dynamic underwater channel conditions from a clean ocean to turbid harbor seawater.

Due to the variability of underwater channels, a robust long-reach UOWC link must be designed against a link loss of roughly up to 100 dB. Meanwhile, the link must be able to tolerate the dynamic changing underwater channels without breaking off. Although linear detectors including APDs have shown their abilities to detect multi-Gbps optical signals transmitted by LD sources, their sensitivities are typically limited by thermal noise [6]. On the other hand, photon-counting detectors can achieve very high sensitivities with moderate data rates on the Mbps scale. In this paper, we propose a reliable long-reach UOWC scheme using LED and PMT, whose concept is illustrated in Figure 1. Due to the advantages of its large light beam, compact structure, low cost, and low power consumption, LEDs are proposed as viable candidates to provide a transmission data rate of several Mbps or even up to hundreds of Mbps for implementing an alignment-released UOWC system. In the demonstration, a commercial LED transmitter is modulated by a predesigned return-to-zero on–off keying (RZ-OOK) with half power semi-angle of 1.25° [7]. With increasing transmission distance z, the receiving radius D_r at the detection area increases, which significantly relaxes the alignment requirement. To achieve high receiver sensitivity and long distance, a typical and practically implemented photocounting receiver is used, which is the PMT combined with a pulse-holding circuit to detect photo-level signals. The received photoelectric current is characterized by a series of discrete rectangular pulses with certain width, whose number satisfies a Poisson distribution. In the demonstration, we propose a digital adaptive DT algorithm for signal recovery. The value of DT is adjusted as a function of the received signal level to achieve the minimum BER value.

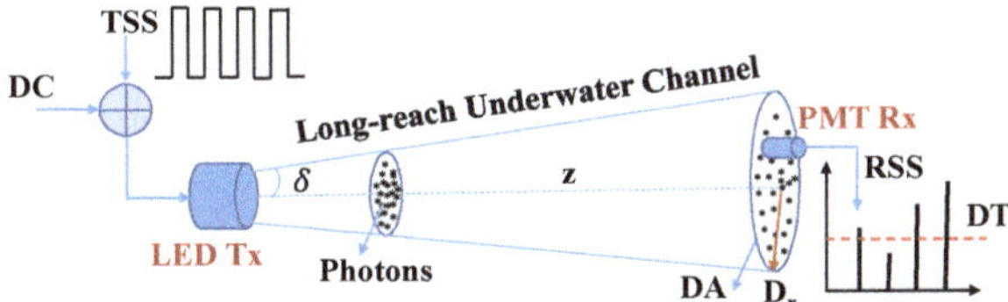

Figure 1. Proposed concept of long-reach UOWC using LED and PMT. DC: direct current, TSS: transmitted signal sequence, DA: detection area, RSS: received signal sequence, DT: decision threshold. δ is the divergence half-angle. z is the transmission distance. D_r is the radius at the detection area.

3. System Model

3.1. LED Transmitter

In our experiment, a commercial low-cost LED at a peak wavelength of 448 nm was employed as the transmitter. The path loss of light caused by water absorption and scattering can be dominated by the Beer–Lambert law.

$$P_r = \eta P_t L_{ch} = \eta P_t e^{-cz} = \eta P_t e^{-(a+b)z}, \tag{1}$$

where η is electrical-to-optical conversion efficiency of LED, a and b represent the coefficients of absorption and scattering, respectively, c is the total loss due to both effects, and z is the underwater transmission distance. L_{ch} is the channel loss, given by $\exp(-cz)$. P_t and P_r are the transmitted electrical power and received optical power (ROP), respectively.

The radiation pattern $I(\phi)$ of the LED obeys the Lambertian model, defined as

$$I(\phi) = \eta P_t \frac{(m_1+1)}{2\pi} \cos^{m_1}(\phi), \tag{2}$$

where ϕ is the angle of irradiance, and $\phi = 0$ is the maximum radiation power angle, i.e., the direct state. m_1 is expressed as the Lambertian emission order of the beam directivity, which is related to the half-power angle $\phi_{1/2}$ of the LED, written as

$$m_1 = \frac{-\ln 2}{\ln(\cos \phi_{1/2})}. \tag{3}$$

The detected optical power by the photon counting receiver at the receiving plane A_{eff} through the distance z is defined as follows [8]:

$$P_r = \frac{I(\phi) L_{ch} A_{eff}}{z^2}. \tag{4}$$

3.2. Underwater Channel

In an underwater environment, the transmitted light is greatly influenced by the optical properties of water. Underwater particles can cause energy attenuation and divergence of the beam. In this section, Kopelevich channel modeling is used as a volume scattering function (VSF) to investigate the extinction coefficient of natural water by simulation [9,10]. The specific form of this model is presented in [9].

Absorption coefficient a and scattering coefficient b denote the spectral absorption and scattering rate of unit interval, respectively. In this paper, we consider fulvic acid, humic acid and chlorophyll as the main absorption components of water [11,12], which can be expressed as follows:

$$a(\lambda) = a_w(\lambda) + a_c(\lambda) + a_f(\lambda) + a_h(\lambda), \tag{5}$$

$$a_c(\lambda) = a_c^0(\lambda)\left(C_c/C_c^0\right)^{0.602}, \tag{6}$$

$$a_f(\lambda) = a_f^0 C_f \exp(-k_f \lambda), \tag{7}$$

$$a_h(\lambda) = a_h^0 C_h \exp(-k_h \lambda), \tag{8}$$

where λ indicates the light wavelength, and $a_\omega(\lambda)$, $a_c(\lambda)$, $a_f(\lambda)$, and $a_h(\lambda)$ are the absorption coefficients caused by pure water, chlorophyll, fulvic acid, and humic acid, respectively. The variables of a_c^0, a_f^0, and a_h^0 represent the chlorophyll, fulvic acid, and humic acid characteristic absorption coefficients, respectively [13–15]. The two constant parameters k_f and k_h are 0.0189 m^{-1} and 0.0111 m^{-1}. C_c, C_f, and C_h indicate the concentrations of chlorophyll, fulvic acid, and humic acid in water (C_c^0 = 1 mg/m^3). The values of C_c are given in Table 1. C_f and C_h are expressed as follows:

$$C_f = 1.74098 C_c \exp(0.12327 C_c / C_c^0), \tag{9}$$

$$C_h = 0.19334 C_c \exp(0.12327 C_c / C_c^0). \tag{10}$$

Table 1. Chlorophyll concentration in four types of water (mg/m^3).

	Pure	Clean Ocean	Coastal	Turbid Harbor
C_c	0	0.31	0.83	5.99

We adopt a small and large particle scattering model to get the scattering coefficient of different types of water, which is a weighted summation with a pure water scattering coefficient [16].

$$b(\lambda) = b_w(\lambda) + b_s^0(\lambda) C_s + b_l^0(\lambda) C_l, \tag{11}$$

where $b_\omega(\lambda)$ indicates the scattering coefficient of pure water, $b_s^0(\lambda)$ and $b_l^0(\lambda)$ denote the scattering coefficients caused by small and large suspended particles, respectively [16,17], and C_s and C_l are the concentrations of both types of particles in water.

The extinction coefficient $c(\lambda)$ is the sum of the absorption coefficient and scattering coefficient. VSF is a very important parameter in underwater channel modeling. It indicates the ratio of scattered intensity (solid angle $\Delta\Omega$ centered on θ) to total incident light intensity at a specific scattering angle. θ indicates the scattering angle. In our model, we adopt the Kopelevich model as the VSF. Compared with the traditional Henyey–Greenstein model, the Kopelevich model not only covers small and large particles, but also can be more accurately applied to high turbid water [9].

VSF for underwater application can be expressed by the combination of pure water, small particles, and large particles [16].

$$p(\lambda,\theta) = b_w(\lambda)p_R(\theta) + b_s^0(\lambda)p_s(\theta)C_s + b_l^0(\lambda)p_l(\theta)C_l, \tag{12}$$

where $p_R(\theta)$, $p_s(\theta)$, and $p_l(\theta)$ indicate the probability density functions for pure water, small particles, and large particles, respectively.

For the Kopelevich model, the total seawater scattering coefficient can be modeled as follows [9]:

$$b(\lambda) = 0.0017 \times \left(\frac{550}{\lambda}\right)^{4.3} + 1.34C_s(\frac{550}{\lambda})^{1.7} + 0.312C_l(\frac{550}{\lambda})^{0.3}, \tag{13}$$

where C_s and C_l are the concentrations of small and large particles, respectively.

We set the weight (unit energy) for each photon, and the energy attenuation of the transmitted light beam is equivalent to the change in weight. We define four main parameters at the transmitter: the wavelength λ, the maximum half-divergence angle θ_{max}, the zenith angle θ, and the azimuth angle φ. Initially, each photo is launched into water with the given maximum half-divergence angle θ_{max} and unit weight. The initial departure direction of the photon is determined on the basis of random variables θ and φ. The direction is generated according to $[-\theta_{max}, \theta_{max}]$ for θ and $[0, 2\pi]$ for φ. The direction vector of emitted photons is $(\sin\theta\cos\varphi, \sin\theta\sin\varphi, \cos\theta)$. After traveling at a certain distance called the free path, emitted photons might lose their energy and change their transmission direction due to collision with particles in the underwater medium. Using a probability model, the free path can be expressed as follows [18]:

$$d = -\ln(\xi)/c, \tag{14}$$

where ξ is a random variable which obeys a uniform distribution within (0, 1].

Due to the collision with particles in the underwater medium, emitted photons lose their energy and change their transmission direction. It is assumed that the weights of emitted photons before and after collision are W_{pre} and W_{post}, which satisfy Equation (15) [18].

$$W_{post} = W_{pre}(1 - a/c). \tag{15}$$

Once scattering occurs, the transmission direction of emitted photons is changed. The new direction vector P_2 after collision is dependent on the old direction vector P_1, scattering angle θ, and azimuth angle φ, as shown in Figure 2. Random variable φ satisfies a uniform distribution within $[0, 2\pi]$.

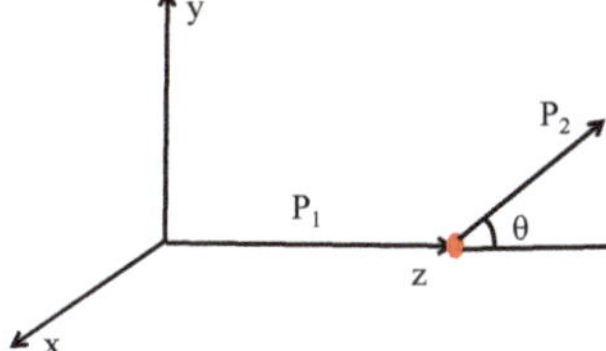

Figure 2. Scattering pattern of emitted photons.

For a single photon, VSF can be considered as the probability density function of the scattering angle. The generating methods of scattering angle for different VSFs are definitely different. As for the Kopelevich model, we use the acceptance–rejection sampling method to get the random scattering angle. According to the old transmission direction vector (ux^i, uy^i, uz^i), the scattering angle θ, and the azimuth angle φ, the transmission direction vector after scattering is represented by (ux^{i+1}, uy^{i+1}, uz^{i+1}) [19].

$$\begin{aligned} u_x^{i+1} &= -u_y^i \sin\theta\cos\phi + u_x^i(\cos\theta + \sin\theta\sin\phi) \\ u_y^{i+1} &= u_x^i \sin\theta\cos\phi + u_y^i(\cos\theta + \sin\theta\sin\phi) \\ u_z^{i+1} &= -({u_x^i}^2 + {u_y^i}^2)\sin\theta\sin\phi/u_z^i + u_z^i\cos\theta \end{aligned} \tag{16}$$

3.3. Photocounting Receiver

After several scattering events, the photons have a chance to be detected by the receiver. Since the solid angle $\Delta\Omega$ of the photon scattering space is small enough, it can be assumed that the VSF among $\Delta\Omega$ is constant. The variable $p(\theta)$ of the scattering direction satisfies

$$\int_0^{\pi} 1/2\pi \int_0^{2\pi} p(\theta)d\theta d\phi = 1. \tag{17}$$

By changing it into the integral of the solid angle, we get

$$\int \frac{p(\theta)}{2\pi\sin\theta}\sin\theta d\theta d\phi = \int \frac{p(\theta)}{2\pi\sin\theta}d\Omega = 1. \tag{18}$$

Thus, the reception probability of the emitted photon is

$$P = \frac{p(\theta)}{2\pi\sin\theta}\Delta\Omega. \tag{19}$$

Considering the conditional probability of free path, the final reception probability becomes

$$P = \frac{p(\theta)}{2\pi\sin\theta}\Delta\Omega \times \exp(-k_s|r_r - r_i|), \tag{20}$$

where r_r is the position of receive window, and r_i is the position where the final scattering before detection happens. In our model, the threshold setting of the photon weight is 10^{-4}, as shown in Table 2. Path loss and impulse response are crucial. We can calculate the path loss by summation of all products of reception probability and receiving photon weights. As for each scattering event, the position prior to scattering is available; thus, the entire path of the photon before detection is recorded. The channel response can be calculated so long as we count the receiving intensity in a given time slot. In summary, we can get the flow chart of the Monte Carlo model as shown in Figure 3. The channel responses of different wavelengths in four types of water are shown in Figure 4. It can be seen from Figure 4a–d that the optimum transmission wavelength is switched from 450 nm (blue) to 595 nm (red) when the water condition is changed from pure to turbid harbor. Moreover, a clear multipath channel characteristic is observed due to heavy scattering as illustrated Figure 4d, which is consistent with the results in [17]. The theoretical analysis and impulse response results under different water conditions guide the design of the experimental system. We can select the optimal wavelength according to the different water conditions to achieve the maximum data rate and the maximum transmission distance.

Table 2. Simulation parameters.

Symbol	Physical Meaning	Value
λ	Incident optical wavelength (unit: nm)	400, 450, 500, 550, and 595
$\theta_{0,max}$	Initial maximum half-divergence angle (random generation within [0, 2π])	8.2°
ξ	Statistical random variable for free path (random generation within (0, 1])	0.6
W	Decision weight at the receiver	>10^{-4}

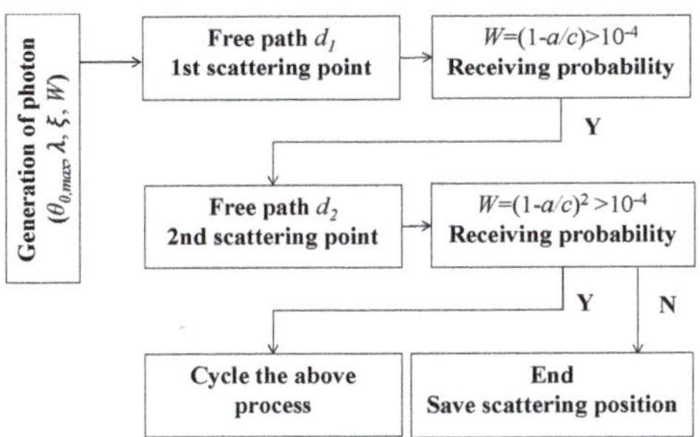

Figure 3. Flow chart of Monte Carlo model for photocounting receiver.

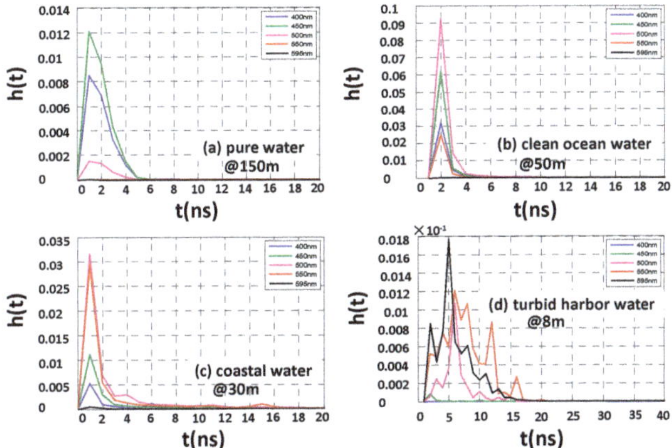

Figure 4. Channel response of different wavelengths in four types of water. The launched wavelengths were set to 400 nm, 450 nm, 500 nm, 550 nm, and 595 nm, respectively. (**a**) pure water (c = 0.056 m^{-1}), (**b**) clean ocean water (c = 0.151 m^{-1}); (**c**) coastal water (c = 0.398 m^{-1}); (**d**) turbid harbor water (c = 2.17 m^{-1}).

4. Experiment and Results

4.1. Experimental Setup and Paraeters

Figure 5 shows a schematic diagram of our experimental UOWC system using a blue LED source and PMT receiver (Hamamatsu, model CR315). An inclination angle of 5° is introduced to the transceiver, which causes huge attenuation to build a non-line-of-sight link. All the signal processing modules are implemented offline by MATLAB. At the transmitter, a pseudo-random bit sequence (PRBS) is generated and then sampled by an arbitrary signal generator (AWG) running at 10 MSa/s (1 Mbps), 20 MSa/s (2 Mbps), 50 MSa/s (5 Mbps), and 100 MSa/s (10 Mbps). Then, the baseband signals combined with a DC bias are injected into the LED. Compared with LD, the LED-based transmitter has no need of strict alignment or high emission power. A real-time oscilloscope is used to convert the analog signal into the digital domain. Simple digital signal processing (DSP) algorithms are applied at the receiving end, such as synchronization, decision, and BER calculation. The data length of each frame is 1151 bits, of which 127 bits are used for synchronization. We use multiple frames of information to increase the number of calculated bits. The number of effective bits used to calculate the BER was 20,718. To avoid synchronization problems, we increased the number of synchronization header bits. Unlike the conventional waveform sampling amplitude demodulation method, the photon-counting pulse signals need to be judged. When the amplitude of the sampled pulse is above the decision threshold voltage (DTV) V_D, one photon is counted. Final decisions on symbol "1" or "0" are made by the counted average values in each symbol. Thus, the BER value can be calculated according to the hard threshold n_{th}. Some key parameters of the proposed UOWC system are summarized and listed in Table 3.

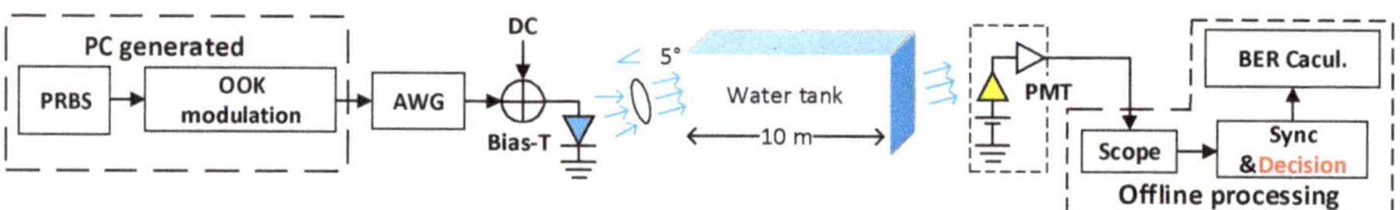

Figure 5. Experimental setup of LED–PMT UOWC system with 5° misalignment between transmitter and receiver. PRBS: pseudo-random bit sequence, AWG: arbitrary signal generator.

Table 3. Key parameters of the proposed UOWC system.

Symbol	Physical Meaning	Value/Unit
m_1	Lambertian order	2.9×10^3
ϕ	Angle of irradiance	5°
$\phi_{1/2}$	Half-power semi-angle of LED	1.25°
z	Transmission distance	10 m
η	E/O conversion efficiency	0.1289
P_t	Transmitted electrical power	1 W
R_b	Transmitted data rate	1/2/5 Mbps

4.2. Attenuation Coefficient Measurement

Water quality significantly impacts the BER performance. The PMT receiver is more sensitive to optical power than other light-sensitive devices such as an APD. Ambient light may annihilate signals. Thus, the experimental system should be thoroughly shaded with black nonreflective material. Our experimental channel was a 10 m long water tank with a volume of 3 m^3. Light absorption and scattering in seawater are caused by inorganic salts and planktonic plants. Some previous studies have shown that a similar effect of aluminum hydroxide or magnesium hydroxide to seawater is observed on the light of particles [20]. In the experiment, we added different concentrations of aluminum hydroxide to pure water to simulate seawater with different degrees of turbidity, i.e., pure seawater, clean seawater, coastal seawater, and harbor seawater, characterized by the parameters of attenuation coefficients.

$$c = \ln \frac{P_c}{P_0} . \frac{1}{z} + c_0. \tag{21}$$

In the experiment, we could not directly measure the relationship between the attenuation coefficient and the aluminum hydroxide concentration due to the presence of an off-angle at the transmitter. A preliminary experiment was carried out using an LD with very narrow divergence angle and a high-sensitivity optical power meter. Because of the reflection and absorption caused by the glass wall, we used Equation (21) to measure the relative attenuation coefficient. The results are shown in Figure 6a. We can see an approximate linear relationship between the aluminum hydroxide concentration and the attenuation coefficient. The parameter c is the measured attenuation coefficient, and c_0 is the attenuation coefficient of pure seawater with a value of 0.056 m^{-1}. The shaded tank was filled with pure water. Then, we added aluminum hydroxide powder to the water at a mass of 3 g each time and measured the ROP as P_c. Figure 6b shows the measured curve of the ROP as a function of the attenuation coefficient varying from 0.2 m^{-1} to 1.3 m^{-1} for different data rates. It can be seen from Figure 3b that the ROP was about −78 dBm for a 2 Mbps data rate at c = 1.3 m^{-1}, which means that a total loss of 99 dB was introduced (launched optical power was 21 dBm). The values of ROP were calculated using the average number of experimentally counted photons according to Equation (22).

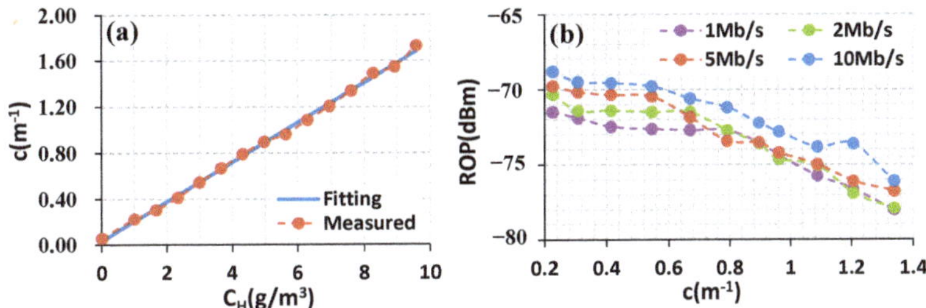

Figure 6. (**a**) Attenuation coefficient as a function of aluminum hydroxide concentration and (**b**) received optical power (ROP) under different water turbidities after 10 m underwater channel.

4.3. Measured BER Performance

In our experiment, we used a Hamamatsu PMT with a spectral response range from 300 nm to 650 nm as the receiver. The quantum efficiency of the PMT was 5%, and the typical dark count was 20 counts/sec. The number of photons counted in symbol "1" was contributed by the signal and the background light, while the photons counted in symbol "0" were caused by the background light and inter-symbol interference. An RZ code with a duty cycle of 0.7 was designed according to Equation (22), since the ROP can be maximized and a clock frequency component is included [21], where ξ is the quantum efficiency of PMT, h is Planck's constant, ν is the frequency of light, T_b is the symbol duration, and n_1 and n_0 are the average numbers of photons contained in symbols "1" and "0".

$$P_{r,PMT} = \frac{1}{\xi} \cdot \frac{7}{20} \cdot \frac{h\nu(n_1 - n_0)}{T_b}. \tag{22}$$

According to the measured results shown in Figure 7, when the number of received photons was less than 20, the measured data followed a relatively strict Poisson distribution since the PMT worked in the linear region. Upon increasing the number of photos to 40, the PMT was subjected to overexposure and worked in the nonlinear region, thus experiencing signal distortion [21]. In this condition, the distribution of the counted photons does not obey a strict Poisson distribution, as shown in Figure 7. The BER value can be calculated using Equation (23), where n_{th} is the hard-decision threshold [5].

$$\mathrm{BER} = \frac{1}{2}\sum_{k=0}^{n_{th}-1} e^{-n_1} \frac{n_1{}^k}{k!} + \frac{1}{2}\sum_{k=n_{th}}^{\infty} e^{-n_0} \frac{n_0{}^k}{k!}, n_{th} = \left[\frac{n_1 - n_0}{\ln n_1 - \ln n_0}\right]. \tag{23}$$

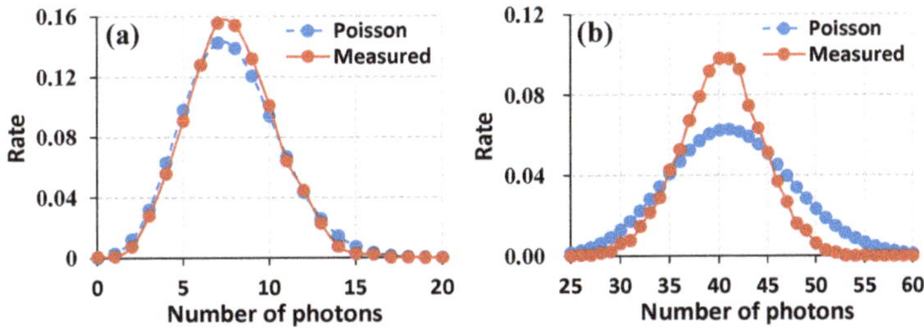

Figure 7. Distribution of the received photons: (**a**) seven photons; (**b**) 40 photons.

We present the measured BER performance under different water conditions in Figure 8. As discussed before, when the number of received photons is increased to around 20 (~−73 dBm), the number of the received photons no longer obeys a Poisson distribution. At this moment, the values of V_D should also be adjusted. In our experiment, the optimal values of V_D were obtained according to the rule of minimizing the BER. As illustrated in Figure 6b, an ROP of −73 dBm corresponded to a 10 m underwater transmission with an attenuation coefficient of 0.8 m^{-1}. When the PMT worked in photon-counting mode ($c > 0.8$ m^{-1}), the number of photons in symbol "1" obeyed a strict Poisson distribution. Thus, the value of DTV V_D was set to 2.5 mV. However, the measured BER performance worsened, especially for 1 Mbps and 2 Mbps data rates, when the attenuation coefficients varied from 0.2 m^{-1} to 0.8 m^{-1} (saturation region of PMT). With the adapted optimal value of V_D = 4.5 mV, error-free transmissions of 1 Mbps and 2 Mbps data rates were successfully achieved. The BER performance enhancement at the 5 Mbps data rate was not significant, because, when increasing the signal baud rate, severe inter-symbol interference was introduced due to the limited bandwidth of LED. Moreover, conclusions can be made according to Figure 8 that the receiver sensitivities of our proposed LED–UOWC systems at 1 Mbps, 2 Mbps, and 5 Mbps data rates were −76 dBm (1.08 m^{-1}), −74 dBm (0.92 m^{-1}), and −70 dBm (0.24 m^{-1}) at the 7% FEC limit of 3.8×10^{-3}, respectively.

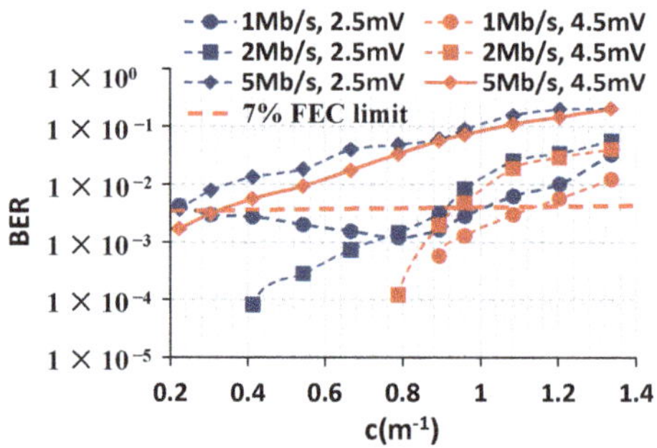

Figure 8. Experimental BER performance under different water turbidities after 10 m.

4.4. The Predicted Performance Based on the Proposed System

As illustrated in Figure 9, we further investigated the proposed system performance under conditions of the first type of seawater (pure, c = 0.056 m^{-1}), the second type (clean, c = 0.151 m^{-1}), and the third type (coastal, c = 0.398 m^{-1}). According to the experimental results illustrated in Figure 4, the required ROP for 2 Mbps at the 7% FEC limit is −74 dBm. Using Equation (4) and the parameters in Table 1, the optical power distribution at the receiving plane within the receiver sensitivity of −74 dBm was established using Lambertian model. Within the receiving radii of 1.28 m, 0.62 m, and 0.29 m, the achievable distances were 83.5 m, 40.5 m, and 19.2 m for the first, second, and third types of seawater, respectively. The maximum transmission distances could be extended to 100 m, 46 m, and 21 m when the receiver was located in the center of the receiving plane, as depicted in Figure 10. With a transmitted electrical power of 10 W, the maximum distances were further increased to 134 m, 60 m, and 27 m. When c exceeded the value of 0.92 m^{-1} (ROP = −74 dBm), as shown in Figure 8, the BER performance for the 2 Mbps signal became worse than the 7% FEC limit. The calculated optical power based on Equation (22) was −74.37 dBm in this condition, which is consistent with the optical power distribution obtained by the Lambertian model, as shown in Figure 9. The experimental 2 Mbps data rate after 10 m could achieve a receiving area of $\pi \times 0.15^2 = 0.07$ m^2. Thus, it is believed that our proposed long-reach UOWC system is capable of achieving an Mbps-scale data rate with an alignment-released configuration.

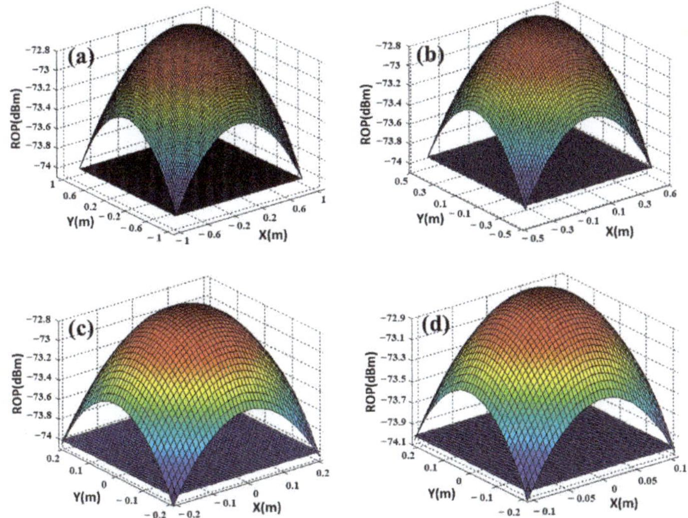

Figure 9. Optical power distribution at the receiving plane within the receiver sensitivity of −74 dBm: (**a**) $c = 0.056\ \text{m}^{-1}$, $z = 83.5$ m, $D_r = 1.28$ m; (**b**) $c = 0.151\ \text{m}^{-1}$, $z = 40.5$ m, $D_r = 0.62$ m; (**c**) $c = 0.398\ \text{m}^{-1}$, $z = 19.2$ m, $D_r = 0.29$ m; (**d**) $c = 0.92\ \text{m}^{-1}$, $z = 10$ m,. $D_r = 0.15$ m.

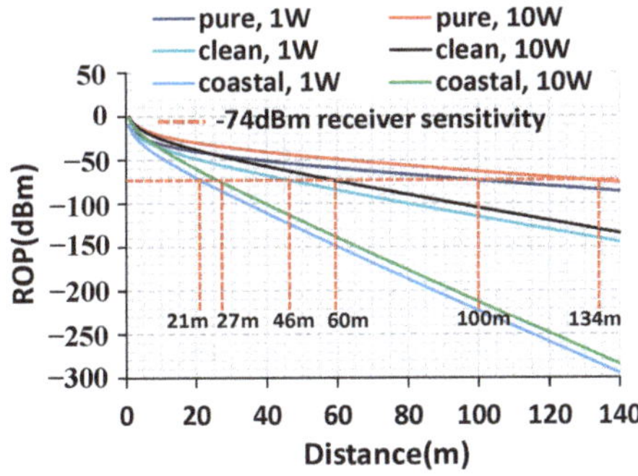

Figure 10. The predicted maximum distances using the proposed system.

5. Discussion

To build a long-range UOWC link or to propagate light through relative turbid water, two factors need to be considered: (i) pointing and alignment, and (ii) multipath interference.

5.1. Pointing and Alignment

To maintain a reliable line-of-sight UOWC link using an LD source after long-distance transmission is very difficult, since the optical beam is quite narrow. At this moment, pointing errors usually occur because of the link misalignment. Using a beam spread function, the link misalignment model for a UOWC system can be expressed as follows [3]:

$$\text{BSF}(L,r) = \text{E}(L,r)e^{-cL} + \int_{0}^{\infty} \text{E}(L,\vartheta)e^{-cL} \times \left\{ \exp\left[\int_{0}^{L} b\widetilde{\beta}(\vartheta(L-z))dz \right] - 1 \right\} J_0(\vartheta r)\vartheta d\vartheta, \quad (24)$$

where $\mathrm{BSF}(L, r)$ is the irradiance distribution at the receiver plane. Employing a LED source with a large beam size corresponds to a large receiving range. Thus, we can get the irradiance distribution more accessibly at the receiver plane.

5.2. Multipath Interference

As illustrated in Figure 4d, a multipath interference effect is produced in an optical turbid harbor underwater channel after 8 m transmission. For a certain data rate, the effect of multipath interference eventually leads to time spreading and waveform distortion, thus decreasing the BER performance due to the inter-symbol interference. Thus, when designing a UOWC system, this issue should be taken into consideration. Fortunately, technologies such as channel equalization [22], adaptive optics, and spatial diversity [23] are capable of suppressing the interference.

6. Conclusions

In this paper, we demonstrated a high-sensitivity long-reach UOWC system using LED and PMT. An experiment was conducted to investigate the BER performance under different water turbidities. Several key factors were taken into consideration during the system design, such as symbol rates, symbol duty cycles, water conditions, PMT characteristics, and decision criteria. With the help of RZ-OOK modulation and a PMT receiver, we experimentally achieved receiver sensitivities of −76 dBm, −74 dBm, and −70 dBm for 1 Mbps, 2 Mbps, and 5 Mbps data rates over a 10 m underwater channel, respectively. More than 100 m distance is achievable for a 2 Mbps data rate in pure seawater at 1 W transmitted power.

Author Contributions: C.L. and Z.L. contributed equally; C.L. and Z.L. performed the investigation and experiment; C.L. performed the analytical calculations and wrote the original draft; D.C. conducted the underwater channel simulation; X.D., F.Y., S.L., and Z.H. discussed the experimental results and revised the manuscript. All authors read and agreed to the published version of the manuscript.

Funding: This work was supported by the National Natural Science Foundation of China (12174002, 11874012), the Anhui Provincial Natural Science Foundation of China (1808085MF186), the China Postdoctoral Science Foundation (2021M690179), the Beijing Postdoctoral research Foundation (2021-ZZ-093), the Innovation project for the Returned Overseas Scholars of Anhui Province (2021LCX011), the Key Research and Development Plan of Anhui Province (202104a05020059), the University Synergy Innovation Program of Anhui Province (GXXT-2020-052), and the Project of State Key Laboratory of Environment-Friendly Energy Materials, Southwest University of Science and Technology (19FKSY0111).

Institutional Review Board Statement: Not applicable.

Informed Consent Statement: Not applicable.

Data Availability Statement: Not applicable.

Conflicts of Interest: The authors declare no conflict of interest.

References

1. Stojanovic, M. Recent advances in high-speed underwater acoustic communications. *IEEE J. Ocean. Eng.* **1996**, *21*, 125–136. [CrossRef]
2. Duntley, S. Light in the sea. *OSA J. Opt. Soc. Am.* **1963**, *53*, 214–233. [CrossRef]
3. Kaushal, H.; Kaddoum, G. Underwater optical wireless communication. *IEEE Access* **2016**, *4*, 1518–1547. [CrossRef]
4. Shen, J.; Wang, J.; Chen, X.; Zhang, C.; Kong, M.; Tong, Z.; Xu, J. Towards power-efficient long-reach underwater wireless optical communication using a multi-pixel photon counter. *OSA Opt. Exp.* **2018**, *26*, 23565–23571. [CrossRef] [PubMed]
5. Rao, H.; Fletcher, A.; Hamilton, S.; Hardy, N.; Moores, J.; Yarnall, T. A burst-mode photon counting receiver with automatic channel estimation and bit rate detection. In Proceedings of the SPIE Conference LASE, San Francisco, CA, USA, 13 April 2016; p. 97390H.
6. Xu, F.; Khalighi, M.; Bourennane, S. Impact of different noise sources on the performance of PIN-and APD-based FSO receivers. In Proceedings of the 11th International Conference on Telecommunications, Graz, Austria, 15–17 June 2011; pp. 211–218.

7. Wang, P.; Li, C.; Xu, Z. A cost-efficient real-time 25 Mb/s system for LED-UOWC: Design, channel coding, FPGA implementation, and characterization. *IEEE/OSA J. Lightw. Technol.* **2018**, *36*, 2627–2637. [CrossRef]
8. Kahn, J.-M.; Barry, J.-R. Wireless infrared communication. *Proc. IEEE* **1997**, *85*, 265–298. [CrossRef]
9. Kopelevich, V. Small-parametric model of the optical properties of seawater. In *Ocean Optics, I: Physical Ocean Optics*; Oxford University Press: Moscow, Russia, 1983; pp. 208–234.
10. Haltrin, V.; Khalturin, V. Propagation of light in sea depth. In *Optical Remote Sensing of the Sea and the Influence of the Atmosphere*; Academy of Sciences of GDR: Berlin, Germany, 1985; Chapter 2, pp. 20–62.
11. Haltrin, V.-I.; Kattawar, G. Self-consistent solutions to the equation of transfer with elastic and inelastic scattering in oceanic optics: I. Model. *OSA Appl. Opt.* **1993**, *32*, 5356–5367. [CrossRef] [PubMed]
12. Haltrin, V.-I. Chlorophyll-based model of seawater optical properties. *OSA Appl. Opt.* **1999**, *38*, 6826–6832. [CrossRef] [PubMed]
13. Prieur, L.; Sathyendranath, S. An optical classification of coastal and oceanic waters based on the specific spectral absorption curves of phytoplankton pigments, dissolved organic matter, and other particulate materials. *Limnol. Oceanogr.* **1981**, *26*, 671–689. [CrossRef]
14. Carder, K.-L.; Stewart, R.-G.; Harvey, G.-R.; Ortner, P.-B. Marine humic and fulvic acids: Their effects on remote sensing of ocean chlorophyll. *Limnol. Oceanogr.* **1989**, *34*, 68–81. [CrossRef]
15. Hawes, S.-K.; Carder, K.-L.; Harvey, G.-R. Quantum fluorescence efficiencies of fulvic and humic acids: Effect on ocean color and fluorometric detection. In Proceedings of the SPIE in Ocean Optics XI, San Diego, CA, USA, 31 December 1992; Volume 1750, pp. 212–223.
16. Haltrin, V.-I. One-parameter model of seawater optical properties. In Proceedings of the Ocean Optics XIV CD-ROM, Kailua-Kona, HI, USA, 10–13 November 1998; pp. 1–7.
17. Gabriel, C.; Khalighi, M.-A.; Bourennane, S.; Léon, P.; Rigaud, V. Monte-Carlo-based channel characterization for underwater optical communication systems. *J. Opt. Commun. Netw.* **2013**, *5*, 1–12. [CrossRef]
18. Ishimaru, A. *Wave Propagation and Scattering in Random Media*; IEEE Press: Piscataway, NJ, USA, 1997.
19. Bohren, C.-F.; Huffman, D.-R. *Absorption and Scattering of Light by Small Particles*; Wiley: Hoboken, NJ, USA, 1988.
20. Pontbriand, C.; Farr, N.; Ware, J.; Preisig, J.; Popenoe, H. Diffuse high-bandwidth optical communications. In Proceedings of the IEEE OCEANS, Quebec City, QC, Canada, 15–18 September 2008; pp. 15–18.
21. Ning, J.; Gao, G.; Zhang, J.; Peng, H.; Guo, Y. Adaptive receiver control for reliable high-speed underwater wireless optical communication with photomultiplier tube receiver. *IEEE Photonics J.* **2021**, *13*, 1–7. [CrossRef]
22. Jaruwatanadilok, S. Underwater wireless optical communication channel modeling and performance evaluation using vector radiative transfer theory. *IEEE J. Sel. Areas Commun.* **2008**, *26*, 1620–1627. [CrossRef]
23. Jamali, M.-V.; Salehi, J.-A.; Akhoundi, F. Performance studies of underwater wireless optical communication systems with spatial diversity: MIMO scheme. *IEEE Trans. Commun.* **2016**, *65*, 1176–1192. [CrossRef]

 *photonics* 

Communication

Dual-Branch Pre-Distorted Enhanced ADO-OFDM for Full-Duplex Underwater Optical Wireless Communication System

Zixian Wei [1], Yibin Li [1], Zhaoming Wang [1], Junbin Fang [2] and Hongyan Fu [1,*]

[1] Tsinghua Shenzhen International Graduate School and Tsinghua-Berkeley Shenzhen Institute (TBSI), Tsinghua University, Shenzhen 518055, China; weizx17@tsinghua.org.cn (Z.W.); liyibin21@mails.tsinghua.edu.cn (Y.L.); wangzm19@mails.tsinghua.edu.cn (Z.W.)
[2] Department of Optoelectronic Engineering, Jinan University, Guangzhou 510632, China; tjunbinfang@jnu.edu.cn
* Correspondence: hyfu@sz.tsinghua.edu.cn

Abstract: In this paper, dual-branch pre-distorted enhanced asymmetrically clipped direct current (DC) biased optical orthogonal frequency division multiplexing (PEADO-OFDM) for underwater optical wireless communication (UOWC) is firstly proposed and simulated. The performances of PEADO-OFDM on the underwater optical channel model (UOCM) are analyzed and further compared with the typical ADO-OFDM. Using the Monte Carlo method for the modeling of UOCM, we adopt a double-gamma function to represent three different water qualities including clear, coastal and harbor waters. The full-duplex architecture enables the removal of Hermitian symmetry (HS) from conventional optical OFDM and can increase the spectral efficiency at the cost of hardware complexity. A new PEADO-OFDM transmitter is also proposed to reduce the complexity of the transmitter. The simulation results exhibit that our proposed dual-branch PEADO-OFDM scheme outperforms the typical ADO-OFDM scheme in spectral efficiency, bit error rate (BER) and stability over the underwater channels of three different water qualities.

Keywords: pre-distorted enhanced; underwater optical wireless communication (UOWC); ADO-OFDM; gamma–gamma function; full-duplex

Citation: Wei, Z.; Li, Y.; Wang, Z.; Fang, J.; Fu, H. Dual-Branch Pre-Distorted Enhanced ADO-OFDM for Full-Duplex Underwater Optical Wireless Communication System. *Photonics* **2021**, *8*, 368. https://doi.org/10.3390/photonics8090368

Received: 17 July 2021
Accepted: 27 August 2021
Published: 1 September 2021

Publisher's Note: MDPI stays neutral with regard to jurisdictional claims in published maps and institutional affiliations.

1. Introduction

High-speed underwater communication plays an increasingly important role in oceanographic research, information transfer and marine development [1]. With the application of autonomous underwater vehicles (AUVs), underwater wireless sensor networks (UWSNs) and remotely operated vehicles (ROVs), underwater optical wireless communication (UOWC) with advantages of high bandwidth, cost-effectiveness and low latency becomes a promising wireless communication technology for short/middle reach data exchange compared with acoustic and radio frequency (RF) communications [2,3]. However, seawater, as an enormous and complex physical, chemical and biological system, contains dissolved substances, suspensions and various kinds of active organisms. With the natural characteristics of inhomogeneities of seawater, the transmission light is strongly attenuated because of the absorption and scattering effects that hinder our construction of a precise underwater optical channel model (UOCM) [4,5]. M. Doniec et al. analyzed the spatial distribution of light energy using the volume scattering function [6], and then the Henyey–Greenstein function-based Monte Carlo method was proved to be an effective means [7]. Tang et al. proposed a double-gamma function based on the mentioned Monte Carlo method, which can describe the UOCM relatively precisely and the impulse responses are calculated [8]. The impulse response calculated from the double-gamma function and Monte Carlo simulation has been shown to have a better fitting effect on real underwater models.

On the other hand, orthogonal frequency division multiplexing (OFDM) is widely used in OWC due to its strong capability to resist inter-symbol interference (ISI). However, the transmitted signals in an OWC system should be non-negative and real-valued due to the intensity modulation and direct detection while the conventional OFDM signals are bipolar complex-valued, which requires the modification of conventional OFDM. Therefore, Hermitian symmetry is commonly utilized before the inverse fast Fourier transform (IFFT) to satisfy the real-valued property. To obtain non-negative signals, various optical OFDM schemes are applied. Direct current biased optical OFDM (DCO-OFDM) adds a DC bias [9], and asymmetrically clipped optical OFDM (ACO-OFDM) utilizes the odd-subcarrier modulation and zero clipping [10]. However, DCO-OFDM and ACO-OFDM suffer from the problems of low power efficiency and low spectral efficiency, respectively. Asymmetrically clipped DC-biased optical OFDM (ADO-OFDM) is the performance tradeoff between DCO-OFDM and ACO-OFDM considering power and spectral efficiency [11–13]. In a conventional ADO-OFDM scheme, the transmitted signals are generated by superimposing the ACO-OFDM signals and DCO-OFDM signals. The zero clipping of the ACO-OFDM branch will introduce clipping noise into the even subcarriers in the transmission of DCO-OFDM signals, bringing difficulties to the demodulation of the DCO-OFDM signals. At the receiver, the ACO-OFDM signals must be demodulated first and subtracted from the reconstructed ADO-OFDM signals, and then the DCO-OFDM signals can be demodulated correctly, which results in high-complexity, delay and error propagation issues in demodulation [14]. The pre-distorted enhanced (PE) operation is applied in PEADO-OFDM to decrease the mutual interference and alleviate the issues caused by the clipping noise of the ACO-OFDM branch [15]. PEADO-OFDM eliminates the inter-carrier interference (ICI) between the ACO-OFDM and the DCO-OFDM branches at the transmitter and is applied as the modulation scheme for downlink in a visible light communication (VLC) system. However, the PE operation requires an additional FFT operation at the transmitter, inducing a more complicated transmitter. Compared with Huang's works in ref. [15], we actually proposed a modified scheme in a practical full-duplex UOWC architecture and a low-complexity scheme of a transmitter, apart from transferring the compensation scheme in the ACO branch from a typical indoor model to various underwater channels. The performance comparisons between four optical OFDM schemes from the perspective of spectral efficiency, power efficiency and detection complexity are shown in Table 1.

In this work, we establish a double-gamma UOCM with three kinds of water qualities to observe the performance of the PEADO-OFDM signal in the UOWC system. The full-duplex architecture increases the capacity to exchange information and the stability of alignment in real deployment. By increasing hardware cost, we propose the dual-branch PEADO-OFDM scheme to remove the Hermitian symmetry (HS) operation in the OFDM scheme, which improves the spectral efficiency. A new PEADO-OFDM transmitter with low complexity is further proposed to substitute an absolute operation for the additional FFT operation at the transmitter. For the first time, we apply PEADO-OFDM in the UOCM by further comparing it with a traditional ADO-OFDM scheme in the aspects of bit error rate (BER), stability and spectral efficiency.

Table 1. General performance comparisons between four optical O-OFDM schemes.

Optical OFDM Scheme	Spectral Efficiency	Power Efficiency	Detection Complexity
DCO-OFDM	high	low	low
ACO-OFDM	low	high	low
ADO-OFDM	high	medium	high
PEADO-OFDM	high	medium	low

2. Channel Model and UOWC System

2.1. Underwater Optical Channel Model

When the light signal propagates in the seawater, the photon suffers from absorption and scattering by the interactions with seawater. The absorption process will reduce the energy of the photon while the scattering process will change the transmit direction of the photon. The absorption coefficient a(λ) and scattering coefficient b(λ) are utilized to evaluate the effects of absorption and scattering, respectively, which vary with water type and wavelength λ. The total energy loss due to absorption and scattering can be described by the attenuation coefficient c(λ), which is given by:

$$c(\lambda) = a(\lambda) + b(\lambda), \tag{1}$$

The a(λ), b(λ) and c(λ) for clear, coastal and harbor water used in this paper are listed in Table 2.

Table 2. The values of a(λ), b(λ) and c(λ) for clear, coastal and harbor water.

Water Type	a (m^{-1})	b (m^{-1})
Clear Water	0.114	0.037
Coastal Water	0.179	0.219
Harbor Water	0.295	1.875

Referring to the close-form expression of the gamma–gamma function given in [8], the impulse response of UOCM is:

$$h(t) = C_1 \Delta t e^{-C_2 \Delta t} + C_3 \Delta t e^{-C_4 \Delta t} \tag{2}$$

where Δt represents the duration of the impulse response.

By using the nonlinear least square criterion:

$$(C_1, C_2, C_3, C_4) = \text{arg}min\left(\int [h(t) - h_{mc}(t)]^2 dt\right) \tag{3}$$

where $h(t)$ and $h_{mc}(t)$ are the impulse response of double-gamma functions and Monte Carlo simulation results, respectively, and the operator arg min $(\cdot)$ is used to return the argument of the minimum.

In the simulation, the Monte Carlo simulations are firstly carried out to obtain the Monte Carlo impulse responses for three kinds of water. Then, Equation (3) can be realized by the curve fitting approach in MATLAB to compute the parameter sets (C_1, C_2, C_3, C_4). By this method, we can build up various impulse responses of clear, coastal and harbor waters. The normalized impulse responses of clear, coastal and harbor waters are shown in Figure 1. We can observe that the time delay spread of the harbor water channel is much larger than that of the other two, which means it will cause more serious ISI.

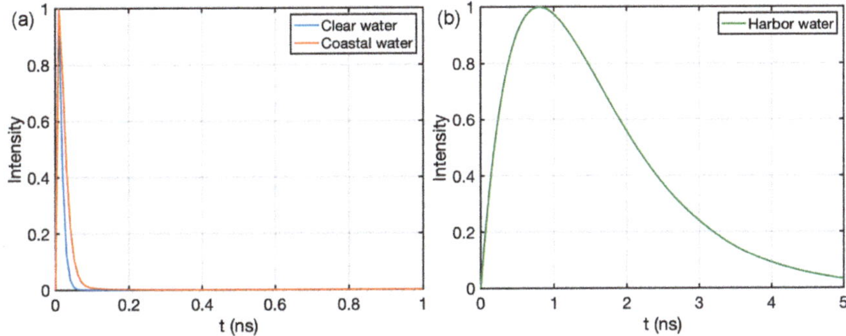

Figure 1. The normalized impulse response in (**a**) clear, (**a**) coastal and (**b**) harbor water using double-gamma functions.

2.2. Full-Duplex Architecture

As shown in Figure 2, the proposed full-duplex UOWC system consists of transceivers, which combine a blue laser diode (LD), a green LD, and two photo-detectors (PDs) within an independent unit. By using this method, blue and green LDs can be used to load the real and imaginary branches of inverse fast Fourier transform (IFFT), respectively. This results in the removal of HS in the optical OFDM scheme. In Figure 2d, the diagonal arrangement of a pair of light sources and a pair of photo-detectors can increase the stability for real applications because the beams can be captured more easily at the receiver side. When both sides adopt this design, a high-speed full-duplex UOWC system will be able to support the requirements of data interaction, simultaneously. For the transmitter and receiver signal processing shown in Figure 2a,b, the details will be introduced in the next section.

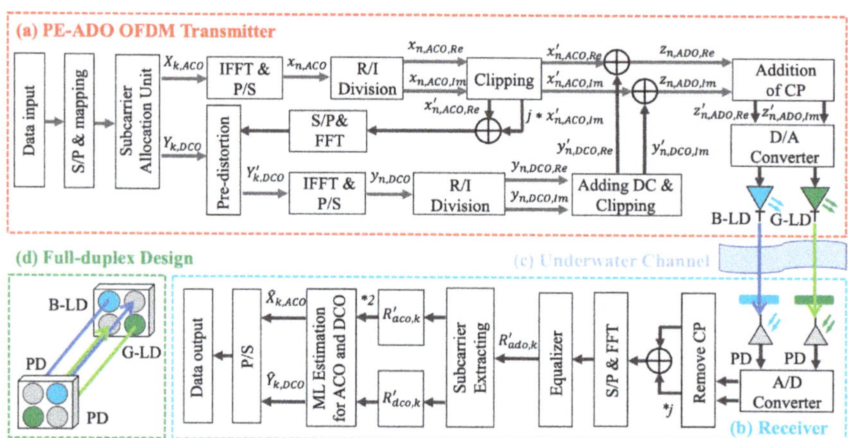

Figure 2. The full-duplex architecture of UOWC system. (**a**) The dual-branch PEADO-OFDM-based transmitter with blue/green LDs. (**b**) The dual-branch PEADO-OFDM-based receiver with double PDs. (**c**) The UOCM. (**d**) The scheme of full-duplex design.

3. Proposed PEADO-OFDM Scheme and Full-Duplex Communication

3.1. The PEADO-OFDM Scheme

PEADO-OFDM is a modified ADO-OFDM scheme with subcarrier allocation, in which the ACO-OFDM branch occupies only a part of odd subcarriers and the DCO-OFDM branch occupies the remaining subcarriers except the 0_th subcarrier, which improves the flexibility of the system. Therefore, the numbers of subcarriers carrying information assigned to the ACO-OFDM branch and DCO-OFDM branch are not fixed anymore.

In the PEADO-OFDM scheme, the clipping noise from the ACO-OFDM branch will be eliminated at the transmitter. For the ACO-OFDM branch, the time-domain clipped ACO-OFDM signals can be obtained through the traditional ACO-OFDM operations. The noise generated from the clipping operation only falls on the even subcarrier. So, after performing the fast Fourier transform (FFT) operation to the clipped ACO-OFDM signals, the clipping noise can be obtained by extracting the data of the even subcarrier, which will be used as the pre-distorted signals. For the DCO-OFDM branch, before the IFFT operation, the original DCO-OFDM symbols on the even subcarriers should subtract the pre-distorted signals first, which is referred to as the pre-distortion process. Then, the traditional DCO-OFDM operations will be performed to obtain the time-domain DCO-OFDM signals.

After superimposing the ACO-OFDM signals and the DCO-OFDM signals in the time domain, the clipping noise that falls on the even subcarriers can be eliminated through pre-distortion, which brings independent demodulation of the ACO-OFDM and DCO-

OFDM branches, leading to a reduction in the processing latency, complexity and error propagation at the receiver.

3.2. The Proposed PEADO-OFDM Transmitter with Low Complexity

The PEADO-OFDM scheme removes the clipping noise at the transmitter side but requires an additional FFT operation. Therefore, we propose another PEADO-OFDM transmitter with low complexity by utilizing the internal characteristics of clipped ACO-OFDM signals.

In ACO-OFDM, the input symbols of IFFT can be represented as:

$$X = [0, X_1, 0, X_3, \ldots, 0, X_{N-1}], \tag{4}$$

where N is the size of IFFT. We denote the output signals of IFFT and the clipped ACO-OFDM signals as x_n and $x_{n,c}$, respectively. $x_{n,c}$ can be described by x_n, which is given by:

$$x_{n,c} = \frac{1}{2}(x_n + |x_n|),\ 0 \leq n < N, \tag{5}$$

Applying the FFT operation to both sides of Equation (4), we can obtain:

$$X_c = \frac{1}{2}X + \frac{1}{2}FFT(|x_n|), \tag{6}$$

where X_c represents the FFT of clipped ACO-OFDM signals $x_{n,c}$. In addition, X_c can be divided into even and odd parts, i.e.,:

$$X_c = X_{odd,c} + X_{even,c}, \tag{7}$$

$$X_{odd,c} = \begin{cases} 0, & if\ k\ is\ even, \\ X_c, & if\ k\ is\ odd, \end{cases} \tag{8}$$

$$X_{even,c} = \begin{cases} X_c, & if\ k\ is\ even, \\ 0, & if\ k\ is\ odd, \end{cases} \tag{9}$$

where $X_{odd,c}$ and $X_{even,c}$ represent the odd part and even part of the X_c, respectively.

It is obvious that $X_{even,c}$ represents the frequency domain form of the clipping noise because the clipping noise only falls on the even subcarrier. For ACO-OFDM signals, it has been proven that after clipping, the symbols on the odd subcarriers are only half of the original symbols in the frequency domain. Thus, $X_{odd,c}$ can be rewritten as:

$$X_{odd,c} = \frac{1}{2}X, \tag{10}$$

Combining Equations (6), (7) and (10), it is easy to derive that:

$$X_{even,c} = \frac{1}{2}FFT(|x_n|), \tag{11}$$

Equation (11) reveals that the frequency domain form of the clipping noise $X_{even,c}$ and the absolute value of the output signals of IFFT are a pair of Fourier transforms. Therefore, we can obtain the clipping noise by employing an absolute operation in the time domain rather than an FFT operation, resulting in lower complexity. The obtained time-domain clipping noise can be used as pre-distortion signals, which will be subtracted by the DCO-OFDM branch after the IFFT operation. The diagram of new PEADO-OFDM transmitter with a dual-branch structure is shown in Figure 3.

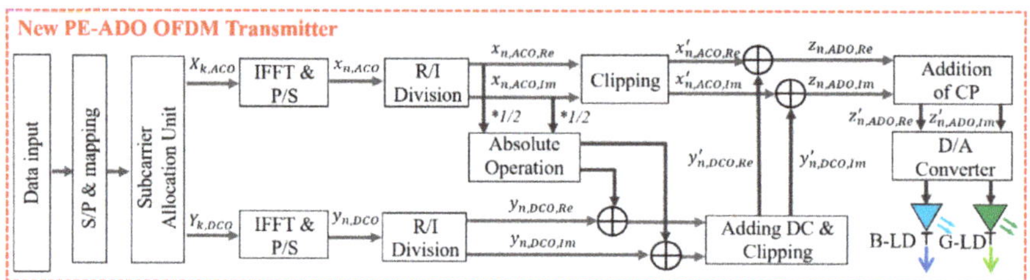

Figure 3. A new PEADO-OFDM transmitter with a dual-branch structure.

3.3. Dual-Branch PEADO-OFDM Structure

In O-OFDM, HS is performed before the IFFT to obtain the real-value signals, where the second half of the subcarrier information is the conjugate symmetry of the first half, which greatly reduces the spectral efficiency. Thus, PEADO-OFDM is proposed without HS, so the time-domain signals after the IFFT are complex-value. Then, the real and imaginary branches of the obtained signals will be divided and transmitted independently. For the ACO-OFDM branch, after the real-imaginary division, the negative signals of both the real branch and the imaginary branch will be clipped. For the DCO-OFDM branch, the clipped real and imaginary branches of the ACO-OFDM signals should be combined to the complex-value signals and converted to the pre-distortion signals through the FFT. Then, the operation of pre-distortion, IFFT, S/P and real-imaginary division are performed sequentially. The DC biases are added to both the real and imaginary branches of the pre-distorted DCO-OFDM signals with all the remaining negative signals clipped to zero, which are:

$$B_{DC,Re} = \mu_{Re}\sqrt{E\left\{y_{n,DCO,Re}^2\right\}}, \tag{12}$$

$$\text{and } B_{DC,Im} = \mu_{Im}\sqrt{E\left\{y_{n,DCO,Im}^2\right\}}, \tag{13}$$

where $y_{n,DCO,Re}$ and $y_{n,DCO,Im}$ represent the real and imaginary branches of the pre-distorted DCO-OFDM signals, respectively, and μ_{Re} and μ_{Im} represent the bias index of the real and imaginary branches, which can be defined as $\beta_{Re} = 10log_{10}(1+\mu_{Re}^2)$ $[dB]$ and $\beta_{Im} = 10log_{10}(1+\mu_{Im}^2)$ $[dB]$, respectively. Hence, by superimposing the real branch from the ACO-OFDM and the DCO-OFDM and the same superimposing operation at the imaginary branch, the real and imaginary branches of the transmitted signals are obtained as:

$$z_{n,ADC,Re} = x'_{n,ACO,Re} + y'_{n,DCO,Re}, \tag{14}$$

$$\text{and } z_{n,ADO,Im} = x'_{n,ACO,Im} + y'_{n,DCO,Im}. \tag{15}$$

Without HS operation, the number of subcarriers conveying information is $N-1$, while in the HS-based O-OFDM scheme, the number of effective subcarriers is $N/2-1$. Therefore, the spectral efficiency of the proposed transmission scheme is significantly improved.

4. Results and Discussions

PEADO-OFDM and ADO-OFDM are simulated and compared in the proposed full-duplex UOWC system with parameters listed in Table 3. In the simulation, a random binary sequence (RBS) is firstly generated and mapped into 4-QAM symbols. Then, the new proposed PEADO-OFDM transmitter structure is applied to generate the PEADO-OFDM signals. The signal powers of both PEADO-OFDM and ADO-OFD schemes are normalized to unity. An additive white Gaussian noise model is utilized in the simulation to simulate different Eb/N0 between signal and noise. At the receiver, the output binary data from the demodulation process will be compared with the transmitted binary sequence to evaluate the system BER.

Table 3. Simulation parameters for dual-branch full-duplex UOWC system.

Simulation Parameters	Symbol	Value
The number of total subcarriers	N	2048
The constellation order	M	4
The bias index of the real branch	β_{Re}	10 dB
The bias index of the imaginary branch	β_{Im}	10 dB
Link communication distance	L	10 m
The duration of channel impulse response	ΔT	5 ns
Bit rate	bps	1×10^9

We denote the number of subcarriers assigned to the ACO-OFDM branch as N_{ACO}. Here, N is the number of total subcarriers, which is divided into odd subcarriers and even subcarriers for ACO-OFDM and DCO-OFDM, respectively, and the number of subcarriers assigned to the ACO-OFDM branch is fixed to $N/2$ in traditional ADO-OFDM. In PEADO-OFDM, the ACO-OFDM branch only occupies a portion of odd subcarriers so that N_{ACO} can be flexibly adjusted to different sizes, i.e., $N/2$, $N/4$ and $N/8$. Figure 4 shows the BER comparison of PEADO-OFDM and ADO-OFDM with the proposed scheme in three types of water, where N_{ACO} is set to 1024, which means that all the odd subcarriers are assigned to the ACO-OFDM branch. Figure 4a,b show that the BER performances in the clear and coastal waters are almost identical. However, in the harbor water, as shown in Figure 4c, the BER performances of both PEADO-OFDM and ADO-OFDM become worse. This result corresponds to the time delay spread shown by the impulse response in Figure 1. Compared with traditional ADO-OFDM in various UOCMs as exhibited in Figure 4, the BER performance of the ACO-OFDM branch in PEADO-OFDM is not changed obviously, while the DCO-OFDM branch in PEADO-OFDM is improved significantly. The PE operation will not affect the demodulation of the ACO-OFDM branch and the DCO-OFDM branch can be demodulated independently without depending on the demodulation of the ACO-OFDM branch. Therefore, the overall BER performance of PEADO-OFDM is optimized.

In addition, the BER performances of the DCO-OFDM branch and ACO-OFDM branch with different N_{ACO} are simulated in the clear water, and the simulation results are given in Figure 5a,b, in which N_{ACO} is set to 1024, 512 and 256. It can be seen from the Figure 5a that with the increase in N_{ACO}, the BER performance of the DCO-OFDM branch in PEADO-OFDM will be improved and the required Eb/N0 for a BER of 10^{-3} is about 14.80 dB, 15.95 dB and 16.37 dB, achieving about 2.62 dB, 1.29 dB and 0.78 dB gains compared with that in ADO-OFDM when $N_{ACO} = 1024$, 512 and 256, correspondingly. The BER performance of the DCO-OFDM branch depends on the channel state and noise when the bias index β is large enough. The larger the N_{ACO} is, the more negative ACO-OFDM signals will be clipped, resulting in lower power of the transmitted signals. Therefore, the noise will lower with the same signal-to-noise ratio, leading to a lower BER.

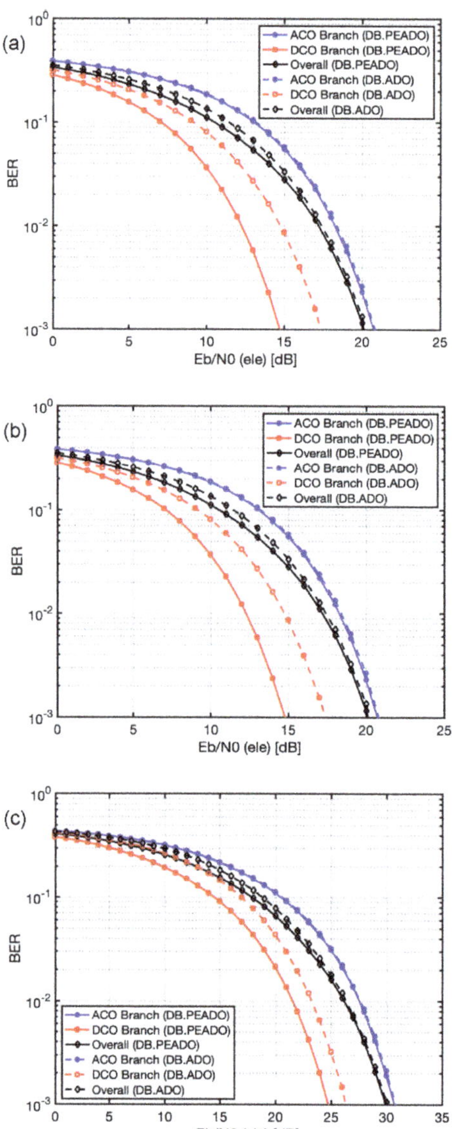

Figure 4. BER performance comparisons of dual-branch ADO-OFDM and PEADO-OFDM in (**a**) clear, (**b**) coastal and harbor water channels. (**c**) the BER performances of both PEADO-OFDM and ADO-OFDM become worse.

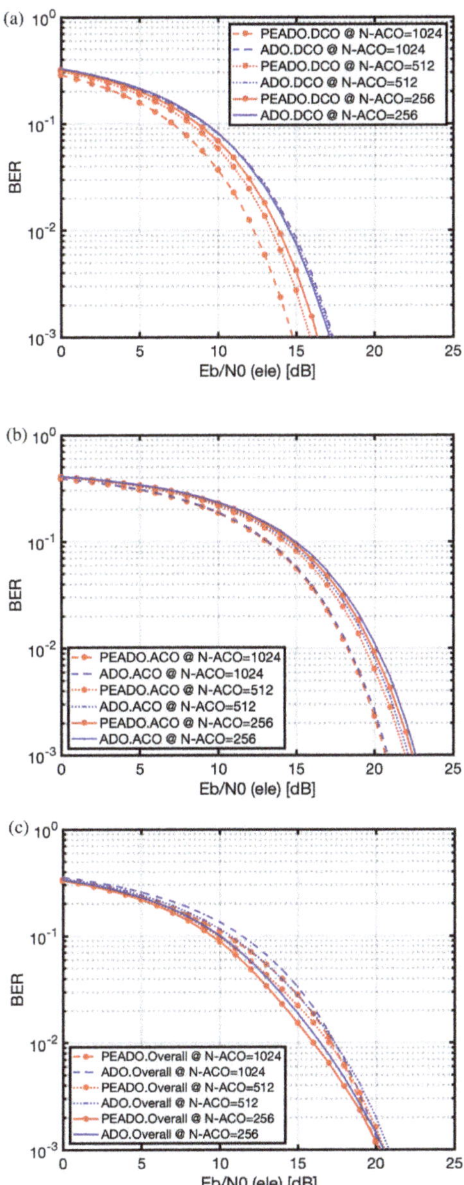

Figure 5. BER performance comparison of (**a**) the DCO-OFDM branch, (**b**) the ACO-OFDM branch and (**c**) the overall dual-branch ADO-OFDM and dual-branch PEADO-OFDM with different N_{ACO} in coastal water channel.

For the ACO-OFDM branch in the PEADO-OFDM and ADO-OFDM schemes, the difference in BER performance becomes obvious when N_{ACO} decreases for the reason that the power of PEADO-OFDM signals is lower than ADO-OFDM signals and the difference is increasing with the decrease in N_{ACO}. In Figure 5c, the overall BER performances of the PEADO-OFDM and ADO-OFDM schemes with different N_{ACO} in the clear water are performed. For any N_{ACO}, the PEADO-OFDM scheme has a better overall BER performance than the conventional ADO-OFDM scheme. To obtain an overall BER of 10^{-3}, the PEADO-OFDM scheme achieves about 0.16 dB, 0.29 dB and 0.28 dB gains compared with the enhanced ADO-OFDM scheme when $N_{ACO} = 1024$, 512 and 256, correspondingly. For the other UOCM channels, the same analysis can obtain similar results.

The proposed modified PEADO-OFDM scheme for the UOWC system not only achieves a lower BER performance on communications, but also allows for easier system scaling. For example, it can be further extended to a multi-channel parallel communication system by using a multi-wavelength transmitter with a blue/green range or spatial division. Such a UOWC system design with the proposed highly efficient scheme proves the communication potential of blue/green LDs, which will play an important role in future high-speed UOWC, integrated systems for underwater drones and underwater information interaction applications.

5. Conclusions

In this work, a novel full-duplex architecture UOWC system adopting the dual-branch PEADO-OFDM scheme is investigated. The UOCM with various typical water qualities based on the double-gamma function and Monte Carlo parameter fitting method is introduced for the evaluation. The dual-branch PEADO-OFDM scheme with real and imaginary branches is introduced firstly, and then an improved transmission scheme combining PEADO-OFDM and UOCM is proposed, in which the HS is evitable. The real and imaginary branches of the transmitted time-domain signals can be transmitted via blue and green LDs, respectively. An improved PEADO-OFDM transmitter with low complexity is also proposed. Compared with the traditional ADO-OFDM scheme, our simulation results show that the proposed dual-branch PEADO-OFDM method can improve the spectrum efficiency and the performance of anti-ISI and BER.

Author Contributions: Conceptualization, Z.W. (Zixian Wei); methodology, Z.W. (Zixian Wei); validation, Z.W. (Zixian Wei) and Y.L.; formal analysis, Y.L.; investigation, Y.L.; resources, Z.W. (Zixian Wei); data curation, Z.W. (Zixian Wei); writing—original draft preparation, Z.W. (Zixian Wei), Y.L. and Z.W. (Zhaoming Wang); writing—review and editing, H.F.; visualization, Y.L.; supervision, J.F. and H.F.; project administration, H.F.; funding acquisition, J.F. and H.F. All authors have read and agreed to the published version of the manuscript.

Funding: This research was funded by the Science, Technology and Innovation Commission of Shenzhen Municipality (JCYJ20180507183815699, WDZC20200820160650001); Guangdong Basic and Applied Basic Research Foundation (2021A1515011450); National Natural Science Foundation of China (61771222); and The Fundamental Research Funds for the Central Universities (21620439).

Conflicts of Interest: The authors declare no conflict of interest.

References

1. Zeng, Z.Q.; Fu, S.; Zhang, H.H.; Dong, Y.H.; Cheng, J.-L. A survey of underwater optical wireless communications. *IEEE Commun. Surv. Tutor.* **2016**, *19*, 204–238. [CrossRef]
2. Saeed, N.; Celik, A.; Al-Naffouri, T.Y.; Alouini, M.S. Underwater optical wireless communications, networking, and localization: A survey. *Ad Hoc Netw.* **2019**, *94*, 101935. [CrossRef]
3. Cossu, G. Recent achievements on underwater optical wireless communication. *Chin. Opt. Lett.* **2019**, *17*, 100009. [CrossRef]
4. Jasman, F.; Green, R.J. Monte Carlo simulation for underwater optical wireless communications. In Proceedings of the 2013 2nd International Workshop on Optical Wireless Communications (IWOW), Newcastle Upon Tyne, UK, 21 October 2013; pp. 113–117.
5. Sahu, S.K.; Shanmugam, P. A study on the effect of scattering properties of marine particles on underwater optical wireless communication channel characteristics. In Proceedings of the OCEANS 2017-Aberdeen, Aberdeen, UK, 19–22 June 2017; pp. 1–7.
6. Gabriel, C.; Khalighi, M.A.; Bourennane, S.; Léon, P.; Rigaud, V. Monte-Carlo-based channel characterization for underwater optical communication systems. *J. Opt. Commun. Netw.* **2013**, *5*, 1–12. [CrossRef]
7. Haltrin, V.I. One-parameter two-term Henyey-Greenstein phase function for light scattering in seawater. *Appl. Opt.* **2002**, *41*, 1022–1028. [CrossRef] [PubMed]
8. Tang, S.J.; Dong, Y.H.; Zhang, X.D. Impulse response modeling for underwater wireless optical communication links. *IEEE Trans. Commun.* **2013**, *62*, 226–234. [CrossRef]
9. Carruthers, J.B.; Kahn, J.M. Multiple-subcarrier modulation for nondirected wireless infrared communication. *IEEE J. Sel. Areas Commun.* **1996**, *14*, 538–546. [CrossRef]

10. Armstrong, J.; Lowery, A.J. Power efficient optical OFDM. *Electron. Lett.* **2006**, *42*, 370–372. [CrossRef]
11. Dissanayake, S.D.; Panta, K.; Armstrong, J. A novel technique to simultaneously transmit ACO-OFDM and DCO-OFDM in IM/DD systems. In Proceedings of the 2011 IEEE GLOBECOM Workshops (GC Wkshps), Houston, TX, USA, 5–9 December 2011; pp. 782–786.
12. Dissanayake, S.D.; Armstrong, J. Comparison of aco-ofdm, dco-ofdm and ado-ofdm in im/dd systems. *J. Lightwave Technol.* **2013**, *31*, 1063–1072. [CrossRef]
13. Xuan, H.; Yang, F.; Song, J. Novel heterogeneous attocell network based on the enhanced ADO-OFDM for VLC. *IEEE Commun. Lett.* **2018**, *23*, 40–43.
14. Na, Z.; Wang, Y.; Xiong, M.; Liu, X.; Xia, J. Modeling and throughput analysis of an ADO-OFDM based relay-assisted VLC system for 5G networks. *IEEE Access* **2018**, *6*, 17586–17594. [CrossRef]
15. Huang, X.; Yang, F.; Pan, C.; Song, J. Pre-Distorted Enhanced ADO-OFDM for Hybrid VLC Networks: A Mutual-Interference-Free Approach. *IEEE Photonics J.* **2020**, *12*, 7901512. [CrossRef]

Article

Low-Complexity Sampling Frequency Offset Estimation and Compensation Scheme for OFDM-Based UWOC System

Hu Li, Tianyang Chen, Zhijie Wang, Bingyao Cao, Yingchun Li and Junjie Zhang *

Key Laboratory of Specialty Fiber Optics and Optical Access Networks, Joint International Research Laboratory of Specialty Fiber Optics and Advanced Communication, Shanghai University, Shanghai 200444, China; lihu@shu.edu.cn (H.L.); tyral_chen@shu.edu.cn (T.C.); zhijie_wang@shu.edu.cn (Z.W.); caobingyao@shu.edu.cn (B.C.); liyingchun@shu.edu.cn (Y.L.)

* Correspondence: zjj@staff.shu.edu.cn

Abstract: In this paper, a simple sampling frequency offset (SFO) estimation and compensation scheme based on two phase-conjugated pilots is proposed and experimentally demonstrated in an OFDM-based underwater wireless optical communication (UWOC) system. The phase shift is obtained by simple multiplication for phase-conjugated pilots, and the results are averaged to perform more accurate phase estimation. The experimental results show that the estimation offset is limited within ±3 ppm when the SFO ranges from −1000 ppm to +1000 ppm over a 1 m tap water channel. Moreover, with the help of the proposed scheme, up to ±300 ppm SFO can be well-compensated with error vector magnitude (EVM) penalties below 1 dB after 1 m underwater transmission. In addition, the results demonstrate that, compared with the ideal case without SFO, our proposed SFO compensation scheme can provide nearly negligible bit error rate (BER) penalties in saltwater with the highly scattering property.

Keywords: underwater wireless optical communication; orthogonal frequency-division multiplexing; sampling frequency offset

Citation: Li, H.; Chen, T.; Wang, Z.; Cao, B.; Li, Y.; Zhang, J. Low-Complexity Sampling Frequency Offset Estimation and Compensation Scheme for OFDM-Based UWOC System. *Photonics* **2022**, *9*, 216. https://doi.org/10.3390/photonics9040216

Received: 16 February 2022
Accepted: 23 March 2022
Published: 25 March 2022

1. Introduction

In recent years, underwater wireless optical communication (UWOC) has been considered a promising technology for future underwater high-speed data transmission [1–3]. Compared with underwater acoustic or radio frequency (RF) communication systems that exhibit either low bandwidth or high attenuation, UWOC can offer a large bandwidth with low propagation loss [4,5]. Because blue-green (400–550 nm) visible light fits within the low absorption window of seawater, blue-green LD/LED-based UWOC can achieve relatively long-distance and high-speed underwater communication at the same time and is expected to play a significant role in various ocean applications [6,7]. Since LD/LED has the fast-switching feature for amplitude modulation, intensity modulation with direct detection (IMDD) is usually applied in a cost-effective UWOC system [8]. To improve the spectral efficiency (SE) of high-speed UWOC systems, orthogonal frequency division multiplexing (OFDM) with high SE and flexible bandwidth has attracted much attention [9–11].

However, one disadvantage of OFDM is its sensitivity to frequency offset [12,13]. Although carrier frequency offset (CFO) does not exist in the real-valued IMDD OFDM systems, sampling frequency offset (SFO) between transceivers inevitably exists due to the clock mismatch of DAC and ADC [14]. Since the transceivers are physically separated, a standard ±200 ppm SFO should be compensated for practical optical transmission systems [15]. Generally, the effects of SFO on received OFDM signals are marked by amplitude attenuation, subcarrier phase rotation, inter-symbol interference (ISI) and inter-carrier interference (ICI) [15,16]. These will seriously deteriorate the system performance. Fortunately, the amplitude attenuation can be compensated for by channel equalization, and the ICI can be regarded as additional noise when the SFO is small. If the number of data-carrying OFDM symbols in each OFDM frame is appropriate and the optimal cyclic prefix/suffix (CP/CS) per OFDM symbol is chosen [16], the SFO-induced ISI can also be avoided. Therefore, the SFO compensation is mainly to correct the SFO-induced subcarrier phase rotation [17].

Several methods have been intensively investigated for SFO estimation and compensation, which can be categorized into the following two types. One is the hardware scheme [18,19], which realizes sampling clock frequency synchronization by adjusting the voltage-controlled oscillator with feedback/feedforward information. This requires high-precision, stable voltage-controlled oscillators and additional analog circuits. Another way of mitigating the SFO effect is digital signal processing (DSP). Depending on the utilization of assistant information, it can also be divided into blind and data-aided estimation methods. An overhead-free SFO compensation scheme based on inter symbol differential detection (ISDD) is proposed in [20,21]. However, this method will cause residual phase noise. Then, the fourth-power algorithm is used for SFO estimation in [22,23]. The application of the method is limited since it is suitable for the OFDM symbols with phases of $\pm\pi/4$ or $\pm 3\pi/4$. The data-aided methods estimate the SFO by using the pilot or training sequence (TS). Then the SFO compensation is completed in time or frequency domain [24–30]. The fourth-order piecewise polynomial interpolation is the most popular time-domain compensation method [24,25], which can alleviate ICI at the expense of additional DSP computing resources. The pilot-aided SFO estimation method can compensate up to ± 1000 ppm SFO by correcting the SFO-induced phase rotation in the frequency domain [26,27]. It is noteworthy that the methods will inevitably reduce SE. In [28–30], the SFO estimation and compensation scheme based on TSs is proposed. The TSs can also be used for synchronization and channel estimation, which improve the SE. Nevertheless, the rigorous OFDM structure limits the compensation range of SFO to less than ± 200 ppm.

In this brief, to balance the compensation range and spectral efficiency, a simple pilot-aided SFO estimation and compensation scheme is proposed. The SFO compensation performance is experimentally investigated in an LD-based UWOC link with a 3-dB bandwidth of approximately 5.36 MHz. The main contributions of this literature can be summarized as follows:

(1) High accuracy and wide SFO compensation range. Up to ± 1000 ppm SFO can be well-estimated with an accuracy of less than ± 3 ppm after 1 m underwater transmission. Moreover, an estimation deviation within ± 2 ppm can be achieved for the SFO below ± 400 ppm;
(2) Low computational complexity. Unlike traditional pilot-aided schemes utilizing the least square (LS) method to fit the SFO slope curve, the proposed scheme utilizes two phase-conjugated pilots for SFO estimation in each OFDM frame and an averaging processor is designed to improve estimation accuracy. These cause the decrease in operational complexity.

2. The Principle of SFO Estimation and Compensation

The OFDM frame consists of one frame head and multiple data-carrying OFDM symbols. The frame head is composed of several training sequences. As illustrated in Figure 1, we assume that the phase-conjugated pilots are inserted after the m-th symbol and the n-th symbol on the k-th subcarrier. On the receiver side, the training sequences are used for symbol synchronization and channel estimation. When the system works in an asynchronous way, the m-th OFDM symbol of the k-th subcarrier in the frequency domain can be approximately written as:

$$R_{m,k} = S_{m,k} \times H_k e^{j\phi_{m,k}} + W_{m,k}, \qquad k = 0, 1, 2, \ldots, N-1 \tag{1}$$

where $S_{m,k}$ and $R_{m,k}$ are the transmitted and received frequency domain signal on the m-th OFDM symbol of the k-th subcarrier, H_k is the channel response on the k-th subcarrier, $W_{m,k}$ represents the noise, $\phi_{m,k} = 2\pi m k N_T \Delta / N$ is the SFO-induced phase rotation for the m-th

OFDM symbol of the k-th subcarrier, N is the FFT size, N_T is the length of the OFDM symbol including CP, the SFO between transmitter and receiver is denoted by $\Delta = (f_t - f_r)/f_r$, f_t and f_r represent the sampling frequency in the DAC and ADC, respectively.

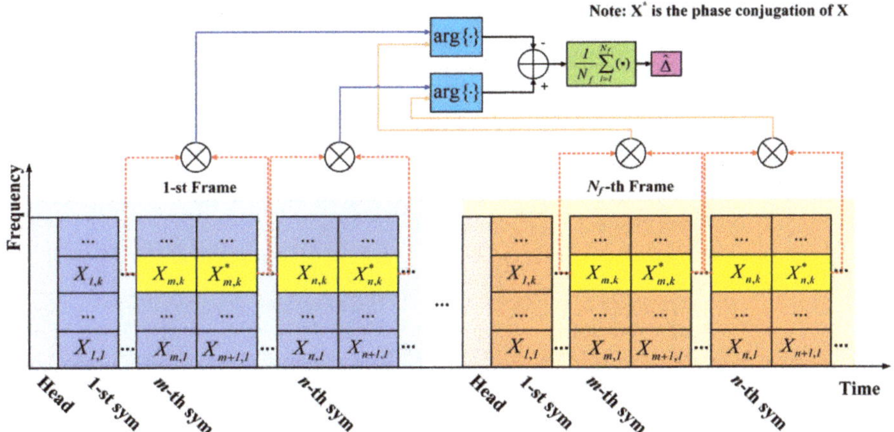

Figure 1. Principle of the SFO estimation based on the phase-conjugated pilots.

After zero-forcing equalization, the equalized signal is expressed as:

$$\hat{R}_{m,k} = S_{m,k}e^{j(\phi_{m,k}+\theta_k)} + W\prime_{m,k} \tag{2}$$

where θ_k is the common phase error caused by inaccurate channel estimation. The OFDM symbols from the same subcarrier index are generally considered to have the same common phase error θ_k [22]. Thus, the product of a pair of phase-conjugated OFDM symbols is followed as:

$$\hat{R}_{m,k} \times \hat{R}_{m+1,k} = \left|S_{m,k}\right|^2 e^{j(\phi_{m,k}+\phi_{m+1,k}+2\theta_k)} + N_{m,k} \tag{3}$$

where $N_{m,k}$ can be considered as additive noise.

Then, the superposition value of phase rotation on the phase-conjugated symbols can be given by:

$$\hat{\varphi}_{m,k} = arg\{\hat{R}_{m,k} \times \hat{R}_{m+1,k}\} = 2\pi N_T k\Delta(2m+1)/N + 2\theta_k + \omega_k \tag{4}$$

where $arg\{\cdot\}$ stands for the angle operation, ω_k represents the phase noise. As shown in Figure 1, a subtraction operation is performed to suppress the common phase error θ_k, the difference of superposition values between two groups of phase-conjugated symbols in the i-th OFDM frame can be expressed as:

$$\hat{\varphi}_{i,k} = \hat{\varphi}_{n,k} - \hat{\varphi}_{m,k} = 4\pi N_T k \times (n-m)\Delta/N + \omega_k' \tag{5}$$

To further reduce the influence of the enhanced phase noise ω_k' and improve the estimation precision, the phase difference $\hat{\varphi}_k$ is obtained by the averaging processor of N_f OFDM frames and can be expressed by:

$$\hat{\varphi}_k = \frac{1}{N_f}\sum_{i=1}^{N_f} \hat{\varphi}_{i,k} \tag{6}$$

In this way, the SFO can be accurately estimated as follows:

$$\hat{\Delta} = \hat{\varphi}_k \times N/(4\pi N_T k \times (n-m)) \tag{7}$$

Finally, the m-th OFDM symbol on the k-th subcarrier should be multiplied by a compensation factor $exp(-j2\pi kmN_T\hat{\Delta}/N)$ to correct the SFO-induced subcarrier phase rotation.

3. Experimental Setup

The experimental setup of the UWOC system based on direct current-biased optical OFDM (DCO-OFDM) is established and demonstrated in Figure 2. The key parameters of the system are listed in Table 1. On the transmitter side, the digital OFDM signal is generated offline with DSP algorithms in MATLAB. Firstly, the generated pseudo-random binary sequence (PRBS) of $2^{17}-1$ is mapped into m-QAM symbols and two phase-conjugated pilots are inserted for SFO estimation. Then, Hermitian symmetry and inverse fast Fourier transform (IFFT) guarantee real signal transmission. The CP and CS are added for each OFDM symbol, and two training sequences (TSs) are added to realize symbol synchronization and channel equalization. Subsequently, the generated digital OFDM signal is loaded into the ROM of a Xilinx AX545 FPGA board and sampled by an 8-bit DAC (AD9280) with a 10 MS/s sampling rate. A variable electrical attenuator (VEA) is added to adjust the amplitude of the OFDM signal. The output analog signal is boosted by an electrical amplifier (AMP-OPA657) and is combined with direct current (DC) bias via a Bias-Tee to drive the blue LD (OSRAM, PL 450B), which can convert an electrical signal into an optical signal for underwater transmission.

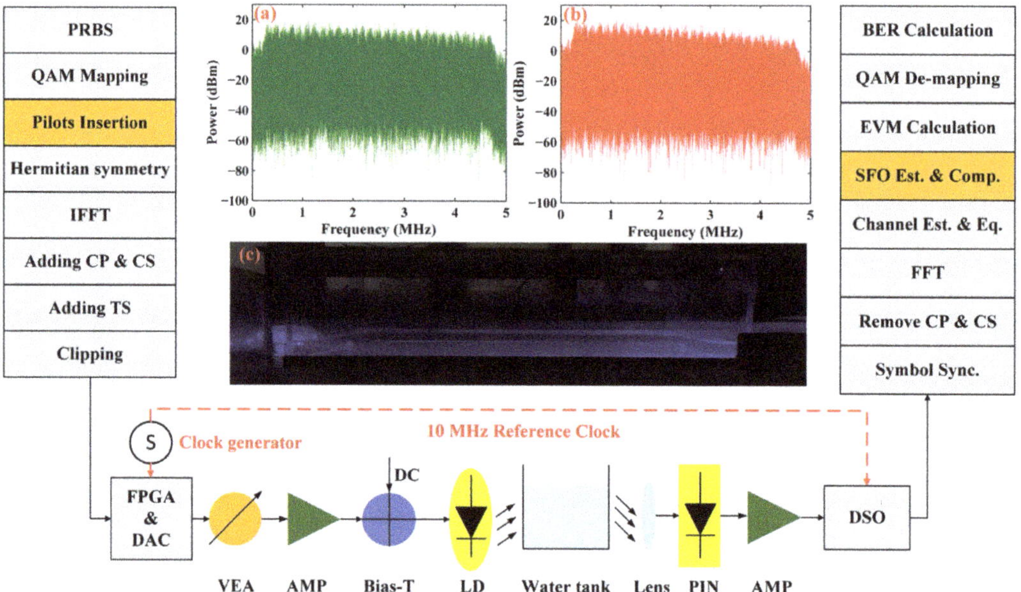

Figure 2. Experimental setup of the UWOC system based on DCO-OFDM. (**a**) The electrical spectra of the transmitted signal; (**b**) the electrical spectra of the received signal; (**c**) the transmission light path of the system.

Table 1. Key parameters of the UWOC system.

Parameters	Values
Physical dimensions of the tank	1 m × 0.5 m × 0.5 m
Laser wavelength	450 nm
LD output power	30 mW
The divergence angles in the parallel/vertical direction	6.5°/22.5°
DAC/ADC resolutions	8/10 bits
DAC/ADC sampling rate	10/50 MS/s
Underwater transmission distance	0.5/1 m
IFFT/FFT size	64
CP and CS length	16
Data-carrying subcarriers	From 3 to 30
Number of TS per OFDM frame	2
Number of OFDM symbol	100

On the receiver side, according to the divergence angle of LD and the attenuation coefficient of tap water in [31], the estimated receiving optical power after 1 m underwater transmission is approximately −22.59 dBm. To further improve the receiving optical power, the output beam is concentrated through the convex lens and fed into a commercial PIN (Hamamatsu, S10784). Then, the detected electrical signal is amplified by the amplifier (AMP-OPA656) and then captured by a digital storage oscilloscope (KEYSIGHT, DSO-S 804A, America). The DSO collects 5,000,000 sampling points each time for MATLAB offline processing. The main DSP flow sequentially includes symbol synchronization, CP and CS removal, 64-point FFT, channel estimation and equalization, SFO estimation and compensation, EVM calculation, QAM de-mapping and BER calculation. Due to the limitation of the receiving amplifier circuit, the 3 dB bandwidth of the system is 5.36 MHz. To investigate the SFO effect, the DAC and ADC are clocked by the same 10 MHz reference clock, which can avoid the difference between reference clocks. Different SFOs can be obtained by changing the frequency of the external clock source for the DAC.

4. Experimental Results and Discussion

4.1. EVM Penalty in Terms of SFO Error

To investigate the effects of SFO estimation accuracy on EVM performance, we artificially set the SFO from −10 to 10 ppm to simulate the SFO error, and the corresponding experiments are carried out in a 1 m tap water channel. Figure 3 shows the measured EVM performance as a function of the SFO error, where the 16-QAM constellations are inserted when the SFO errors are ±3 ppm and ±10 ppm, respectively. As shown in Figure 3, with the increase in the SFO error, the EVM performance gradually declines. For the SFOs with opposite values, the EVM penalties are essentially the same due to the adding of CP and CS in each OFDM symbol. When the SFO error is up to ±10 ppm, the EVM performance has deteriorated significantly. Compared with the error-free case, the corresponding EVM penalties are approximately 2.5 dB. When the SFO errors are dipped to ±3 ppm, the EVM penalties are only 0.3 dB, and the phase rotations of the constellations are very slight, which has a negligible effect on the system performance. This is also supported by our previous work in real-time optical OFDM systems [32]. Therefore, we can take the estimated offset <±3 ppm as our accuracy target because it has little impact on the system performance.

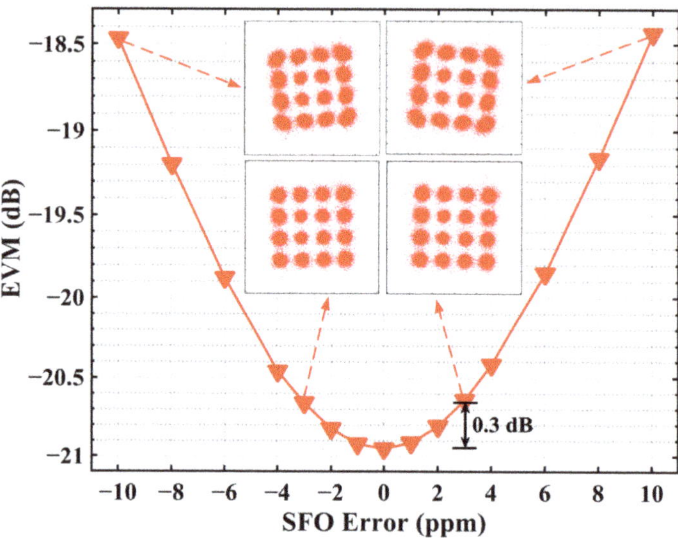

Figure 3. EVM performance versus SFO estimation error.

4.2. Influence of Frame Number and Pilot Interval on Estimation Accuracy

In order to further identify the optimal pilot interval and the convergence of multi-frame averaging, the comparison of estimation accuracy is implemented after electric back-to-back (EBTB) and 1 m tap water channel transmission when the SFO is 600 ppm. The pilot interval is defined as *n-m* in Equation (7). Figure 4a shows the estimated offset obtained from the average result of 120 frames at different pilot intervals. As shown in Figure 4a, when the pilot interval is 4, the estimated offset is too large and the impact of SFO cannot be well-alleviated. The reason is that the phase difference in Equation (5) is small and the estimated SFO value is more vulnerable to noise. Furthermore, the target accuracy can be successfully achieved as the pilot interval is up to 34. In such a case, the curve of estimated offset tends to be flat, and the further increase in the pilot interval will not improve the estimation accuracy. Moreover, since the signal after EBTB transmission is less affected by noise, the estimation accuracy of EBTB is more accurate and the estimated offset is less than ±1 ppm. Figure 4b demonstrates the corresponding estimated offset versus the frame number when the pilot interval is 34. It shows clearly that the SFO estimated offsets can all converge very fast, and the required frame numbers are approximately 40. The fast convergence of SFO estimation can further depress the computational complexity. In the following discussion, the pilot interval is fixed at 34 and the phase-conjugated pilots are inserted on the 22nd subcarrier in each frame. This is because the phase shift of high subcarriers rises faster as the symbol index increases. The large phase difference can achieve more precise estimation accuracy [27].

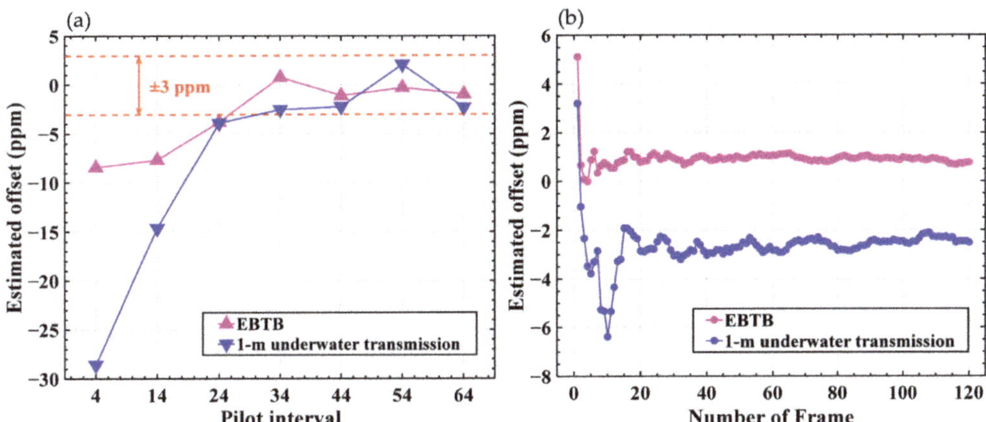

Figure 4. (**a**) The estimated offset at different pilot intervals; (**b**) The convergence performance versus frame number when the pilot interval is 34.

4.3. SFO Estimation Accuracy and Compensation Range

Following the convergence analysis, we investigate the estimation accuracy of the proposed scheme under a wide SFO range. Figure 5 shows the estimated SFO value and offset through a 1 m tap water channel in the IMDD-OFDM system. It can be seen that when the SFO varies from −1000 ppm to +1000 ppm, the estimated SFO values are very close to the real SFO, and the estimated offset is within ±3 ppm. In particular, when the SFO is less than ±400 ppm, an estimation deviation within ±2 ppm can be achieved. The results demonstrate that the proposed SFO estimation technique is capable of achieving high-accuracy SFO estimation in an asynchronous OFDM-based UWOC system.

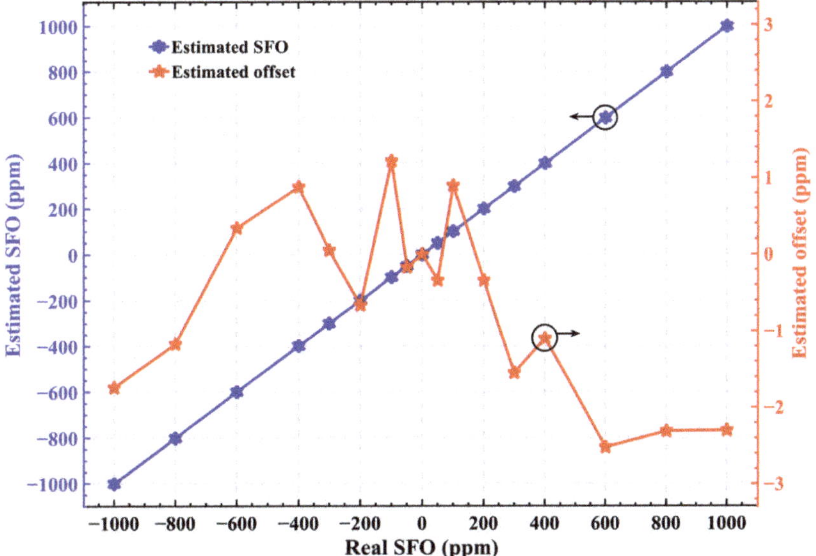

Figure 5. Estimated SFO value and estimation offset versus real SFO.

Figure 6 shows the EVM performance of a system over different SFO after 1 m underwater transmission, where the 16-QAM constellations are also inserted in the presence of ±50 ppm and ±600 ppm SFO. It can be seen that the EVM performance without SFO compensation deteriorates rapidly with the increase in SFO. As shown in Figure 6a–d, the phase rotations can be observed in 16-QAM constellations without SFO compensation, and the constellations are hard to distinguish. In addition, the directions of phase rotation for the negative and positive SFO value are opposite and the rotation angle is proportional to the value of SFO. After using the proposed SFO compensation scheme, the system performance can be significantly improved. From Figure 6e–h, the constellations become much clearer because the phase rotations are corrected with SFO compensation. Compared with the SFO-free case, even if the SFO is up to ±1000 ppm, the EVM penalties after SFO compensation are less than 3 dB. Moreover, when the SFO is less than ±300 ppm, the EVM with compensation is very close to the case without SFO and the EVM penalty is below 1 dB.

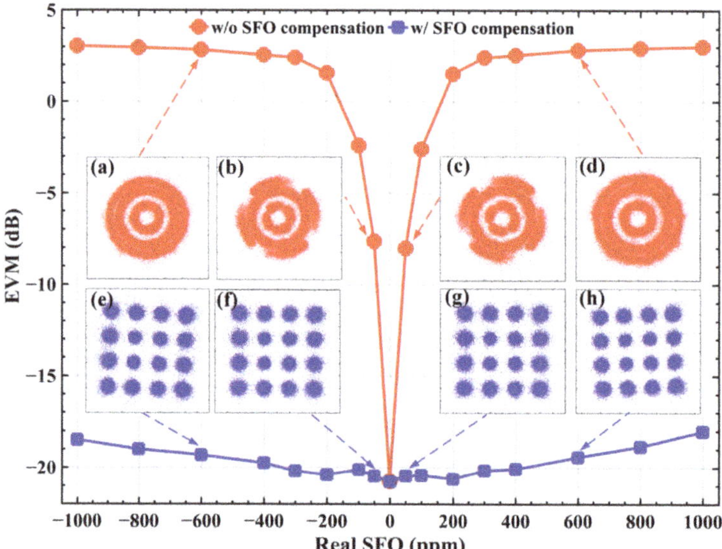

Figure 6. EVM versus sampling clock frequency offset. Subfigures show the 16-QAM constellations (**a**–**d**) without SFO compensation; (**e**–**h**) with SFO compensation.

4.4. SFO Compensation Performance in Different Underwater Environments

Broken surface waves and rain produce air bubbles in the ocean. To evaluate the influence of air bubbles on the compensation performance of UWOC, we put the oxygen pump in the water tank to generate the bubbles and disturbances. The physical device diagram is shown in Figure 7. The exhaust volume is set to 16 L/min and the experiments are carried out under a 1 m tap water link. The EVM performance after SFO compensation is shown in Table 2. The results show that the EVM penalties are similar to the cases without bubbles. Compared with the SFO-free case, the EVM at 200 ppm has no loss after compensation. The EVM penalty is 2 dB in the presence of 1000 ppm SFO. It can prove that the air bubbles have little impact on SFO compensation performance, which shows the proposed SFO estimation and compensation method is robust.

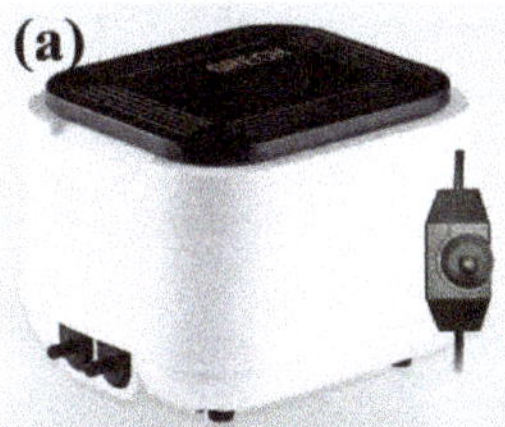

Figure 7. Physical device diagram. (**a**) Oxygen pump; (**b**) Water surface with bubble volume of 16 L/min.

Table 2. The EVM performance of different SFO.

SFO (ppm)	**0**	**200**	**1000**
EVM (dB)	−21.4	−21.9	−19.4

Light propagation undergoes serious reduction due to the absorption and scattering in seawater [33]. Therefore, we add sea salt in the 0.5 m link to assess the system performance. Since a longer communication distance always means a lower received optical power and SNR, we move the condenser lens away from the direction perpendicular to the incident light to reduce the SNR of the receiver, which can show the potential of the UWOC system at low SNR or long-distance transmission. The BER performance over the different salinity of water is presented in Figure 8 and the corresponding receiving optical power (ROP) is listed in Table 3. It is worth noting that the transform model of BER and SNR/EVM based on Gaussian noise is still applicable in the underwater channel [11]. It can be found that the concentrations of sea salt-added water are dominated for the signal degradation when the SFO is 0 ppm. With the increase in salinity, the signal attenuation is more serious due to the scattering effect, which results in the deterioration of BER performance. In the presence of SFO, the main reason for the deterioration in system performance becomes the SFO-induced phase rotation. The BER is approximately 0.4 without SFO compensation. With the help of the proposed SFO compensation method, even if the SFO is up to 1000 ppm, the BER performance has been greatly improved and is very close to the ideal case without SFO. The constellation diagrams corresponding to 1000 ppm, 200 ppm and 0 ppm are inserted in Figure 8a–c, respectively. It can be seen from the constellation diagrams that up to 1000 ppm SFO can be well-compensated with slight BER penalty.

4.5. Algorithm Complexity Comparisons

Table 4 summarizes the complexity comparisons between the proposed SFO estimation method and some conventional SFO estimation algorithms. Since the SFO value is obtained by averaging multiple SFO estimation values in these schemes, and the computational complexity is mainly determined by a single SFO estimation, Table 4 lists the complexity of one SFO estimation. In Table 4, N_d represents the number of data-carrying subcarriers and M is the number of the pilot symbols. In [26], the phase-shift slope on each OFDM symbol can be acquired by the least square (LS) method after extracting the phases on M pilot subcarriers, and the results of each OFDM symbol are averaged to obtain the accurate SFO value. In [30], the training sequences at the frame head and frame tail for SFO estimation are used, and the LS method is also applied to improve the estimation accuracy where the computational complexity depends on the number of data-carrying subcarriers. A blind SFO estimation scheme for QPSK and 16-QAM modulation is proposed in [22] by extracting two symbols in each frame to perform a fourth-power algorithm, which needs 4 multipliers. For the proposed SFO estimation method, we only need 2 multipliers in each frame for SFO estimation. Therefore, compared with the other SFO estimation and compensation methods, the proposed scheme has lower computational complexity.

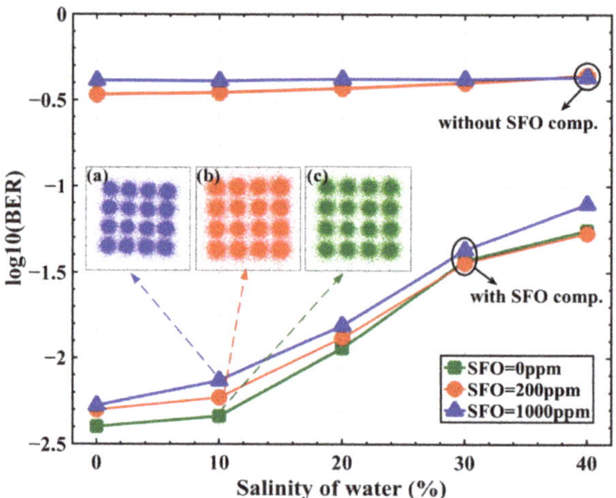

Figure 8. BER performance versus water salinity under different SFO. Subfigures show the 16-QAM constellations when the SFO is (**a**) 1000 ppm; (**b**) 200 ppm; (**c**) 0 ppm.

Table 3. The receiving optical power at different salinities.

Salinity	0%	10%	20%	30%	40%
ROP (dBm)	−14.5	−15.22	−16.12	−17.59	−18.82

Table 4. Complexity comparisons of SFO estimation.

Algorithm	Multiply/Divide	Add/Subtract	arg{·}
[26]	$M/2+1$	$M/2-1$	M
[30]	N_d+1	N_d-1	N_d
[22]	4	1	2
Proposed	2	1	2

5. Conclusions

In this paper, a simple SFO estimation and compensation scheme is proposed for the asynchronous OFDM-based UWOC system. The coarse phase shift is obtained by the product of two phase-conjugated pilots, and then an averaging processor of multi frames is used for accurate SFO estimation. The experimental results demonstrate that the proposed scheme can achieve an estimation accuracy within ±3 ppm under the SFO range of ±1000 ppm. A standard ±200 ppm SFO can be well-compensated with negligible EVM penalties after 1 m underwater transmission. Afterwards, the robustness is verified in the water with bubbles and sea salt, which shows the proposed algorithm has great application potential in a real underwater environment. Moreover, it can be concluded that our proposed scheme has low computational complexity and, thus, is a very promising solution for real-time OFDM-based UWOC systems. Our algorithm can also be extended to high-speed UWOC systems, and the corresponding experiments will be carried out the future work.

Author Contributions: Conceptualization, H.L.; methodology, H.L.; software, H.L. and J.Z.; validation, H.L. and T.C.; formal analysis, J.Z.; investigation, H.L. and J.Z.; resources, J.Z.; writing—original draft preparation, H.L.; writing—review and editing, H.L., J.Z., T.C., Z.W., B.C. and Y.L.; supervision, J.Z. All authors have read and agreed to the published version of the manuscript.

Funding: This work was supported in part by National Key Research and Development Program of China (2021YFB2900800), the Science and Technology Commission of Shanghai Municipality (Project No. 20511102400, 20ZR1420900) and 111 project (D20031).

Conflicts of Interest: The authors declare no conflict of interest.

References

1. Shen, C.; Guo, Y.; Oubei, H.M.; Ng, T.K.; Liu, G.; Park, K.H.; Ho, K.T.; Alouini, M.S.; Ooi, B.S. 20-meter underwater wireless optical communication link with 1.5 Gbps data rate. *Opt. Express* **2016**, *24*, 25502–25509. [CrossRef] [PubMed]
2. Chi, N.; Haas, H.; Kavehrad, M.; Little, T.D.; Huang, X.L. Visible light communications: Demand factors, benefits and opportunities [Guest Editorial]. *IEEE Wirel. Commun.* **2015**, *22*, 5–7. [CrossRef]
3. Chi, N.; Zhou, Y.; Wei, Y.; Hu, F. Visible light communication in 6G: Advances, challenges, and prospects. *IEEE Veh. Technol. Mag.* **2020**, *15*, 93–102. [CrossRef]
4. Wang, F.; Liu, Y.; Jiang, F.; Chi, N. High speed underwater visible light communication system based on LED employing maximum ratio combination with multi-PIN reception. *Opt. Commun.* **2018**, *425*, 106–112. [CrossRef]
5. Ho, C.M.; Lu, C.K.; Lu, H.H.; Huang, S.J.; Cheng, M.T.; Yang, Z.Y.; Lin, X.Y. A 10m/10Gbps underwater wireless laser transmission system. In Proceedings of the Optical Fiber Communication Conference, Optical Society of America, Th3C-3, Los Angeles, CA, USA, 19–23 March 2017.
6. Kaushal, H.; Kaddoum, G. Underwater optical wireless communication. *IEEE Access* **2016**, *4*, 1518–1547. [CrossRef]
7. Zeng, Z.; Fu, S.; Zhang, H.; Dong, Y.; Cheng, J. A survey of underwater optical wireless communications. *IEEE Commun. Surv. Tutor.* **2016**, *19*, 204–238. [CrossRef]
8. Chowdhury, M.Z.; Hossan, M.T.; Islam, A.; Jang, Y.M. A comparative survey of optical wireless technologies: Architectures and applications. *IEEE Access* **2018**, *6*, 9819–9840. [CrossRef]
9. Oubei, H.M.; Duran, J.R.; Janjua, B.; Wang, H.Y.; Tsai, C.T.; Chi, Y.C.; Ng, T.K.; Kuo, H.C.; He, J.H.; Alouini, M.S.; et al. 4.8 Gbit/s 16-QAM-OFDM transmission based on compact 450-nm laser for underwater wireless optical communication. *Opt. Express* **2015**, *23*, 23302–23309. [CrossRef]
10. Chen, Y.; Kong, M.; Ali, T.; Wang, J.; Sarwar, R.; Han, J.; Guo, C.; Sun, B.; Deng, N.; Xu, J. 26 m/5.5 Gbps air-water optical wireless communication based on an OFDM-modulated 520-nm laser diode. *Opt. Express* **2017**, *25*, 14760–14765. [CrossRef]
11. Zou, P.; Zhao, Y.; Hu, F.; Chi, N. Underwater visible light communication at 3.24 Gb/s using novel two-dimensional bit allocation. *Opt. Express* **2020**, *28*, 11319–11338. [CrossRef]
12. Wu, S.; Liu, P.; Bar-Ness, Y. Phase noise estimation and mitigation for OFDM systems. *IEEE Trans. Wirel. Commun.* **2006**, *5*, 3616–3625. [CrossRef]
13. Schmidl, T.M.; Cox, D.C. Robust frequency and timing synchronization for OFDM. *IEEE Trans. Commun.* **1997**, *45*, 1613–1621. [CrossRef]
14. Zhou, Z.; He, J.; Ma, J.; Chen, M. Experimental demonstration of an SFO-robustness scheme with fast OFDM for IMDD passive optical network systems. *J. Lightwave Technol.* **2020**, *38*, 5608–5616. [CrossRef]
15. Chen, M.; Zhou, H.; Zheng, Z.; Deng, R.; Chen, Q.; Peng, M.; Tang, X. OLT-centralized sampling frequency offset compensation scheme for OFDM-PON. *Opt. Express* **2017**, *25*, 19508–19516. [CrossRef] [PubMed]
16. Chen, M.; Liu, G.; Zhou, H.; Chen, Q.; He, J. Inter-symbol differential detection-enabled sampling frequency offset compensation for DDO-OFDM. *IEEE Photonics Technol. Lett.* **2018**, *30*, 2095–2098. [CrossRef]
17. Xi, D.; Chen, M.; Zhang, L.; Wang, L.; Chen, G.; Zhou, H. Digital interpolation-based SFO compensation in OCT precoding-enabled NHS-OFDM transmission systems. *Opt. Fiber Technol.* **2021**, *66*, 102642. [CrossRef]
18. Jin, X.Q.; Tang, J.M. Optical OFDM synchronization with symbol timing offset and sampling clock offset compensation in real-time IMDD systems. *IEEE Photonics J.* **2011**, *3*, 187–196. [CrossRef]
19. Giddings, R.P.; Tang, J.M. Experimental demonstration and optimisation of a synchronous clock recovery technique for real-time end-to-end optical OFDM transmission at 11.25Gb/s over 25km SSMF. *Opt. Express* **2011**, *19*, 2831–2845. [CrossRef]
20. Liu, Y.; He, J.; Chen, M.; Xiao, Y.; Cheng, Y. 64APSK constellation scheme for short-reach DMT with ISDD enabled SFO compensation. *Opt. Commun.* **2020**, *467*, 125689. [CrossRef]
21. Xi, D.; Chen, M.; Tang, X.; Zhang, L.; Wang, L.; Chen, G.; Liu, G.; Zhou, H. Sampling frequency offset compensation in non-Hermitian symmetric optical OFDM. *Opt. Commun.* **2020**, *472*, 126048. [CrossRef]
22. Wen, H.; Geng, K.; Chen, Q.; Deng, R.; Chen, M.; Zong, T.; Liu, F. A Simple Blind Sampling Frequency Offset Estimation Scheme for Short-Reach DD-OFDM Systems. *IEEE Photonics J.* **2020**, *12*, 1–10. [CrossRef]
23. Luo, Z.; Chen, M.; Chen, Q.; Wen, H. Simple blind estimation of sampling frequency offset for real-time asynchronous DDO-OFDM system. *Opt. Commun.* **2021**, *490*, 126890. [CrossRef]
24. Hu, Q.; Jin, X.; Xu, Z. Compensation of sampling frequency offset with digital interpolation for OFDM-based visible light communication systems. *J. Lightwave Technol.* **2018**, *36*, 5488–5497. [CrossRef]
25. Hu, Q.; Jin, X.; Liu, W.; Guo, D.; Jin, M.; Xu, Z. Comparison of interpolation-based sampling frequency offset compensation schemes for practical OFDM-VLC systems. *Opt. Express* **2020**, *28*, 2337–2348. [CrossRef]

26. Chen, M.; He, J.; Tang, J.; Chen, L. Pilot-aided sampling frequency offset estimation and compensation using DSP technique in DD-OOFDM systems. *Opt. Fiber Technol.* **2014**, *20*, 268–273. [CrossRef]
27. Zhang, Z.; Zhang, Q.W.; Li, Y.C.; Song, Y.X.; Zhang, J.J.; Chen, J. A single pilot subcarrier based sampling frequency offset estimation and compensation algorithm for optical IMDD OFDM systems. *IEEE Photonics J.* **2016**, *8*, 1–9. [CrossRef]
28. Deng, R.; He, J.; Chen, M.; Wei, Y.; Shi, J.; Chen, L. Real-time VLLC-OFDM HD-SDI video transmission system with TS-based SFO estimation. In Proceedings of the 2017 Optical Fiber Communications Conference and Exhibition (OFC), Los Angeles, CA, USA, 19–23 March 2017.
29. Zhang, M.; Bi, M.; Xin, H.; Li, L.; Fu, Y.; He, H.; Hu, W. Low complexity and high accuracy sampling frequency offset estimation and compensation algorithm in IMDD-OFDM system. In Proceedings of the 2017 16th International Conference on Optical Communications and Networks (ICOCN), Wuzhen, China, 7–10 August 2017.
30. Ma, J.; He, J.; Chen, M.; Zhou, Z.; Liu, Y.; Xiao, Y.; Cheng, Y. Cost-effective SFO compensation scheme based on TSs for OFDM-PON. *J. Opt. Commun. Netw.* **2019**, *11*, 299–306. [CrossRef]
31. Oubei, H.; Durán, J.; Janjua, B.; Wang, H.; Tsai, C.; Chi, Y.; Ng, T.; Kuo, H.; He, J.; Alouini, M.; et al. Wireless optical transmission of 450 nm, 3.2 Gbit/s 16-QAM-OFDM signals over 6.6 m underwater channel. In Proceedings of the CLEO: Science and Innovations, San Jose, CA, USA, 5–10 June 2016.
32. Zhang, J.J.; Tang, Z.H.; Giddings, R.; Wu, W.; Wang, W.L.; Cao, B.Y.; Zhang, Q.W.; Tang, J.M. Stage-Dependent DSP operation range clipping-induced bit resolution reductions of full parallel 64-Point FFTs incorporated in FPGA-Based optical OFDM receivers. *J. Lightwave Technol.* **2016**, *34*, 3752–3760. [CrossRef]
33. Tian, P.; Chen, H.; Wang, P.; Liu, X.; Chen, X.; Zhou, G.; Zhang, S.; Lu, J.; Qiu, P.; Qian, Z.; et al. Absorption and scattering effects of Maalox, chlorophyll, and sea salt on a micro-LED-based underwater wireless optical communication. *Chin. Opt. Lett.* **2019**, *17*, 100010. [CrossRef]

Article

An Indoor Visible Light Positioning System for Multi-Cell Networks

Roger Alexander Martínez-Ciro [1,*], Francisco Eugenio López-Giraldo [1], José Martín Luna-Rivera [2] and Atziry Magaly Ramírez-Aguilera [3]

1 Facultad de Ingenierías, Instituto Tectonológico Metropolitano (ITM), Calle 54A No. 30-01, Barrio Boston, Medellin 050012, Colombia; franciscolopez@itm.edu.co
2 Facultad de Ciencias, Universidad Autónoma de San Luis Potosí (UASLP), Av. Chapultepec 1570, San Luis Potosi 78295, Mexico; mlr@fciencias.uaslp.mx
3 Unidad Académica de Ingeniería Eléctrica, Universidad Autónoma de Zacatecas (UAZ), Campus Siglo XXI, Zacatecas 98160, Mexico; atziry.ra@uaz.edu.mx
* Correspondence: rogermartinez@itm.edu.co; Tel.: +57-460-0727 (ext. 5586)

Abstract: Indoor positioning systems based on visible light communication (VLC) using white light-emitting diodes (WLEDs) have been widely studied in the literature. In this paper, we present an indoor visible-light positioning (VLP) system based on red–green–blue (RGB) LEDs and a frequency division multiplexing (FDM) scheme. This system combines the functions of an FDM scheme at the transmitters (RGB LEDs) and a received signal strength (RSS) technique to estimate the receiver position. The contribution of this work is two-fold. First, a new VLP system with RGB LEDs is proposed for a multi-cell network. Here, the RGB LEDs allow the exploitation of the chromatic space to transmit the VLP information. In addition, the VLC receiver leverages the responsivity of a single photodiode for estimating the FDM signals in RGB lighting channels. A second contribution is the derivation of an expression to calculate the optical power received by the photodiode for each incident RGB light. To this end, we consider a VLC channel model that includes both line-of-sight (LOS) and non-line-of-sight (NLOS) components. The fast Fourier transform (FFT) estimates the powers and frequencies of the received FDM signal. The receiver uses these optical signal powers in the RSS-based localization application to calculate the Euclidean distances and the frequencies for the RGB LED position. Subsequently, the receiver's location is estimated using the Euclidean distances and RGB LED positions via a trilateration algorithm. Finally, Monte Carlo simulations are performed to evaluate the error performance of the proposed VLP system in a multi-cell scenario. The results show a high positioning accuracy performance for different color points. The average positioning error for all chromatic points was less than 2.2 cm. These results suggest that the analyzed VLP system could be used in application scenarios where white light balance or luminaire color planning are also the goals.

Keywords: visible light communication (VLC); visible light positioning (VLP); free-space communication; RGB LED

Citation: Martínez-Ciro, R.A.; López-Giraldo, F.E.; Luna-Rivera, J.M.; Ramírez-Aguilera, A.M. An Indoor Visible Light Positioning System for Multi-Cell Networks. *Photonics* **2022**, *9*, 146. https://doi.org/10.3390/photonics9030146

Received: 13 December 2021
Accepted: 7 February 2022
Published: 1 March 2022

1. Introduction

Indoor positioning systems (IPS) have been widely studied in the literature for different applications such as indoor navigation in museums and exhibition centers, tracking people or objects in indoor scenarios, robot movement control, location-based advertisement distribution in stores, etc. [1]. Some existing technologies have been used to provide indoor localization services, including Bluetooth, Wi-Fi, infrared ray, radio frequency identification, ultra-wideband, ZigBee, ultrasonic, and more recently, visible light communication (VLC). In particular, visible light positioning (VLP) is a VLC-based technology that has attracted more attention due to its high accuracy and low-cost implementation [2]. In recent years, VLC-based localization systems have already been proposed for application in healthcare facilities, indoor public spaces, shopping centers, underground mines, the Internet of Things (IoT), etc. [2–7]. Various methods can be adopted to build a VLP system, such as angle of arrival (AOA), time of arrival (TOA), and received signal strength (RSS). Although TOA requires strict synchronization between transmitting and receiving, AOA needs higher hardware requirements. On the other hand, RSS-based methods have attracted extensive attention for their high positioning accuracy, low cost, and simplicity in hardware. In this way, this paper adopts an RSS-based indoor localization solution [8–10].

In principle, a light detector-based VLP system uses a photodiode sensor to capture the optical signal from WLED lamps and utilizes the RSS information of WLED lamps for positioning calculation with a trilateration algorithm [11]. The adoption of WLEDs in lighting applications has yielded many VLP solutions in the literature. The advantages of WLEDs include a long lifetime, small size, low power consumption, and high efficiency [12,13]. A VLP solution based on WLEDs can be classified according to the number of LED lights used for positioning, i.e., a single LED or multiple LEDs. In 2013, Kim et al. proposed an RSS-based VLP system using a radio frequency (RF) carrier allocation technique with three WLEDs. The results of the experiments for a single-cell showed position error estimates of about 2.4 cm on average in an indoor space of $60 \times 60 \times 60$ cm^3 [14]. In 2016, Hsu et al. proposed an indoor visible light positioning experiment that combines the LED's ID positioning, RSS, radio frequency carrier allocation, and a solar cell as an optical receiver. This VLP system also uses three WLEDs in a single cell, which achieved centimeter-level positioning accuracy [15]. Another experimental VLP solution is presented in [16], based on linear interpolation, RF carrier, and three white LEDs. The results showed that the achieved positioning error is lower than 5 cm. The work of Xu et al. [17] used multiple photodiodes to help WLED position calculation with projective geometry and RSS indications. Their VLP system achieved a positioning error of 13 cm with an architecture based on two photodiodes and two WLEDs. In [18], Cai et al. used a particle swarm optimization (PSO) algorithm to perform a three-dimensional coordinate estimation in an indoor environment of $0.9 \times 0.9 \times 1.5$ m^3 with four WLEDs and one photodiode. Nevertheless, it is not advantageous to use PSO in a real-time VLP system with multiple cells due to the iterations required for the localization problem. In [19], Huang et al. employ received signal strength for a two-dimensional VLP system with a positioning accuracy of about 8 cm in a small $200 \times 60 \times 60$ cm^3 space. On the other hand, there are only few research works that have proposed a VLP architecture for a multi-cell area based on white LEDs [14–16,20–23]. For example, in [24], Little et al. proposed a multi-cell lighting testbed for a VLP system based on the RSS technique and radio frequency allocation. The testbed was constructed with 15 WLEDs in an indoor environment of $1.8 \times 3.9 \times 1.47$ m^3 with an average accuracy of 50 cm. However, this work did not provide simulation or experimental results. In addition, when the application scenario has a large coverage area, the complexity of the implementation could be high due to the multiple transmitter LEDs needed in the VLP system. In [25], a visible light positioning system is proposed for indoor Internet of Things. The simulation system is conducted with 4 cells and 16 WLEDs in a room size of $10 \times 10 \times 3$ m^3. A filter bank of multicarrier-based subcarrier multiplexing (FBNC-SCM) techniques was exploited to provide a high-speed data rate and high-accuracy positioning. However, Tx and Rx modulation and demodulation processes are very complex due to the multiple sub-processes such as the synchronization of all WLEDs, band and low-pass filters, inverse fast Fourier transform, quadrature amplitude modulation and others.

On the other hand, there are few existing works on multi-color transmission channels for VLP systems. Trichromatic WLEDs based on red–green–blue (RGB) LEDs have become promising in the VLP system because they offer the possibility to perform wavelength division multiplexing (WDM) and color shift keying (CSK) [26,27]. In [26], Vieira et al. proposed a VLP based on RGB LEDs with WDM, a double PIN photodetector with two UV light biased gates, the RSS technique, and the multilateration method. Additionally, an extra ultraviolet LED was added to the system for error control in the synchronization process of the transmitter and receiver. The VLP system performs different filtering processes and decodes encoded signals for recovering the transmitted data information. However, Vieira et al. [26] only presents a VLP architecture for one cell where the complexity is affected due to multiple processes. Furthermore, it does not provide an explicit expression to estimate the Euclidean distances of the VLP system based on RGB LEDs. Moreover, the work in [26] does not report the result of the evaluation of the localization error's performance.

In this paper, we propose for the first time an indoor VLP solution for a multi-cell network using RGB LEDs. The positioning system combines a frequency division multiplexing (FDM) scheme with an RSS-based trilateration method within a network with K cells. Each cell consists of three RGB LED transmitters and one photodiode detector located at the mobile user as the optical receiver. This method resembles the VLC system using color-shift keying (CSK) modulation, as proposed in the standard IEEE 802.15.7 [27–30]. Some important features of the CSK scheme are adapted to the localization problem in the VLP system of this work. Furthermore, we enable FDM signals on the CSK symbols to transmit the identification (ID) of the luminaries but also to mitigate inter-cell interference. A total of seven chromatic points are used in the VLC transmitter configuration. The expression of the Euclidean distance for the 7-CSK constellation is calculated according to the optical channel response and the power spectral density in the received signal. For the analysis of the VLP system, we adopted the VLC channel with line-of-sight (LOS) and non-line-of-sight (NLOS) components. The evaluation of the error performance is given as a function of the chromatic point transmitted by the RGB LED. Finally, the properties of the proposed positioning system are investigated using Monte Carlo simulations.

The sections to come are organized as follows. Section 2 presents the conventional VLP system model based on WLEDs. The proposed VLP system model is then introduced in Section 3 to include RGB LEDs. After that, the results are presented in Section 4, followed by the conclusions in Section 5.

2. VLP System Model Based on WLEDs

This section gives a brief overview of the operating principles of the existing VLP system models using WLEDs [14–16]. Although some research works in the literature consider multiple cells within the VLP architecture [14–16,20], they confine the simulation or experimental evaluations to the single-cell scenario. Therefore, let us first review the basic VLP model using WLEDs. Figure 1 presents a conventional single-cell architecture of the VLP system based on WLEDs for a two-dimensional (2D) location system [14,15,20].

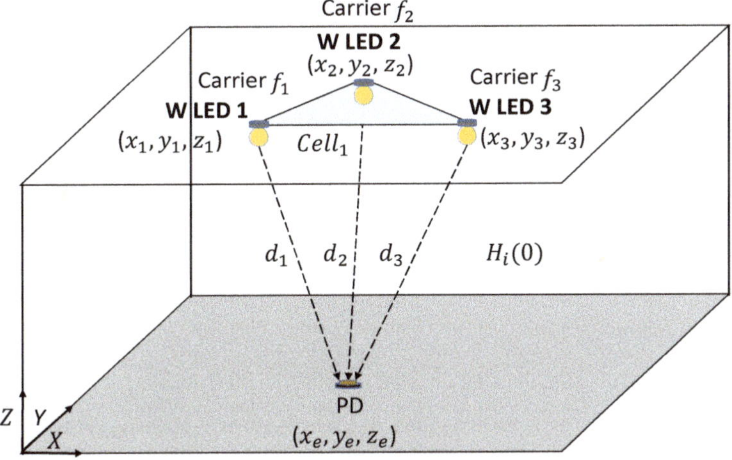

Figure 1. Architecture and geometric model of the VLP for one cell based on three WLEDs.

In this architecture, three WLEDs placed on the ceiling in a triangular configuration provide coverage for the cell. The principle of VLP is that each WLED transmits a unique signal that allows users to calculate their position. An optical receiver, such as a photodiode (PD), can use this information and a trilateration algorithm to calculate the location of the user.

The application of the trilateration method requires the signal of at least three WLEDs to calculate a 2D position of the user. It estimates the Euclidean distances d_i between the PD and each WLED position, which mathematically can be expressed as [14,15,20]:

$$d_i = \sqrt{(x_e - x_i)^2 + (y_e - y_i)^2 + (z_e - z_i)^2}, \tag{1}$$

where $(x_i,\ y_i,\ z_i)$ is the coordinate of the i-th WLEDs for $i = 1,2,3$, and $(x_e,\ y_e,\ z_e)$ is the PD position. The height of each WLED on the ceiling is the same; that is, $z_1 = z_2 = z_3 = Z$ and $z_e = 0$, where $(X,\ Y,\ Z)$ are the workspace dimensions. Then, the estimated 2D position, $\mathbf{X} = [x_e\, y_e]^T$, can be calculated by following a system of two linear equations. Such a system can be rewritten in a matrix form as follows [14,15]:

$$\mathbf{AX} = \mathbf{B}, \tag{2}$$

where

$$\mathbf{A} = \begin{bmatrix} x_2 - x_1 & y_2 - y_1 \\ x_3 - x_1 & y_3 - y_1 \end{bmatrix},\quad \mathbf{X} = \begin{bmatrix} x_e \\ y_e \end{bmatrix},\quad \text{and } \mathbf{B} = \begin{bmatrix} \frac{(d_1^2 - d_2^2 + x_2^2 + y_2^2 - x_1^2 - y_1^2)}{2} \\ \frac{(d_1^2 - d_3^2 + x_3^2 + y_3^2 - x_1^2 - y_1^2)}{2} \end{bmatrix}.$$

As previously mentioned in Section 1, several approaches can be used to calculate the Euclidean distances between the PD and the WLEDs, such as RSS, AOA, or TOA [14–16,20]. For the study of the VLP system proposed in this work, we take as reference the WLED-based VLP system presented by Constanzo et al. [20]. However, it should be noted that their model is useful for a VLP system with one cell. Therefore, in Constanzo et al., the Euclidean distances d_i are derived using the RSS method [20]. The FDM signals are adopted to divide the total bandwidth into a series of frequency sub-bands corresponding to each optical signal transmitted by WLEDs, P_{Topt}, with line-of sight (LOS). The light sensor transforms the incident optical power into a photovoltage, $x_i(t)$, conditioned and processed by electronic devices. The calculation of d_i between a WLED and the light sensor can be calculated as follows [20]:

$$d_i = \left(\frac{h_i^m (m+1) A \zeta P_{Topt} T(\psi) g(\psi)}{2\pi P_{ri}} \right)^{\frac{1}{m+2}}, \tag{3}$$

where h_i^m is the height between the i-th WLED and the PD receiver, with m as the order of the Lambertian radiation; A is the effective area of the photodiode; ζ is the calibrating factor; $T(\psi)$ is the gain of the optical filter; $g(\psi)$ is the receiver's optical concentrator gain; ψ is the angle of incidence with respect to the axis normal to the receiver surface; and P_{ri} is the receiver's optical power. Observe that m is related to $\phi/2$, which is the transmitter semi-angle, by $m = -ln2/ln(cos(\phi/2))$. In [20], they proposed to use the power spectral density of the photovoltage signal to estimate the received optical power values, as shown in the equation below (4).

$$Pr_i = \int_{f_i - \frac{f_i}{Q_i}}^{f_i + \frac{f_i}{Q_i}} PSD[x_i(t)]df, \tag{4}$$

where $PSD[\cdot]$ defines the power spectral density of the signal, and f_i and Q_i, for $i = 1,2,3$, are the carrier frequencies and the quality factors of peak filters, respectively.

Finally, the $(x_i,\ y_i,\ z_i)$ position of the luminaires is estimated by identifying binary codes or the frequency of the carrier signals associated with each ith transmitter [14–16,20].

3. Proposed VLP System Model Based on RGB LEDs

3.1. System Model

The conceptual multi-cell architecture of the proposed VLP system is presented in Figure 2. The coverage area of the indoor scenario is divided into K small cells, $cell_k$, for $k = 1, 2, 3 \ldots K$, each consisting of three RGB LEDs. Figure 3 shows the proposed method of the VLP system for a multi-cell network based on luminaire (L) RGB LEDs. This system includes three stages: (1) transmission protocol, (2) optical receiver, and (3) localization process. The first stage consists of the color coding based on the CIE 1931 chromaticity space, the FDM+DC technique, RGB LEDs as VLC transmitters, and optical channels based on the Lambertian model. The optical receiver is based on the Thorlabs' PDA36A light detector model, where we exploit the sensitivity response of the photodiode for estimating the RGB light power as a function of the photocurrent. This process is similar to those utilized in [27,31]. Then, the PDA36A converts the photocurrent into a voltage signal with a transimpedance amplifier. Finally, in the localization process, the FFT is applied to the voltage signal. After that, we suggest a mathematical procedure to calculate the Euclidean distance between the RGB LED emitter and the optical receiver. Next, the trilateration algorithm is used to estimate the receiver position.

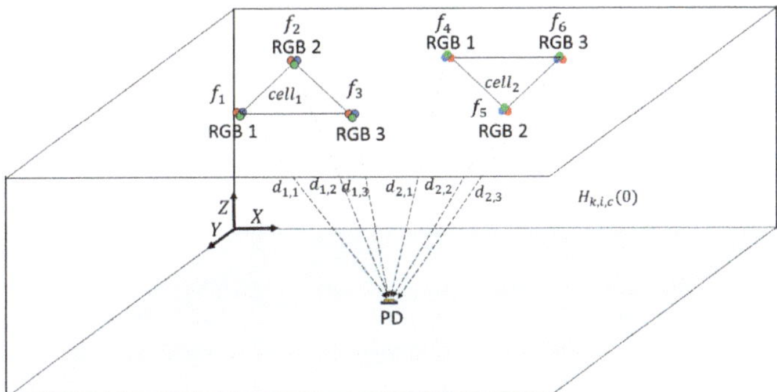

Figure 2. Muti-cell architecture of the proposed VLP system based on RGB LEDs.

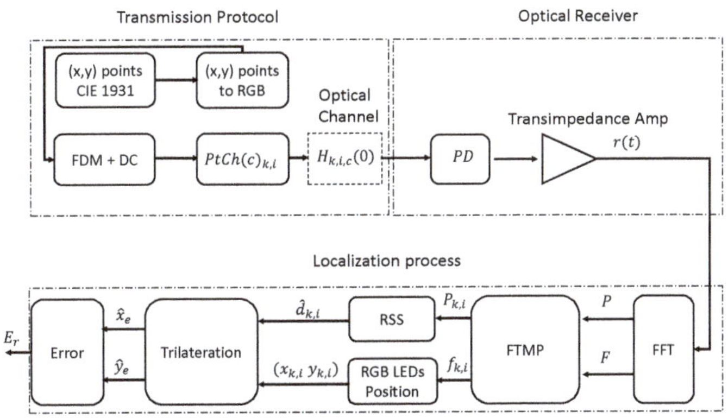

Figure 3. Proposed method for the RGB LED-based VLP system.

3.2. Transmission Protocol

In this section, we use the notation of $PtCh(c)_{k,i}$ to mention the transmitted optical power of some c channels for c = red (R), green (G) or blue (B), of the RGB LED i in $cell_k$. Note that all channels of an RGB LED transmit the same carrier signal. This is to take advantage of the maximum power emitted by each RGB channel. This is explained in detail in this section.

In every $cell_k$ of the network, each RGB LED will transmit a VLC signal with a unique identifier through a specific carrier frequency. Our approach allows for exploring the CIE 1931 RGB color space [32] to enable symbol mapping and constellation design in the same fashion as in color-shift keying (CSK) schemes [27]. Figure 4 illustrates an example of the color space constellation diagram for the 7-CSK modulation scheme with a triangular region denoted by the "RGB" vertices.

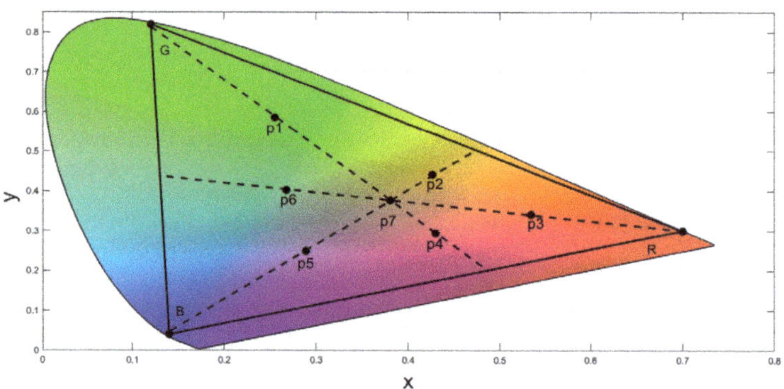

Figure 4. CIE 1931 chromatic points to explore in the VLP system based on RGB LEDs.

The transmitter of the VLP system is composed of multiple processes. The first process is the selection of the chromatic point on CIE 1931 space. This constellation point generates normalized optical powers for each RGB light channel. Then, these power levels are used to configure the FDM scheme and the direct current (DC) generation block. The signals generated by these systems (FDM + DC) modulate the bias current of the RGB LEDs. The FDM signal encodes the identifiers (ID) of each RGB LED through frequencies, and the DC signal configures each RGB LED to yield positive optical signals. This mechanism ensures that the RGB LED transmits the ID while maintaining color balance.

Specifically, the transmission protocol is defined as follows. First, we select one chromatic point on CIE 1931 color space (see Figure 4) for transmitter configuration. The chromatic point is represented by the normalized optical power vector $[P_r,\ P_g,\ P_b]^T$ through the Equation (5). Each chromatic point $(x_p,\ y_p)$ could be explored in the transmission of the carrier signals of the VLP system. Therefore, any constellation point (from P_1 to P_7) can be represented by the vector of optical powers

$$\mathbf{P} = \begin{bmatrix} P_r \\ P_g \\ P_b \end{bmatrix} = \begin{bmatrix} x_r & x_g & x_b \\ y_r & y_g & y_b \\ 1 & 1 & 1 \end{bmatrix}^{-1} \begin{bmatrix} x_p \\ y_p \\ 1 \end{bmatrix}, \quad (5)$$

where the ordered pairs $(x_r,\ y_r)$, $(x_g,\ y_g)$ and $(x_b,\ y_b)$ represent the color values $(x,\ y)$ associated with the peak wavelength of the sources of red, green, and blue light, respectively [27]. The optical power signals $[P_r\ P_g\ P_b]^T$ of ith RGB LEDs in each $cell_k$ are transmitted through the free space channel $H_{k,i,c}(0)$.

Using an FDM technique, the ith RGB LED in $cell_k$ will transmit its identification signal $PtCh(c)_{k,i}$ in a specific RF carrier wave with a modulation frequency $f_{k,i}$ through each particular channel c. Since the carrier wave exhibits signal variations from negative to positive values, a DC bias should be applied. Therefore, the carrier signals transmitted by each channel of the RGB LED are shown in Equation (6):

$$PtCh(c)_{k,i} = \frac{P_c PLED_c sin(2\pi f_{k,i}t) + 1)}{2}, \tag{6}$$

where P_c is the normalized optical power for channel c, and $PLED_c$ corresponds to the maximum emission optical power of the c channels of the LEDs. Notice that these optical powers will depend on the characteristics of the RGB LED used. The chromaticity of the light emitted by RGB LEDs is established through the aforementioned process. Following electrical-to-optical conversion, the optical signal is propagated through free space, where the channel DC gain of the visible light communication system, $H_{k,i,c}(0)$, is composed of an LOS and non-LOS (NLOS) component, such that $H_{k,i,c}(0) = H_{k,i,c}(LOS) + H_{k,i,c}(NLOS)$ [33]. This can be expressed by (7) [33]:

$$H_{k,i,c}(0) = \frac{(m+1)}{2\pi d_{k,i}^2} Acos_{k,i}^m(\phi)cos(\psi)T(\psi)g(\psi) + \frac{A}{A_{room}} \frac{\langle \rho_c \rangle}{1 - \langle \rho_c \rangle}, \tag{7}$$

where A_{room} is the room surface, and $\langle \rho_c \rangle$ expresses the average reflectivity of a given room. According to the definition of $\langle \rho_c \rangle$, we can adopt the $\langle \rho_r \rangle$, $\langle \rho_g \rangle$ and $\langle \rho_b \rangle$ values proposed in [34]. On the other hand, we use the definition in [20,35] for the LOS channel. They assume $cos(\psi) = cos(\phi)$ is also equal to $h_{k,i}/d_{k,i}$ for horizontal orientation. Equation (7) can be rewritten as:

$$H_{k,i,c}(0) = \frac{(m+1)}{2\pi d_{k,i}^{m+3}} Ah^{m+1}T(\psi)g(\psi) + \frac{A}{A_{room}} \frac{\langle \rho_c \rangle}{1 - \langle \rho_c \rangle}. \tag{8}$$

3.3. Optical Receiver

We make use of only one photodetector to convert the optical signals received from the RGB channels, to an electrical photocurrent [30,31]. Then, the photocurrent signal is used for the PD positioning problem, considering RF identification and the RSS technique. Next, the receiver optical power $PrCh(c)$ for c channels in the $cell_k$ can be determined as follows:

$$PrCh(c) = \sum_{k=1}^{K} \sum_{i=1}^{3} PtCh(c)_{k,i} H_{k,i,c}(0). \tag{9}$$

We use a single photodiode as an optical receiver of the RGB signals, in a similar fashion to the CSK-based VLC system proposed in [27,31] for low complexity receivers. However, the photodiode transforms each component of the RGB light into a photocurrent signal $r(t)$ corrupted by additive Gaussian noise $n(t)$ (AWGN), as shown in Equation (10):

$$r(t) = G_{amp} S PrCh(c) R(\lambda_c) + n(t), \tag{10}$$

where G_{amp} corresponds to the transimpedance gain; S is the scale factor; and the scalars $R(\lambda_R)$, $R(\lambda_G)$ and $R(\lambda_B)$ determine the photodiode responsivity associated with each red, green and blue wavelength, respectively [36]. Note that, the receiver optical power $PrCh(c)$ is made up of several carrier signals on different RGB light channels. Such a scheme takes advantage of the FDM structure by avoiding the use of multiple photodiodes with RGB filters in the receiver. Therefore, the localization process requires decomposing the FDM signal at the receiver by applying the fast Fourier transform (FFT) on the RGB optical power signal received at the PD. This process allows for estimating the power-spectral density (PSD) of the received power signal, as shown in Equation (11):

$$PSD[PrCh(c)R(\lambda_c)] = PSD\left[\frac{r(t)}{G_{amp}S}\right]. \tag{11}$$

3.4. Localization

In this section, we describe the localization stage for the VLP system based on RGB LEDs. This process is mainly divided into two phases. At first, the Euclidean distances between the RGB LED transmitters and the optical receiver are computed by processing the FFT of the voltage signal generated with the light sensor and using an RSS-based method. A simple frequency identification algorithm is used for the RGB LED position estimation. The second phase performs the trilateration algorithm that uses the Euclidean distance and the RGB LED position parameters to estimate the receiver location.

Figure 5 shows an example of the received RGB power vector for a VLP system with two cells. In this example, we use six carrier frequencies to modulate the RGB LEDs with $Pr_{k,i}$ as the maximum value of the received optical power for the R, G and B channels of the $cell_k$ with $k = 1,2$. Considering Equation (11), we can apply the integral to the PSD signal in the interval $f_{k,i} \pm \Delta$ to calculate $Pr_{k,i}$, as it is shown in Equation (12):

$$Pr_{k,i} = R(\lambda_c)PtCh(c)_{k,i}H_{k,i,c}(0) = max\left(\int_{f_{k,i}-\Delta}^{f_{k,i}+\Delta} PSD\left[\frac{r(t)}{G_{amp}S}\right]df\right), \tag{12}$$

where $f_{k,i}$ represents the carrier frequency of luminaire i in cell k, and $\pm\Delta$ is the frequency limits used to delimit the area of peak powers $Pr_{k,i}$. It should be noted that only three optical signals are needed for the two-dimensional positioning problem. Consequently, a basic search algorithm is carried out on the signal $Pr_{k,i}$ to identify the three-maximum powers (TMP) and the associated frequencies. The Euclidean distance $\hat{d}_{k,i}$ is then estimated by measuring the received signal strength (RSS). Note that $Pr_{k,i}$ and $\hat{d}_{k,i}$ are scalar values. Therefore, we only consider the root mean square (RMS) values of the optical signal transmitted by each channel c, with $Ptrms(c)_{k,i} = RMS(PtCh(c)_{k,i})$. In addition, it should be noted that each RGB LED transmits the same carrier signal through the c channels. Therefore, the total contribution of the RMS optical powers emitted by the c channels was considered as a summation. Consequently, and substituting Equation (8) in (12), the Euclidean distance is estimated as shown in Equation (13):

$$\hat{d}_{k,i} = \sqrt[m+3]{\frac{(m+1)\sum_{c=R}^{B}\left(R(\lambda_c)Ptrms(c)_{k,i}\right)Ah^{m+1}T(\psi)g(\psi)}{2\pi\left(Pr_{k,i}-\sum_{c=R}^{B}\left(R(\lambda_c)Ptrms(c)_{k,i}H_{k,i,c}(NLOS)\right)\right)}}. \tag{13}$$

On the other hand, the frequencies $f_{k,i}$ are used to identify the coordinate $(x_{k,i}, y_{k,i})$ of luminaire i of the $cell_k$. Finally, we perform the trilateration algorithm with $\hat{d}_{k,i}$ and $(x_{k,i}, y_{k,i})$ to estimate the position of PD $(\hat{x}_e, \hat{y}_e)$ using Equation (2).

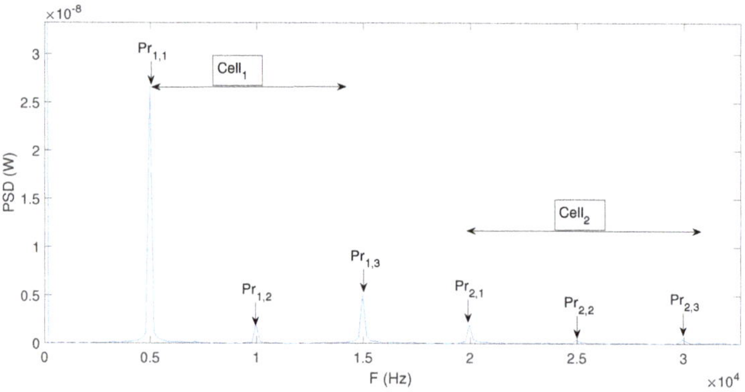

Figure 5. Power spectral density on the RGB optical power vector for VLP-based RGB LEDs

4. Simulation Results and Discussion

In this section, for the various color points addressed in Section 3, we evaluate the indoor positioning performance under the LOS and NLOS VLC channel model. We use the Monte Carlo method to evaluate the positioning error of the proposed RGB LED- based VLP system. The simulation results are presented using different mobile locations and chromatic points for the RGB LEDs. The chromaticity coordinates on the CIE 1931 of the red, green, and blue LEDs were (0.70, 0.30), (0.19, 0.78), and (0.09, 0.13). We explored a total of seven color points p_1 to p_7 for the RGB LED configuration. Each chromatic point was evaluated in the VLP system. The specific chromatic point (x_p, y_p) and the normalized optical power vector $[P_r\ P_g\ P_b]^T$ are shown in Table 1. It should be noted that, each optical power vector satisfies the rule $P_r + P_g + P_b = 1$ for the CIE 1931 standard [32]. The simulated VLP system considers two cells with a total of six RGB LEDs, as it is shown in Figure 2. This architecture allows us to validate the proposed method presented in Figure 3. We highlight that this architecture can be extended for multiple cells depending on the area of the application scenario.

Table 1. Color coordinates and normalized optical power vector.

Chromatic Point	(x_p, y_p)	$[P_r\ P_g\ P_b]^T$
p_1	(0.2549, 0.5849)	$[0.2274\ 0.6228\ 0.1498]^T$
p_2	(0.4264, 0.4410)	$[0.5236\ 0.3396\ 0.1369]^T$
p_3	(0.5349, 0.3411)	$[0.7105\ 0.1492\ 0.1403]^T$
p_4	(0.4310, 0.2940)	$[0.5233\ 0.1512\ 0.3255]^T$
p_5	(0.2892, 0.2490)	$[0.2728\ 0.1770\ 0.5502]^T$
p_6	(0.2676, 0.4024)	$[0.2416\ 0.3841\ 0.3743]^T$
p_7	(0.3804, 0.3769)	$[0.4395\ 0.2854\ 0.2751]^T$

The parameters used in the simulations of the VLP system are summarized in Table 2. The ceiling is fixed at a height of 2.2 m from the receiver (photodiode), assuming that in real applications, the user could carry the receiver at a height (z_e) of 0.8 m from the floor. We divide the space in the X–Y plane into a 4 × 7 grid. Each point of the grid indicates a test spot and the real coordinates (x_e, y_e) of the receiver. For the VLC channel model, we use the definition in [33] for the NLOS channel and we can adopt the average reflectivity proposed in [34]. However, the reflection coefficients presented in [34] were estimated considering the power spectrum of a WLED as an emission source. Consequently, the values for $\langle \rho_r \rangle$, $\langle \rho_g \rangle$ and $\langle \rho_b \rangle$ of the RGB LED adopted in this article are approximate values of the *R*, *G* and *B* components of the WLED proposed by [34]. Next, we estimated the received RF power with the power spectral density according to Equation (11), where the maximum value of the received optical power for *R*, *G* and *B* channels were computed with Equation (12). After solving the Equations (13) and (2), the positioning result can be obtained. Please note that the positioning error is then defined as the Euclidean distance between the actual coordinate (x_e, y_e) and the estimated mobile receiver position $(\hat{x}_e, \hat{y}_e)$ on all the test points.

Figures 6 and 7 show the results of the positioning errors obtained for all chromatic points $(p_1–p_7)$. All chromatic points showed similar localization error performance. It can be seen that the lowest positioning errors were obtained for p_6 (see Figure 7b) (on average 1.96 cm), and the highest errors for p_5 (see Figure 7a) (on average 2.13 cm). This small variation in localization error could be related to the effect of the NLOS component and photodetector responsivity for some wavelengths on the performance of the VLP system.

Table 2. Simulation parameters.

Parameters	Value	Parameters	Value
Dimension space	$X = 6$ $Y = 3$ $Z = 3$ m	RGB LED optical power for channel	3.333 W
RGB 1 coordinates	(0.75, 0.75, 3) m	$f_{1,1}$	5 kHz
RGB 2 coordinates	(1.50, 2.25, 3) m	$f_{1,2}$	10 kHz
RGB 3 coordinates	(2.25, 0.75, 3) m	$f_{1,3}$	15 kHz
RGB 4 coordinates	(3.75, 2.25, 3) m	$f_{2,1}$	20 kHz
RGB 5 coordinates	(4.50, 0.75, 3) m	$f_{2,2}$	25 kHz
RGB 6 coordinates	(5.25, 2.25, 3) m	$f_{2,3}$	30 kHz
RGB LED half-power angle	$\phi/2 = 60°$	$R(\lambda_R)$	0.41 A/W
Detector area	$A = 13^{-6}$ m^2	$R(\lambda_G)$	0.2 A/W
G_{amp}	0.75^6 V/A	$R(\lambda_B)$	0.12 A/W
Scale factor	$S = 0.5$	Noise (RMS)	340^{-6} V
$T(\psi)$	1	$g(\psi)$	1
Sampling rate	96,000 kHz	FFT points	1500
Δ	300 Hz	$\langle \rho_r \rangle$	0.0733
$\langle \rho_g \rangle$	0.0450	$\langle \rho_b \rangle$	0.0558
A_{room}	90 m^2	m	1

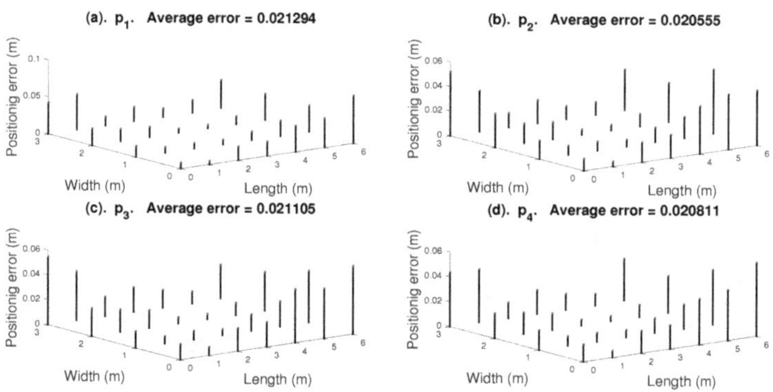

Figure 6. Positioning error for chromatic points p_1 to p_4.

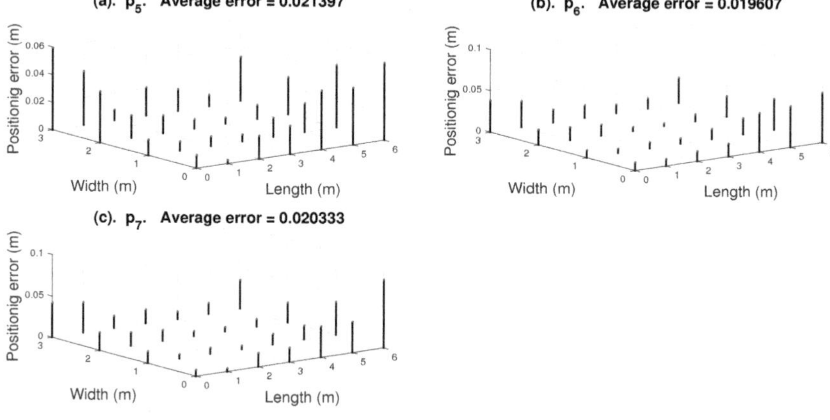

Figure 7. Positioning error for chromatic points p_5 to p_7.

The performance for some of the chromatic points showed larger errors at the sides or corners of the room. This could be caused by the low optical power provided by the RGB channels on the surface of the photodiode. However, the results of the maximum average localization error of 2.13 cm show a good performance of the VLP system based on RGB LEDs. These results validate the proposed method and the derived equations in the parameter estimation process. We remark that any chromatic point in the CIE 1931 space could be exploited in the design of the VLP system based on RGB LEDs. Nevertheless, the color balance should be set according to the lighting requirements of the application scenario. For example, the p_7 color point (see Figure 7c) can be used if the application scenario requires white light balance. Otherwise, the chromaticity of the light contributions by points p_1 to p_6 can be useful in other scenarios such as museums, where the color planning of the luminaires can be configured for some artwork or specimens.

The RGB LED-based VLP system showed good error performance as compared to existing WLED-based VLP architectures [14,16,17,19,24]. Additionally, we note that in VLP architecture based on RGB LEDs, the signal-to-noise ratio could be positively impacted by transmitting the same carrier signal over the RGB channels. Our proposed VLP system also allows us to explore the color temperature of the RGB luminaire, considering the CIE 1931 standard, which is not possible with WLED because the color temperature is fixed. However, it is important to note that the VLP architecture based on RGB LEDs requires more hardware components than that based on WLEDs. On the other hand, our VLP system proposes cells based on three RGB LEDs, where the mathematical expression to estimate the Euclidean distances were derived. Our system requires fewer components than the VLP architecture proposed by [26] because they employed an additional ultraviolet LED in the VLP system based on RGB LEDs. Additionally, they do not provide an explicit expression for estimating the Euclidean distances.

5. Conclusions

In this work, we presented a novel design of a VLP system based on RGB LEDs considering multiple cells and a frequency division multiplexing scheme. We proposed indoor positioning adopting VLC based on the LOS and NLOS environment, which was investigated by simulation. We propose a new method to estimate the position of the mobile user receiver within a multiple-cell coverage network, suggesting different possible sets of chromatic points in CIE 1931 space for the configuration of the transmitters. This system adopted the RSS and trilateration techniques that combine frequency identification for the localization problem. The proposed design achieves a simpler and more flexible transmitter position identification than using frequency identification combined with the modulation technique. The resulting VLP system is able to perform lighting system chromaticity control, visible light communication and indoor positioning simultaneously. A proof-of-concept example was developed to test this approach. We emulated a practical indoor scenario by setting the height of the room at 3 m with a distance of 2.2 m between the RGB LEDs, on the ceiling of the room, and the photodiode receiver. In addition, numerical simulations have been carried out to validate the analytical expression derived for the Euclidean distance vector. These results showed good agreement between the experimental simulation and the theoretical analysis. We observed a high positioning accuracy performance for the chromatic point p_5. However, the average positioning error for all chromatic points was less than 2.2 cm. The proposed VLP system would be useful for various indoor location-based applications. The chromatic point p_7 could be recommended for an application scenario in which the lighting requires white light balance. More importantly, this VLP-based RGB LED maintains compatibility with the use of the CIE 1931 standard.

Author Contributions: Conceptualization, J.M.L.-R. and R.A.M.-C.; formal analysis, J.M.L.-R. and R.A.M.-C.; investigation, F.E.L.-G., R.A.M.-C. and A.M.R.-A.; software, R.A.M.-C.; writing–original draft preparation, R.A.M.-C. and F.E.L.-G.; writing–review and editing, F.E.L.-G., A.M.R.-A. and J.M.L.-R. All authors have read and agreed to the published version of the manuscript.

Funding: This research received no external funding.

Institutional Review Board Statement: Not applicable.

Informed Consent Statement: Not applicable.

Data Availability Statement: Not applicable.

Acknowledgments: This research and the APC were funded by Instituto Tecnológico Metropolitano [grant PCI 20203], Colombia.

Conflicts of Interest: The authors declare no conflict of interest.

References

1. Zafari, F.; Gkelias, A.; Leung, K.K. A Survey of Indoor Localization Systems and Technologies. *IEEE Commun. Surv. Tutor.* **2019**, *21*, 2568–2599. [CrossRef]
2. Zhuang, Y.; Cao, Y.; Wu, Y.; Member, S.; Thompson, J.; Haas, H. A Survey of Positioning Systems Using Visible LED Lights. *IEEE Commun. Surv. Tutor.* **2018**, *20*, 1963–1988. [CrossRef]
3. Yasir, M.; Ho, S.; Vellambi, B. Indoor Position Tracking Using Multiple Optical Receivers. *J. Lightwave Technol.* **2016**, *34*, 1166–1176. [CrossRef]
4. Rahman, A.B.M.M.; Li, T.; Wang, Y. Recent Advances in Indoor Localization via Visible Lights: A Survey. *Sensors* **2020**, *20*, 1382. [CrossRef] [PubMed]
5. Chen, P.; Pang, M.; Che, D.; Yin, Y.; Hu, D.; Gao, S. A Survey on Visible Light Positioning from Software Algorithms to Hardware. *Wirel. Commun. Mob. Comput.* **2021**, *2021*, 9739577. [CrossRef]
6. Almadani, Y.; Ijaz, M.; Rajbhandari, S.; Raza, U.; Adebisi, B. Applications of Visible Light Communication for Distance Estimation: a Short Survey. In Proceedings of the 2019 IEEE Jordan International Joint Conference on Electrical Engineering and Information Technology (JEEIT), Amman, Jordan, 9–11 April 2019; pp. 261–265. [CrossRef]
7. Chaudhary, N.; Alves, L.N.; Ghassemblooy, Z. Current Trends on Visible Light Positioning Techniques. In Proceedings of the 2019 2nd West Asian Colloquium on Optical Wireless Communications (WACOWC), Tehran, Iran, 27–28 April 2019; pp. 100–105. [CrossRef]
8. Stevens, N.; Steendam, H. Magnitude of the Distance Estimation Bias in Received Signal Strength Visible Light Positioning. *IEEE Commun. Lett.* **2018**, *22*, 2250–2253. [CrossRef]
9. Guo, X.; Ansari, N.; Hu, F.; Shao, Y.; Elikplim, N.R.; Li, L. A survey on fusion-based indoor positioning. *IEEE Commun. Surv. Tutor.* **2020**, *22*, 566–594. [CrossRef]
10. Luo, J.; Fan, L.; Li, H. Indoor Positioning Systems Based on Visible Light Communication: State of the Art. *IEEE Commun. Surv. Tutor.* **2017**, *19*, 2871–2893. [CrossRef]
11. Karunatilaka, D.; Zafar, F.; Kalavally, V.; Parthiban, R. LED Based Indoor Visible Light Communications: State of the Art. *IEEE Commun. Surv. Tutor.* **2015**, *17*, 1649–1678. [CrossRef]
12. Do, T.H.; Yoo, M. An in-depth survey of visible light communication based positioning systems. *Sensors* **2016**, *16*, 678. [CrossRef]
13. Popoola, O.; Sinanović, S.; Popoola, W.; Ramirez, R. Optical Boundaries for LED-Based Indoor Positioning System. *Computation* **2019**, *7*, 7. [CrossRef]
14. Kim, H.S.; Kim, D.R.; Yang, S.H.; Son, Y.H.; Han, S.K.; Liao, X.; Lin, K.; Chen, Y. An Indoor Visible Light Communication Positioning System Using a RF Carrier Allocation Technique. *J. Lightwave Technol.* **2013**, *31*, 131–144. [CrossRef]
15. Hsu, C.W.; Wu, J.T.; Wang, H.Y.; Chow, C.W.; Lee, C.H.; Chu, M.T.; Yeh, C.H. Visible Light Positioning and Lighting Based on Identity Positioning and RF Carrier Allocation Technique Using a Solar Cell Receiver. *IEEE Photonics J.* **2016**, *12*, 1–7. [CrossRef]
16. Wu, Y.C.; Hsu, K.L.; Liu, Y.; Hong, C.Y.; Chow, C.W.; Yeh, C.H.; Liao, X.L.; Lin, K.H.; Chen, Y.Y. Using Linear Interpolation to Reduce the Training Samples for Regression Based Visible Light Positioning System. *IEEE Photonics J.* **2020**, *12*, 1–5. [CrossRef]
17. Xu, W.; Wang, J.; Shen, H; Zhang, H.; You, X. Indoor Positioning for Multiphotodiode Device Using Visible-Light Communications. *IEEE Photonics J.* **2016**, *8*, 1–11.
18. Cai, Y.; Guan, W.; Wu, Y.; Xie, C.; Chen, Y.; Fang, L. Indoor High Precision Three-Dimensional Positioning System Based on Visible Light Communication Using Particle Swarm Optimization. *IEEE Photonics J.* **2017**, *9*, 1–20. [CrossRef]
19. Huang, N.; Gong, C.; Luo, J.; Xu, Z. Design and Demonstration of Robust Visible Light Positioning Based on Received Signal Strength. *J. Lightwave Technol.* **2020**, *38*, 5695–5707. [CrossRef]
20. Costanzo, A.; Loscri, V. Error compensation in indoor positioning systems based on software defined visible light communication. *Phys. Commun.* **2019**, *34*, 235–245. [CrossRef]
21. Chen, Y.; Zheng, H.; Liu, H.; Han, Z.; Ren, Z. Indoor High Precision Three-Dimensional Positioning System Based on Visible Light Communication Using Improved Hybrid Bat Algorithm. *IEEE Photonics J.* **2020**, *12*, 1–13. [CrossRef]
22. Alam, F.; Parr, B.; Mander, S. Visible Light Positioning Based on Calibrated Propagation Model. *IEEE Sens. Lett.* **2019**, *3*, 1–4. [CrossRef]
23. Anastou, A.C.; Delibasis, K.K.; Boulogeorgos, A.A.; Sandalidis, H.G.; Vavoulas, A.; Tasoulis, S.K. A Low Complexity Indoor Visible Light Positioning Method. *IEEE Access* **2021**, *9*, 57658–57673. [CrossRef]
24. Little, T.; Rahaim, M.; Abdalla, I.; Lam, E.; Mcallister, R.; Vegni, A. A multi-cell lighting testbed for VLC and VLP. In Proceedings of the 2018 Global LiFi Congress, Paris, France, 8–9 February 2018; Volume 1, pp. 1–6.

25. Yang, H.; Zhong, W.; Chen, C.; Alphones, A.; Du, P.; Zhang, S.; Xie, X. Coordinated Resource Allocation-Based Integrated Visible Light Communication and Positioning Systems for Indoor IoT. *IEEE Trans. Wirel. Commun.* **2020**, *19*, 4671–4684. [CrossRef]
26. Vieira, M.A.; Vieira, M.; Louro, P.; Mateus, L.; Vieira, P. Indoor positioning system using a WDM device based on a-SiC:H technology. *J. Lumin.* **2017**, *191*, 135–138. [CrossRef]
27. Martínez Ciro, R.A.; López Giraldo, F.E.; Betancur Perez, A.F.; Luna Rivera, J.M. Design and Implementation of a Multi-Colour Visible Light Communication System Based on a Light-to-Frequency Receiver. *Photonics* **2019**, *6*, 42. [CrossRef]
28. IEEE Computer Society Sponsored by the IEEE Standards Association: IEEE Standard for Local and Metropolitan Area Networks—Part 15.7: Short-Range Wireless Optical Communication Using Visible Light. September 2011. Available online: https://standards.ieee.org/findstds/standard/802.15.7-2011.html (accessed on 27 November 2018).
29. Mendes Matheus, L.E.; Borges Vieira, A.; Vieira, L.F.; Vieira, M.A.; Gnawali, O. Visible Light Communication: Concepts, Applications and Challenges. *IEEE Commun. Surv. Tutor.* **2019**, *21*, 3204–3237. [CrossRef]
30. Martínez Ciro, R.A.; López Giraldo, F.E.; Betancur Perez, A.F.; Luna Rivera, M. Characterization of Light-To-Frequency Converter for Visible Light Communication Systems. *Electronics* **2018**, *7*, 165. [CrossRef]
31. Luna-Rivera, J.M.; Suarez-Rodriguez, C.; Guerra, V.; Perez-Jimenez, R.; Rabadan-Borges, J.; Rufo-Torres, J. Low-complexity colour-shift keying-based visible light communications system. *IET Optoelectron.* **2015**, *9*, 191–198. [CrossRef]
32. CIE-1931, International Commission on Illumination, Colorimetry—Part 3: CIE Tristimulus Values, ISO 11664-3:2012(E)/CIE S014-3/E:2011. Available online: http://www.cie.co.at/publications/colorimetry-part-3-cie-tristimulus-values (accessed on 20 October 2020).
33. Zhang, X.; Duan, J.; Fu, Y.; Shi, A. Theoretical accuracy analysis of indoor visible light communication positioning system based on Received Signal Strength Indicator. *J. Lightwave Technol.* **2014**, *32*, 4180–4186. [CrossRef]
34. Lee, K.; Park, H.; Barry, J. Indoor channel characteristics for visible light communications. *IEEE Commun. Lett.* **2011**, *15*, 217–219. [CrossRef]
35. Plets, D.; Bastiaens, S.; Martens, L.; Joseph, W. An Analysis of the Impact of LED Tilt on Visible Light Positioning Accuracy. *Electronics* **2019**, *8*, 389. [CrossRef]
36. THORLABS, PDA36A(-EC) Si Switchable Gain Detector. Available online: https://www.thorlabs.com/thorproduct.cfm?partnumber=PDA36A-EC (accessed on 11 January 2021).

 *photonics*

Article

An Optimal Scheme for the Number of Mirrors in Vehicular Visible Light Communication via Mirror Array-Based Intelligent Reflecting Surfaces

Ling Zhan [1,2], Hong Zhao [2], Wenhui Zhang [3] and Jiming Lin [4,*]

[1] Guangxi Key Laboratory of Wireless Wideband Communication and Signal Processing, Guilin University of Electronic Technology, Guilin 541004, China; zhanling@guet.edu.cn
[2] School of Information and Communication, Guilin University of Electronic Technology, Guilin 541004, China; zhaohong@guet.edu.cn
[3] School of Computer Science and Information Security, Guilin University of Electronic Technology, Guilin 541004, China; zhangwh@guet.edu.cn
[4] College of Electronic and Information Engineering, Beibu Gulf University, Qinzhou 535011, China
* Correspondence: linjm@guet.edu.cn

Abstract: The optimization problem of the number of mirrors under energy efficiency (EE) maximization for vehicular visible light communication (VVLC) via mirror array-based intelligent reflecting surface (IRS) is investigated. Under considering that the formulated optimization problem is subject to the real and non-negative of the transmitted signal, the maximum power consumption satisfied luminous ability and eye safety, the minimum achievable rate, and the required bit error ratio (BER), EE is proved to be a unimodal function of the number of mirrors. Then, the binary search-conditional iteration (BSCI) algorithm is proposed for quickly finding the optimal number of mirrors with maximum EE. Numerical results demonstrate that fewer mirrors can obtain the maximum EE, and the computational complexity of the BSCI algorithm is reduced by 10^5 orders of magnitude, compared with the Bubble Sort method.

Keywords: vehicular visible light communication (VVLC); intelligent reflecting surface (IRS); the number of mirrors; energy efficiency (EE)

Citation: Zhan, L.; Zhao, H.; Zhang, W.; Lin, J. An Optimal Scheme for the Number of Mirrors in Vehicular Visible Light Communication via Mirror Array-Based Intelligent Reflecting Surfaces. *Photonics* **2022**, *9*, 129. https://doi.org/10.3390/photonics9030129

Received: 25 January 2022
Accepted: 23 February 2022
Published: 24 February 2022

1. Introduction

Reliable information transmission between vehicles is essential [1–3] in the intelligent transportation system (ITS). Vehicle-to-vehicle (V2V) communication mainly adopts radio frequency (RF) communications currently [4–7]. RF communications are prone to problems, such as lack of spectrum resources, electromagnetic interference, and synchronization limitations when the traffic flow is large and the vehicles are very dense, which brings enormous challenges to reliable V2V communication.

In the visible light communication (VLC) system, the information is sent by the LEDs' high-speed flashing and transmitted through the channel to the receiver [8–10], such as a Photo-Diode (PD) [11], image sensor [12], or high-speed camera [13]. The received optical signal is converted into the electrical signal through photoelectric conversion firstly; then after signal processing, the original information is restored. It can realize the communication while satisfying the luminous ability, which can be used as a technology complementing the RF communications and improve the efficiency of resources, which has the characteristics of rich spectrum resources, high energy efficiency, and greenness.

With the continuous progress of semiconductor technology, LED gradually replaces the traditional light source and becomes an important choice for lamps [14–16], which provides a hardware basis for realizing VLC. When the vehicle is driving on the road, the headlamps or taillights between the front and rear vehicles can be used as the transmitter, and the receiver can be installed on another vehicle, and the light emitted by the LED can reach the receiver directly through the line-of-sight (LOS) link [17–20]. For VLC with the non-line-of-sight (NLOS) link, the road surface can be used as the reflector [21]. The light emitted by the headlamp reaches the road surface firstly and then reaches the receiver through the reflection of the road surface. In this case, the receiver is in the front and the transmitter is in the back, and a certain distance should be maintained to ensure that the reflected light is within the field of view (FOV) of the receiver.

When the headlamps or taillights are used as transmitters, the light emitted by the transmitters cannot reach the receivers which are installed on other vehicles for parallel. According to the propagation characteristics of optics, it also cannot be reached by road reflection. The auxiliary means need to be considered to realize VLC between parallel vehicles.

The intelligent reflecting surface (IRS) [22,23] is a tunable metasurface composed of many low-cost passive reflective elements, which can manipulate the wavelength, polarization, and phase of the incident wave [24,25]. In the RF-based vehicular networks, the metasurfaces can revise the Snell's law that redirecting the radio waves in the desired direction, which solves the problem that the communication is obstructed by strong obstacles and extends coverage in the highly dynamic vehicular environment [26], realizing keyless, secure transmission [27].

In the optical wireless communication, beam steering [28], beam shaping [29], and improving the service level of the link [30] for coherent light using metasurface-based IRS have been studied. For incoherent light (such as visible light), AM Abdelhady et al. [31] install IRS on the wall which reflects the incident light to the receiver by intelligently controlling the phase gradient of each metasurface and the orientation of each mirror in the indoor environment and the results proved that the performance of the mirror array is better than that of the metasurface. For the VLC between parallel vehicles, the mirror array-based IRS can be installed on the transportation infrastructure, and the light from the transmitter is reflected into the receiver by controlling the rotation angle of each mirror, which solves the problem of realizing VLC for parallel vehicles. Compared with the hybrid VLC-WiFi [32], the hardware implementation is simple, and the disadvantages of RF communication are solved.

In wireless communications, EE is defined as the ratio of transmitted bits to energy consumption. It is usually expressed in bits per Joule (bits/J) [33–35]. The higher the EE, the less energy the system expends for the same communication performance. It mainly contains two elements that are achievable rate and power consumption.

- To ensure the effectiveness of the communication system, the achievable rate needs to reach a certain value. Since the transmitted signal is non-negative, real, and limited amplitude, the classical Shannon capacity formula is not suitable for VLC. Researchers have been studied the lower bound of the achievable capacity of the VLC system [36–38], and the achievable rate is proportional to the signal-to-noise ratio (SNR) [39]. Each mirror in the IRS is independently controlled, and the light reaches the receiver through their reflection. The total channel gain equals the sum of channel gain corresponding to each mirror, and the SNR becomes larger with the number of mirrors increasing.So, the achievable rate is not only related to the channel gain corresponding to each mirror, but also to the number of mirrors.
- In the IRS-aided VLC system, the power consumption of the system is mainly included that of the transmitter, receiver, and IRS. The power consumption of the transmitter and receiver mainly includes signal power, DC offset, and the hardware static power consumption [40,41]. The power consumption of the IRS equals the sum of that for each mirror rotating. Therefore, the total power consumption changes depending on the number of mirrors.

Because the achievable rate and power consumption are related to the number of mirrors, the EE is also affected by the number of mirrors in the VLC system via mirror array-based IRS. To get the maximum EE, it is necessary to optimize the number of mirrors. Although the time allocation, power control, and phase matrix are analyzed for EE optimization [42], the influence of the number of mirrors in IRS on EE has not been analyzed, as far as the authors know.

The main contributions of this paper are as follows.

- The VLC system via mirror array-based IRS for parallel vehicles is designed, which provides convenience for parallel vehicles to realize VLC. The right headlamp of the right vehicle is used as the transmitter, the receiver is installed between the two headlamps of the left vehicle, and the IRS is installed on the street light pole. The channel model of the system is analyzed, and the channel gain is calculated.
- The calculation methods of the achievable rate and power consumption are given. According to the system model, the calculation formulas of the SNR and the instantaneous achievable rate are given. Based on reference [40], the total power consumption of the system and the power consumption of each mirror are analyzed. Both the achievable rate and the total power consumption are functions of the number of mirrors $\boldsymbol{N}$, and thus EE is also a function of $\boldsymbol{N}$.
- The number of mirrors optimization problem under the EE maximization is formulated. Considering the non-negative of the transmitted signal, the maximum power consumption satisfied luminous ability and eye safety, the minimum achievable rate, and the required bit error rate (BER), the optimal value of $\boldsymbol{N}$ is found. According to the constraints and the properties of the achievable rate, EE is proved to be a unimodal function.
- The binary search-conditional iterative (BSCI) algorithm is proposed to optimize $\boldsymbol{N}$. According to the constraints of the optimization problem, the range of $\boldsymbol{N}$ is analyzed. The BSCI algorithm is proposed, which has low computational complexity and can quickly find the optimal value of $\boldsymbol{N}$.
- The optimization of $\boldsymbol{N}$ with different minimum achievable rates, noise power, and distance between vehicle and IRS is simulated. Firstly, the influence of the minimum achievable rate on the range of $\boldsymbol{N}$ is analyzed. Then, the optimal value of $\boldsymbol{N}$ is analyzed when the minimum achievable rate is constant and the noise power is different. Finally, the optimal value of $\boldsymbol{N}$ is analyzed when the distance between the vehicle and the IRS changes when the minimum achievable rate and noise power are constant. The theoretical analysis of this paper and the performance of the BSCI algorithm are proved.

Mathematical notations and definitions are presented in Table 1.

The remainder of this paper is organized as follows. In Section 2, the VLC system via mirror array-based IRS for parallel vehicles is designed, and the calculation methods of achievable rate and total power consumption are given. In Section 3, the optimization problem is formulated, and the range of $\boldsymbol{N}$ is analyzed according to the constraints. EE is proved to be a unimodal function, and the BSCI algorithm is proposed. The numerical results of the optimization of $\boldsymbol{N}$ with different minimum achievable rates, noise power, and distances between the vehicle and IRS are provided in Section 4. Finally, the conclusions and future research directions are drawn in Section 5.

Table 1. Mathematical notations and definitions.

Notations	Definitions
x_s	X-coordinate of the transmitter *S* as measured from the upper left corner of the IRS
y_s	Y-coordinate of the transmitter *S* as measured from the IRS along the road
z_s	Z-coordinate of the transmitter *S* as measured from the upper left corner of the IRS
x_d	X-coordinate of the receiver *D* as measured from the upper left corner of the IRS
y_d	Y-coordinate of the receiver *D* as measured from the IRS along the road
h_d	Z-coordinate of *D* as measured from the transmitter *S*
w_m	Width of each mirror
h_m	Height of each mirror
Δw_m	Edge-to-edge inter-mirror separation distances along the x-axis
Δh_m	Edge-to-edge inter-mirror separation distances along the z-axis
n_k	The number of mirrors of each column in the IRS
n_l	The number of mirrors of each row in the IRS
ρ	Mirror reflection efficiency
P_i	Transmitted power
m	Order of Lambertian emission
$\Phi_{1/2}$	Half-power semiangle of an LED
$\theta^S_{R_{i,j}}$	Irradiance angle of the LED from the transmitter *S* to mirror $R_{i,j}$
$\theta^D_{R_{i,j}}$	Incidence angle of the PD from mirror $R_{i,j}$ to the receiver *D*
ϖ	Current-to-light conversion efficiency
A_d	Physical area of the PD
$T_s(\cdot)$	Optical filter gain
$g(\theta)$	Optical concentrator gain
μ	Refractive index
Ψ_c	FOV of the PD
ζ	Efficiency of the transmit power amplifier
I_{DC}	DC-offset
A	Amplitude constraint of the signal
ε	The variance of the signal
η	Responsivity of the PD
N	Total number of mirrors in the IRS
B	VLC system modulation bandwidth
P_{max}	The maximum power threshold
BER_t	The maximum acceptable BER
N_{max}	The maximum number
$\mathbb{E}(\cdot)$	Expectation operator

2. System Model and Analysis

2.1. System Model

The considering scenario is that VLC via mirror array-based IRS for the parallel vehicles in adjacent lanes. The right headlamp (LED light source) of the right vehicle is used as the transmitter, and the PD is installed in the middle of the two headlamps of the left vehicle. The mirror array-based IRS is installed on the street light pole, and the height of the center point is consistent with the headlamps. Figure 1 shows the application scenario of the VLC system via mirror array-based IRS for parallel vehicles.

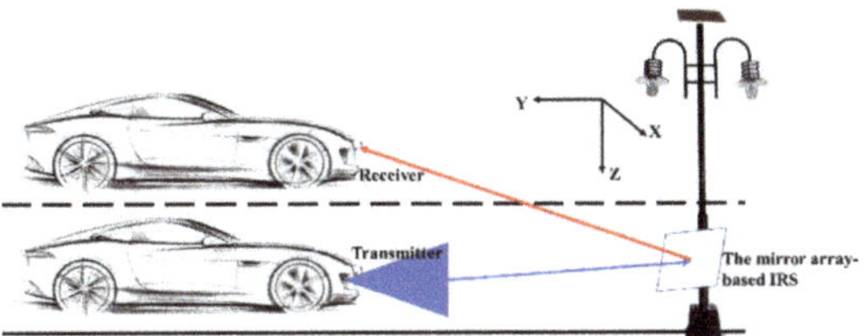

Figure 1. Application scenario of the VLC system via mirror array-based IRS for parallel vehicles.

The model diagram of this scenario is given in Figure 2 for the convenience of analysis.

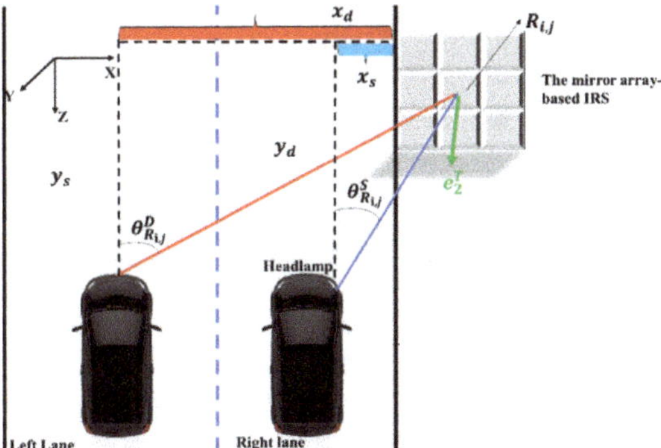

Figure 2. Model of the VLC system via mirror array-based IRS for parallel vehicles.

For the mirror array-based IRS, the rotation angle of each mirror in the IRS can be controlled independently without interfering with others. One of the mirrors is analyzed as an example. We define a Cartesian coordinate system whose origin is at the center of the mirror $R_{i,j}$ $(1 \leq i \leq n_k,\ 1 \leq j \leq n_l)$. The position vector of the transmitter **S** can be expressed as

$$\mathbf{S} = \begin{bmatrix} -(x_s + \frac{w_m}{2} + (j-1)(w_m + \Delta w_m)) \\ y_s \\ -\left(z_s + \frac{h_m}{2} + (i-1)(h_m + \Delta h_m)\right) \end{bmatrix}, \tag{1}$$

The position vector of the PD can be expressed as

$$\mathbf{D} = \begin{bmatrix} -(x_d + \frac{w_m}{2} + (j-1)(w_m + \Delta w_m)) \\ y_d \\ h_d - \left(z_s + \frac{h_m}{2} + (i-1)(h_m + \Delta h_m)\right) \end{bmatrix}, \tag{2}$$

To ensure that the reflected light reaches the receiver, each mirror must be rotated according to the position of the transmitter and receiver to obtain the appropriate angle. The mirror is first arranged via the clockwise rotation of the local z-axis with an angle $\beta_{i,j}$ and the local negative x-axis with an angle $\alpha_{i,j}$. The normal vector direction of the mirror after rotation is expressed as

$$\hat{\mathbf{N}_{\mathrm{i,j}}} = \frac{\hat{\mathbf{R}_{\mathrm{i,j}}\mathbf{S}} + \hat{\mathbf{R}_{\mathrm{i,j}}\mathbf{D}}}{\sqrt{2 + 2\hat{\mathbf{R}_{\mathrm{i,j}}\mathbf{S}}^{\mathrm{T}}\hat{\mathbf{R}_{\mathrm{i,j}}\mathbf{D}}}}, \tag{3}$$

where $\hat{\mathbf{R}_{\mathrm{i,j}}\mathbf{S}} = \frac{\mathbf{S}-\mathbf{R}_{\mathrm{i,j}}}{\|\mathbf{S}-\mathbf{R}_{\mathrm{i,j}}\|_2}$, $\hat{\mathbf{R}_{\mathrm{i,j}}\mathbf{D}} = \frac{\mathbf{D}-\mathbf{R}_{\mathrm{i,j}}}{\|\mathbf{D}-\mathbf{R}_{\mathrm{i,j}}\|_2}$, $\|\cdot\|_2$ denote the ℓ_2-norm, and $(\cdot)^{\mathrm{T}}$ denotes the transpose operator.

The relation between normal vector and rotation angle can be expressed as

$$\hat{N_{i,j}} = \begin{bmatrix} \sin(\beta_{i,j})\cos(\alpha_{i,j}) \\ \cos(\beta_{i,j})\cos(\alpha_{i,j}) \\ \sin(\alpha_{i,j}) \end{bmatrix}. \tag{4}$$

In the actual scene, the distance of light transmission is much larger than the size of the light source, so it can be regarded as a point light source. The irradiance of the point light source after being reflected by the mirror $R_{i,j}$ to the PD can be expressed as [31]

$$E_{i,j} = \frac{\rho(m+1)P_i \cos^m\left(\theta^S_{R_{i,j}}\right)}{2\pi(\|R_{i,j}D\|_2 + \|R_{i,j}S\|_2)^2}\cos\left(\theta^D_{R_{i,j}}\right), \tag{5}$$

where $E_{i,j}$ represents the irradiance at the detector center contributed by the mirror $R_{i,j}$. m is the order of Lambertian emission [43] related to the half-power semiangle of LED $\Phi_{1/2}$ which can be expressed as $m = -ln2/ln(\cos\Phi_{1/2})$.

According to Figure 2, $\cos\left(\theta^S_{\mathbf{R_{i,j}}}\right) = \mathbf{e}_2^T\hat{\mathbf{R}_{i,j}\mathbf{S}} = \mathbf{e}_2^T(\mathbf{S} - \mathbf{R}_{i,j})/\|\mathbf{S} - \mathbf{R}_{i,j}\|_2$, $\cos\left(\theta^D_{\mathbf{R_{i,j}}}\right) = \mathbf{e}_2^T\hat{\mathbf{R}_{i,j}D} = \mathbf{e}_2^T(\mathbf{D} - \mathbf{R}_{i,j})/\|\mathbf{D} - \mathbf{R}_{i,j}\|_2$, and $\mathbf{e}_2^T = [\mathbf{0}, \mathbf{1}, \mathbf{0}]$.

According to the theory of VLC transmission [44], the direct current (DC) gain of the channel can be obtained as

$$H^{IRS}_{i,j} = \frac{\varpi T_s\left(\theta^D_{\mathbf{R_{i,j}}}\right)A_d\rho(m+1)\cos^m\left(\theta^S_{\mathbf{R_{i,j}}}\right)}{2\pi(\|\mathbf{R_{i,j}D}\|_2 + \|\mathbf{R_{i,j}S}\|_2)^2}\cos\left(\theta^D_{\mathbf{R_{i,j}}}\right)g\left(\theta^D_{\mathbf{R_{i,j}}}\right), \tag{6}$$

$g\left(\theta^D_{R_{i,j}}\right)$ can be given as

$$g(\theta) = \begin{cases} \frac{\mu^2}{sin^2(\psi_c)} & 0 \le \theta \le \psi_c \\ 0 & \theta > \psi_c \end{cases}, \tag{7}$$

The total DC gain can be obtained as

$$H^{IRS} = \sum_{i=1}^{n_k}\sum_{j=1}^{n_l} H^{IRS}_{i,j}. \tag{8}$$

2.2. SNR

The transmitted signal of the LED can be expressed as

$$x = \sqrt{\zeta}s + I_{DC}, \tag{9}$$

where s is the input message.

The transmitted signal must be real and non-negative in VLC, and the optical power must be limited to human eye safety and illumination requirement. Generally, we assume that the signal s satisfies the following conditions:

$$-A \le s \le A, \tag{10a}$$

$$\mathbb{E}(s) = 0, \tag{10b}$$

$$\mathbb{E}\left(s^2\right) = \varepsilon, \tag{10c}$$

$$A > 0, \tag{10d}$$

$$\varepsilon > 0, \tag{10e}$$

$$\sqrt{\zeta}s \le I_{DC}, \tag{10f}$$

The total electrical power of the LED driver can be expressed as

$$\mathbb{E}\left(\left(\sqrt{\zeta}s+I_{DC}\right)^2\right)=\mathbb{E}\left(\zeta s^2+2\sqrt{\zeta}sI_{DC}+I_{DC}^2\right)=\zeta\varepsilon+I_{DC}^2, \tag{11}$$

where $P_i=\zeta\varepsilon$ is the power of the signal s.

The total electrical power should be limited, i.e.,

$$\zeta\varepsilon+I_{DC}^2\leq P_{max}. \tag{12}$$

The received signal can be expressed as

$$y=\eta H^{IRS}x+w, \tag{13}$$

where w is the additive white Gaussian noise obeys a distribution $\mathcal{N}(0,\sigma^2)$ with mean zero and variance σ^2.

After removing the constant DC-offset, the SNR γ can be expressed as

$$\gamma=\frac{\left(\eta H^{IRS}\right)^2\cdot P_i}{\sigma^2}, \tag{14}$$

The BER of the optical OOK modulation is given by

$$BER=Q\left(\sqrt{SNR}\right), \tag{15}$$

where

$$Q(x)=\frac{1}{\sqrt{2\pi}}\int_x^\infty e^{-y^2/2}dy. \tag{16}$$

2.3. The Achievable Rate

Because of the non-negative and real-valued amplitude, the classic Shannon capacity formula is not appropriate to VLC. In reference [36], a tight lower bound for dimmable VLC is proposed, so the achievable instantaneous rate can be expressed as

$$R=\frac{1}{2}B\ \log_2\left(1+\frac{e}{2\pi}\gamma\right), \tag{17}$$

where e is the value of the base of natural logarithms.

The Formula (8) can be rewritten as

$$H^{IRS}(N)=\sum_{n=1}^{N}H_n. \tag{18}$$

where $\boldsymbol{N}$ is the total number of mirrors in the IRS. H_n is the channel gain and arranged in decreasing order of magnitude. That is, $H_1=H_{max}$.

The Formula (17) can be rewritten as a function of $\boldsymbol{N}$

$$R(N)=\frac{1}{2}B\ \log_2\left(1+\frac{e}{2\pi}\cdot\frac{\eta^2P_i\left(\sum_{n=1}^{N}H_n\right)^2}{\sigma^2}\right). \tag{19}$$

Assuming that the minimum achievable instantaneous rate of the VLC system is R_{min}, that is

$$\frac{1}{2}B\ \log_2\left(1+\frac{e}{2\pi}\gamma\right)\geq R_{min}, \tag{20}$$

and

$$\gamma \geq \left(2^{\frac{2R_{min}}{B}} - 1\right)\frac{2\pi}{e}. \tag{21}$$

According to Formulas (14) and (21), we can obtain:

$$NH_1 \geq \sum_{n=1}^{N} H_n \geq \sqrt{\frac{\left(2^{\frac{2R_{min}}{B}} - 1\right)\frac{2\pi\sigma^2}{e}}{\eta^2 P_i}}. \tag{22}$$

The minimum number of mirrors required to satisfy the in Equation (22) is

$$N \geq \sqrt{\frac{\left(2^{\frac{2R_{min}}{B}} - 1\right)\frac{2\pi\sigma^2}{e}}{\eta^2 P_i}} / H_1. \tag{23}$$

According to the law of energy conservation, the received power is less than or equal to the transmitted power, we can get $\left(\eta H^{IRS}\right)^2 \leq 1$. So, it must hold $\eta \sum_{n=1}^{N} H_n \leq 1$.

Due to H_n being arranged in decreasing order of magnitude, the sufficient condition $\eta N H_1 \leq 1$ can provide a simple upper-bound of the maximum number of mirrors, that is $N \leq \frac{1}{\eta H_1}$.

When the IRS is installed on traffic infrastructures, its size must be limited in order not to obstruct traffic. When the size of each mirror is fixed, it is assumed that the maximum number of mirrors in the IRS is N_{max}. So $N \leq min\left\{N_{max}, \frac{1}{\eta H_1}\right\}$.

2.4. The Total Power Consumption

In the VLC system via mirror array-based IRS, the total power consumption is composed of the transmit power, the hardware static power consumed in the transmitter and receiver, and IRS. The mirrors do not consume any transmit power since they are passive elements. The total power consumption model is shown in Figure 3.

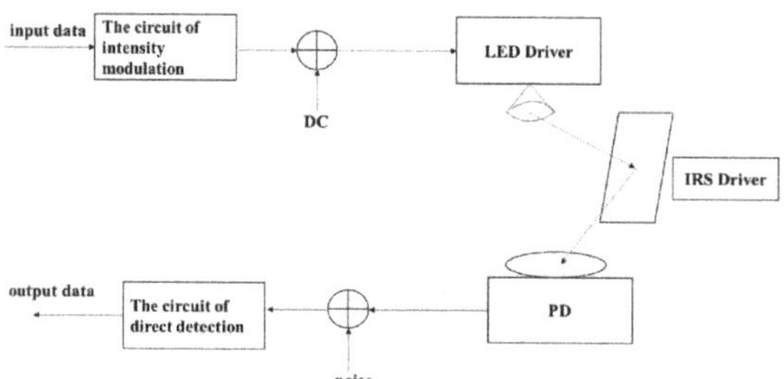

Figure 3. The total power consumption model in the VLC system via mirror array-based IRS.

In the system, the purpose of the mirror is to get a suitable position by rotation and reflect the emitted light to PD. Therefore, the hardware static power of the mirror array-based IRS is mainly used to control the rotation angle of mirrors. According to Formula (4), the rotation angle of the mirror is related to the normal vector. The normal vector depends on the distance between each mirror and the transmitter or receiver. Because the interval of mirrors is much smaller than the distance between mirrors and transmitter or receiver, the difference of mirrors rotation angle in IRS is relatively small. Therefore, the power consumption for each mirror rotation can be regarded as the same.

Based on the above considerations, the total power consumption of IRS-assisted VLC system can be expressed as

$$P_{total} = P_x + P_{hsp} + NP_m = \varepsilon + I_{DC}^2 + P_{hsp} + NP_m, \tag{24}$$

where P_{hsp} is the values of the hardware static power consumed in the transmitter and receiver. P_m is the value of the power consumed in each mirror of IRS.

To analyze the influence of the number of mirrors on EE, the Formula (24) can be rewrit as

$$P_{total} = NP_m + P_{else}, \tag{25}$$

where

$$P_{else} = \varepsilon + I_{DC}^2 + P_{hsp}. \tag{26}$$

3. The Number of Mirrors Optimization

3.1. Problem Formulation

According to the definition of EE, it can be expressed as

$$EE(N) = \frac{R(N)}{P_{total}} = \frac{\frac{1}{2}B\ \log_2\left(1 + \frac{e}{2\pi} \cdot \frac{\eta^2 P_i \left(\sum_{n=1}^{N} H_n\right)^2}{\sigma^2}\right)}{NP_m + P_{else}}. \tag{27}$$

Proposition 1. *Set* $\delta = \frac{e}{2\pi} \cdot \frac{\eta^2 P_i}{\sigma^2}$, $R(N) = \frac{1}{2}B\log_2\left(1 + \delta\left(\sum_{n=1}^{N} H_n\right)^2\right)$. *when* $\delta\left(\sum_{n=1}^{N} H_n\right)^2 \geq 1$, $R(N+1) - R(N) \geq R(N+2) - R(N+1)$.

Proof of Proposition 1.

$$\begin{aligned} R(N+1) - R(N) &= \tfrac{1}{2}B\ \log_2\left(\frac{1+\delta\left(\sum_{n=1}^{N+1} H_n\right)^2}{1+\delta\left(\sum_{n=1}^{N} H_n\right)^2}\right) = \tfrac{1}{2}B\ \log_2\left(\frac{1+\delta\left(H_{N+1}+\sum_{n=1}^{N} H_n\right)^2}{1+\delta\left(\sum_{n=1}^{N} H_n\right)^2}\right) \\ &= \tfrac{1}{2}B\ \log_2\left(\frac{1+\delta\left(\left(\sum_{n=1}^{N} H_n\right)^2+2\left(H_{N+1}\sum_{n=1}^{N} H_n\right)+\left(H_{N+1}\right)^2\right)}{1+\delta\left(\sum_{n=1}^{N} H_n\right)^2}\right) \\ &= \tfrac{1}{2}B\ \log_2\left(1 + \frac{2\delta\left(H_{N+1}\sum_{n=1}^{N} H_n\right)}{1+\delta\left(\sum_{n=1}^{N} H_n\right)^2} + \frac{\delta\left(H_{N+1}\right)^2}{1+\delta\left(\sum_{n=1}^{N} H_n\right)^2}\right). \end{aligned} \tag{28}$$

Similarly,

$$R(N+2) - R(N+1) = \tfrac{1}{2}B\ \log_2\left(1 + \frac{2\delta\left(H_{N+2}\sum_{n=1}^{N+1} H_n\right)}{1+\delta\left(\sum_{n=1}^{N+1} H_n\right)^2} + \frac{\delta\left(H_{N+2}\right)^2}{1+\delta\left(\sum_{n=1}^{N+1} H_n\right)^2}\right). \tag{29}$$

If $R(N+1) - R(N) \geq R(N+2) - R(N+1)$, it holds that

$$\begin{aligned} &\tfrac{1}{2}B\ \log_2\left(1 + \frac{2\delta\left(H_{N+1}\sum_{n=1}^{N} H_n\right)}{1+\delta\left(\sum_{n=1}^{N} H_n\right)^2} + \frac{\delta\left(H_{N+1}\right)^2}{1+\delta\left(\sum_{n=1}^{N} H_n\right)^2}\right) \\ &\geq \tfrac{1}{2}B\ \log_2\left(1 + \frac{2\delta\left(H_{N+2}\sum_{n=1}^{N+1} H_n\right)}{1+\delta\left(\sum_{n=1}^{N+1} H_n\right)^2} + \frac{\delta\left(H_{N+2}\right)^2}{1+\delta\left(\sum_{n=1}^{N+1} H_n\right)^2}\right). \end{aligned} \tag{30}$$

That is

$$\frac{2\left(H_{N+1}\sum_{n=1}^{N} H_n\right)}{1+\delta\left(\sum_{n=1}^{N} H_n\right)^2} + \frac{\left(H_{N+1}\right)^2}{1+\delta\left(\sum_{n=1}^{N} H_n\right)^2} \geq \frac{2\left(H_{N+2}\sum_{n=1}^{N+1} H_n\right)}{1+\delta\left(\sum_{n=1}^{N+1} H_n\right)^2} + \frac{\left(H_{N+2}\right)^2}{1+\delta\left(\sum_{n=1}^{N+1} H_n\right)^2}. \tag{31}$$

Since $H_{N+1} \geq H_{N+2}$ and $1 + \delta\left(\sum_{n=1}^{N} H_n\right)^2 \leq 1 + \delta\left(\sum_{n=1}^{N+1} H_n\right)^2$, so

$$\frac{\left(H_{N+1}\right)^2}{1 + \delta\left(\sum_{n=1}^{N} H_n\right)^2} \geq \frac{\left(H_{N+2}\right)^2}{1 + \delta\left(\sum_{n=1}^{N+1} H_n\right)^2}. \tag{32}$$

Then, for (31) to hold, it is sufficient that

$$\frac{\sum_{n=1}^{N} H_n}{1+\delta\left(\sum_{n=1}^{N} H_n\right)^2} \geq \frac{\sum_{n=1}^{N+1} H_n}{1+\delta\left(\sum_{n=1}^{N+1} H_n\right)^2}. \tag{33}$$

If we set $t = \sum_{n=1}^{N} H_n$, the inequality (33) can be written as an equivalent function

$$f(t) = \frac{t}{1+\delta t^2}. \tag{34}$$

If *f(t)* is a monotonically decreasing function, then

$$\frac{d(f(t))}{dt} = \frac{1}{1+\delta t^2} - \frac{2\delta t^2}{(1+\delta t^2)^2} = \frac{1-\delta t^2}{(1+t^2)^2} \leq 0. \tag{35}$$

So, when $\delta\left(\sum_{n=1}^{N} H_n\right)^2 \geq 1$, the inequality (33) can hold and $R(N+1) - R(N) \geq R(N+2) - R(N+1)$. Hence the proof follows. □

Proposition 2. *$EE(N)$ in (27) is a unimodal function.*

Proof of Proposition 2. Under the previously considered constraints of **N**, $R(N)$ is an increasing function that grows more and more slowly. For the denominator in the Formula (27), $P_m \ll P_{else}$. $NP_m + P_{else}$ increases with increasing of **N**, and the growth rate becomes slower and slower.

For $EE(N)$, its changes are divided into two cases:

1. It keeps increasing with the increasing of **N**. The peak value of EE will not appear within the range of **N**;
2. There exists an N', $EE(N)$ decreases monotonically when $N \geq N'$. At this time, $EE(N') \geq EE(N'+1) \geq EE(N'+2)$.

When $EE(N') \geq EE(N'+1)$, it holds

$$\frac{R(N')}{N'P_m + P_{else}} \geq \frac{R(N'+1)}{(N'+1)P_m + P_{else}}, \tag{36}$$

and

$$N' \leq -\frac{P_{else}}{P_m} + \frac{R(N')}{R(N'+1) - R(N')}. \tag{37}$$

Since $R(N') = -(R(N'+1) - R(N')) + R(N'+1)$, (37) can be rewritten as

$$N' \leq -\frac{P_{else}}{P_m} - 1 + \frac{R(N'+1)}{R(N'+1) - R(N')}. \tag{38}$$

In Proposition 1, we proof that when $\delta\left(\sum_{n=1}^{N} H_n\right)^2 \geq 1$, $R(N+1) - R(N) \geq R(N+2) - R(N+1)$.

So, the in Equation (38) implies

$$N' \leq -\frac{P_{else}}{P_m} - 1 + \frac{R(N'+1)}{R(N'+2) - R(N'+1)}. \tag{39}$$

We have

$$\frac{R(N'+1)}{(N'+1)P_m + P_{else}} \geq \frac{R(N'+2)}{(N'+2)P_m + P_{else}}. \tag{40}$$

It means that $EE(N'+1) \geq EE(N'+2)$.

So, if $EE(N') \geq EE(N'+1)$, it can be proved that $EE(N') \geq EE(N'+1) \geq EE(N'+2) \geq EE(N'+3) \geq EE(N'+\cdots)$. To sum up, $EE(N)$ is either monotonically increasing, or there exists an N', with $EE(N)$ monotonically decreasing when $N \geq N'$. Therefore, $EE(N)$ is a unimodal function, and hence the proof follows.

Our aim is to find the optimal number of mirrors with the maximum EE under the unique constraints of VLC. With the conditions of Equations (10a) to (10f), the optimization problem can be formulated as

$$\max_{N} EE(N) \tag{41}$$

$$\textbf{s.t.} \quad \sqrt{\zeta} s \leq I_{DC}, \tag{42a}$$

$$\zeta\varepsilon + I_{DC}^2 \leq P_{max}, \tag{42b}$$

$$R(N) \geq R_{min}, \tag{42c}$$

$$\delta\left(\sum_{n=1}^{N} H_n\right)^2 \geq 1, \tag{42d}$$

$$N \leq min\left\{N_{max}, \frac{1}{\eta H_1}\right\}, \tag{42e}$$

$$BER \leq BER_t. \tag{42f}$$

where R_{min} is the minimum achievable rate. □

3.2. BSCI Algorithm

Assuming that $N = \{N \in \mathbb{N}^+ | \underline{N} \leq N \leq \overline{N}\}$ ($\mathbb{N}^+$ is the set of positive integers). Equations (42c) to (42f) can be used to obtain the range of $\mathbf{N}$.

When $R(N) \geq R_{min}$, we have

$$\frac{1}{2}B \log_2\left(1+\delta\left(\sum_{n=1}^{N} H_n\right)^2\right) \geq R_{min}. \tag{43}$$

It holds

$$NH_1 \geq \sum_{n=1}^{N} H_n \geq \sqrt{\frac{\left(2^{\frac{2R_{min}}{B}}-1\right)}{\delta}},$$

and

$$N \geq \sqrt{\frac{\left(2^{\frac{2R_{min}}{B}}-1\right)}{\delta H_1^2}}. \tag{44}$$

When $\delta\left(\sum_{n=1}^{N} H_n\right)^2 \geq 1$, it holds

$$NH_1 \geq \sum_{n=1}^{N} H_n \geq \frac{1}{\sqrt{\delta}},$$

and

$$N \geq \frac{1}{H_1\sqrt{\delta}}. \tag{45}$$

when $BER \leq BER_t$, according to (15), we set $BER_t = Q(\sqrt{\gamma_t})$.

So,

$$\frac{\eta^2 P_i \left(\sum_{n=1}^{N} H_n\right)^2}{\sigma^2} \geq \gamma_t, \tag{46}$$

it holds

$$NH_1 \geq \sum_{n=1}^{N} H_n \geq \sigma\sqrt{\gamma_t}/\eta\sqrt{P_i},$$

and

$$N \geq \frac{\sigma\sqrt{\gamma_t}}{\eta H_1 \sqrt{P_i}}. \tag{47}$$

According to the above conditions, we have

$$\underline{N} = \left\{ \underline{N} \in \mathbb{N}^+ \middle| \underline{N} \geq max\left\{ \sqrt{\frac{\left(2^{\frac{2R_{min}}{B}} - 1\right)}{\delta H_1^2}}, \frac{1}{H_1\sqrt{\delta}}, \frac{\sigma\sqrt{\gamma_t}}{\eta H_1 \sqrt{P_i}} \right\} \right\}. \tag{48}$$

According to the constraints of Equation (42e), we have

$$\overline{N} = \left\{ \overline{N} \in \mathbb{N}^+ \middle| \overline{N} \leq min\left\{ N_{max}, \frac{1}{\eta H_1} \right\} \right\}. \tag{49}$$

Under the constraints, $EE(N)$ is divided into three cases to find the maximum value:

1. If $EE(\underline{N}) \geq EE(\underline{N}+1) \geq EE(\underline{N}+2)$, $EE(N)$ decreases monotonically with $\boldsymbol{N}$. $EE(\underline{N})$ is the maximum value of $EE(N)$ and the optimal value of $\boldsymbol{N}$ is $\underline{N}$;
2. If $EE(\overline{N}) \geq EE(\overline{N}-1) \geq EE(\overline{N}-2)$, $EE(N)$ increases monotonically with $\boldsymbol{N}$. The peak value of EE does not appear within this range and the optimal value of $\boldsymbol{N}$ does not exist;
3. If it is not the case of (1) and (2), $EE(N)$ increases first and then decreases with $\boldsymbol{N}$. To reduce the amount of computation, the binary search (Algorithm 1) method is used to find the maximum value of $EE(N)$ as follows.

Step 1: set the iterative range. The starting point is $u = \underline{N}$ and the ending point is $v = \overline{N}$.

Step 2: set $b = \frac{(u+v)}{2}$. If b is not an integer, the largest integer less than b is used to conclusion.

Step 3: if $EE(b) \geq EE(b+1)$, $v = b$. Otherwise, $u = b$.

Step 4: repeat steps 2–3 until $(v - u) \leq 1$. Return $EE(v)$ which is the maximum value of $EE(N)$ and v which is the optimal value of $\boldsymbol{N}$.

Algorithm 1: The Binary Search Method

```
Given N̲, N̄, P_m, P_else, and δ
Calculate R(N) in the range of N = {N ∈ ℕ+ | N̲ ≤ N ≤ N̄} using the Formula (19)
set u = N̲ and v = N̄
while ((v − u) > 1)
    b = floor((u+v)/2)
    if EE(b) ≥ EE(b + 1)
        v = b
else
        u = b
end
end
Return EE(v), v
```

Based on the above analysis, the BSCI algorithm (Algorithm 2) is proposed to find the N_{opt}, which is the optimal value of N, and $EE_{max}(N)$ which is the maximum value of $EE(N)$. The specific steps are as follows.

Step 1: input the parameters of LED, PD, and IRS.

Step 2: calculate the iterative range $N = \{N \in \mathbb{N}^+ | \underline{N} \leq N \leq \overline{N}\}$ according to Formulas (48) and (49).

Step 3: calculate $R(N)$ with the iterative range of N according to Formula (19).
Step 4: conditional iteration.

If $EE(\overline{N}) \geq EE(\overline{N}-1) \geq EE(\overline{N}-2)$, N_{opt} does not exist.
If EE $EE(\underline{N}) \geq EE(\underline{N}+1) \geq EE(\underline{N}+2)$, $N_{opt} = \underline{N}$ and $EE_{max}(N) = EE(\underline{N})$.
If not in the above two cases, N_{opt} and $EE_{max}(N)$ are obtained by using the binary search method.

Step 5: output N' and $EE_{max}(N)$.

Algorithm 2: The BSCI Algorithm

```
Given the parameter values of the LED, PD, and IRS
calculate the iterative range N = {N ∈ N+ | N_ ≤ N ≤ N̄} according to Formulas (48) and (49).
calculate R(N) with the iterative range of N according to Formula (19).
for N = N_ : 1 : N̄
  if EE(N̄) ≥ EE(N̄ − 1) ≥ EE(N̄ − 2)
   N_opt does not exist;
break;
  else if EE(N_) ≥ EE(N_ + 1) ≥ EE(N_ + 2)
            N_opt = N_;
            EE_max(N) = EE(N_);
else
            N_opt and EE_max(N) are obtained by using the binary search method;
end if
end if
end for
output N_opt and EE_max(N)
```

According to the BSCI algorithm, when $EE(N)$ increases or decreases monotonically in the range of **N**, the required result can be obtained only by one conditional decision. When $EE(N)$ increases monotonically first and then decreases monotonically, the inflection point can be found quickly by using the binary search method. Compared with the Bubble Sort method, the amount of computation is greatly reduced and the computational efficiency is improved.

4. Numerical Results

4.1. Simulation Parameters

The main simulation parameters of the IRS-aided VLC system are listed in Table 2. Assume that two parallel vehicles are driving along the centerline of the neighbor lanes and the width of each lane is 3.5 m. The height of the high-beam headlamp is 0.62 m, and the separation between two headlamps is 1.12 m. IRSs are installed on the street light pole, and the height of the center is consistent with the height of the headlamp.

The coordinate values of the transmitter and receiver can be obtained as follows.

$$\begin{aligned} x_s &= \left(\tfrac{3.5}{2} - \tfrac{1.12}{2}\right) = 1.19,\ y_s = 10,\ z_s = -\left(w_m \cdot \tfrac{n_l}{2}\right) = -\left(0.05 * \tfrac{n_l}{2}\right). \\ x_d &= \left(3.5 + \tfrac{3.5}{2}\right) = 5.25,\ y_d = y_s = 10,\ z_d = 0. \end{aligned}$$

For the transmitter signals, set A = 2, $\varepsilon = 1$, and $I_{DC}^2 = 45$ dBm. The noise power is $\sigma^2 = -98$ dBm. The maximum value of the electrical power of the system $P_{max} = 50$ dBm, and the hardware static power consumed in the transmitter and receiver $P_{hsp} = 30$ dBm. The size of each mirror in IRS is 0.01×0.01 m^2, and the spacing between mirrors is zero. To ensure traffic safety and avoid collision with the IRS when driving, we set the IRS to have 60 mirrors in each row and 60 mirrors in each column, so the maximum number of mirrors is $N_{max} = 60 \times 60 = 3600$. The power consumption of each mirror is $P_m = 20$ dBm. The mirror reflection efficiency is 0.8. The modulation mode is optical OOK modulation and the BER_t is 10^{-6}.

Table 2. Main simulation parameters.

Parameter	Value
z_m	0.62 m
(w_m, h_m)	(0.01, 0.01) m
$\Phi_{1/2}$	60 deg.
Ψ_c	35 deg.
A_d	1.0 cm^2
T_s	1.0
μ	1.5
ϖ	0.44 W/A
ρ	0.8
η	0.54 A/W
ζ	1.2
B	20 MHz

4.2. Numerical Results

4.2.1. EE Performance with Different R_{min}

The minimum achievable rate R_{min} can represent the effectiveness of the communication system. According to Formula (48), R_{min} can affect the starting iterative value of $\boldsymbol{N}$. Figure 4 illustrates the EE versus $\boldsymbol{N}$ with different R_{min}.

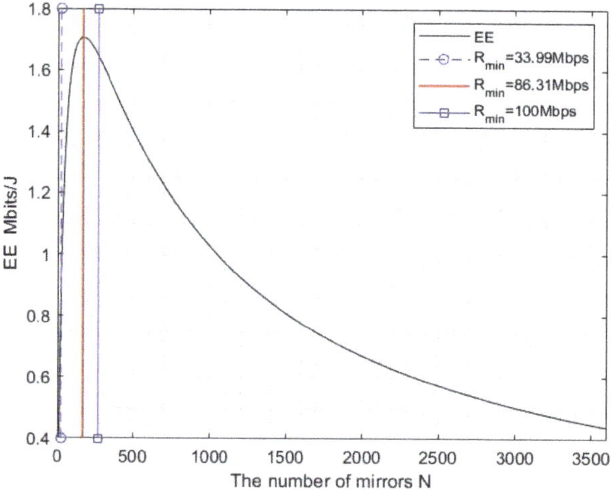

Figure 4. EE versus N with different R_{min}.

As can be seen from Figure 4, EE shows a trend of increasing first and then decreasing monotonically with the increasing of N, itmeans that EE(N) is a unimodal function, which is consistent with the proof in the paper. With the different R_{min}, the iterative range $N = \{N \in \mathbb{N}^+ | \underline{N} \le N \le \overline{N}\}$ changes. When $R_{min} < 33.99$ Mbps, the iterative range of N is independent of R_{min}. This is because $\min\{N \in \mathbb{N}^+ | R(N) \ge R_{min}\} \le \frac{\sigma\sqrt{\gamma_t}}{\eta H_1 \sqrt{P_i}}$, and $\frac{1}{H_1\sqrt{\delta}} \le \frac{\sigma\sqrt{\gamma_t}}{\eta H_1 \sqrt{P_i}}$, so the starting iterative point of N is $\underline{N} = min\Big\{N \in \mathbb{N}^+ \Big| N \ge \frac{\sigma\sqrt{\gamma_t}}{\eta H_1 \sqrt{P_i}}\Big\}$. When $R_{min} \ge 33.99$ Mbps, $\min\{N \in \mathbb{N}^+ | R(N) \ge R_{min}\} \ge \frac{\sigma\sqrt{\gamma_t}}{\eta H_1 \sqrt{P_i}} \ge \frac{1}{H_1\sqrt{\delta}}$, $\underline{N} = \min\{N \in \mathbb{N}^+ | R(N) \ge R_{min}\}$.

When 33.99 Mbps $\le R_{min} \le$ 86.31 Mbps, $EE(N)$ increases first and then decreases monotonically. The binary search method proposed can be used to find N_{opt}. When 86.31 Mbps $\le R_{min} \le R(\overline{N})$, such as $R_{min} = 100$ Mbps in the Figure 4, $EE(N)$ decreases monotonically, so $N_{opt} = \underline{N}$ and $EE_{max}(N) = EE(\underline{N})$. When $R_{min} > R(\overline{N})$, the achievable rate cannot meet the requirements and N_{opt} does not exist.

Since the changing of R_{min} will changes the iterative range of N, the calculation amount will also be different when finding N_{opt}. Table 3 gives N_{opt} and iterations with different R_{min}.

Table 3. N_{opt} and iterations with different R_{min}.

R_{min} (Mbps)	N_{opt}	$EE_{max}(N)$ (Mbits/J)	Iterations of Bubble Sort Method	Iterations of BSCI Algorithm
40	168	1.7049	6363528	12
50	168	1.7049	6313681	12
60	168	1.7049	6242811	12
70	168	1.7049	6144265	12
80	168	1.7049	6004845	12
90	191	1.7004	5812345	1
100	270	1.6417	5546115	1

As can be seen from the Table 3, when $R_{min} = 40$, 50, 60, 70, 80 Mbps, $N_{opt} = 168$. This is consistent with the previous analysis. Compared to the total number of mirrors in the IRS $N_{max} = 3600$, $EE(N)$ can be maximized only using 4.67% of the total number of mirrors in IRS after optimization. The BSCI algorithm needs 12 iterations to find N_{opt}, which reduces the amount of computation by 10^5 orders of magnitude, compared with the Bubble Sort method. When $R_{min} = 90$, 100 Mbps, $EE(N)$ decreases monotonically and N_{opt} increases gradually. The BSCI algorithm only needs one iteration to find N_{opt}. However, the Bubble Sort method still needs a lot of computation. Therefore, the BSCI algorithm is more efficient.

4.2.2. EE Performance with Different σ^2

For VVLC, the noise will affect the EE performance, especially the background light noise. To facilitate comparison, we set $R_{min} = 40$ Mbps. Figure 5 illustrates the EE versus N with different noise power σ^2.

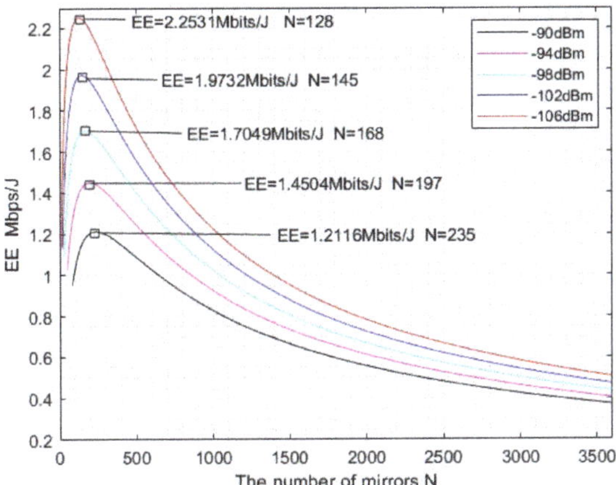

Figure 5. EE versus N with different noise power σ^2.

As can be seen from Figure 5, $EE(N)$ increases first and then decreases with different noise power σ^2, which is a unimodal function. When the number of mirrors in IRS is constant, the EE becomes larger with the smaller noise power. When the number of mirrors is fixed, the denominator in Formula (27) is the same and the lower noise power causes SNR to increase, resulting in the continuous increase of $R(N)$, thus $EE(N)$ also increases.

With the noise power becoming lower, the N_{opt} is smaller and $EE_{max}(N)$ is larger. According to Formula (41), the $EE(N)$ is the largest corresponding to N_{opt}. It can be seen from Formula (27) that when the increasing speed of the numerator is greater than that of the denominator, the $EE(N)$ keeps increasing, otherwise, the $EE(N)$ keeps decreasing. When the noise power is smaller, the SNR and the $R(N)$ is larger. Taking the growth ratio of the adjacent $R(N)$ as an example, the denominator in the ratio $R(N+1)/R(N)$ will also become larger, and the ratio will be less than that at high noise power at this time. Therefore, the numerator grows faster than that of the denominator, easily obtaining the maximum $EE(N)$. Taking $\sigma^2 = -106$ dBm as an example, the EE has the maximum value only using 128 mirrors. The remaining mirrors in the IRS can be used to support the VLC of multiple vehicles, which improves the utilization of mirrors in the IRS.

However, the lower noise power makes the starting iterative point of **N** smaller when calculating the maximum EE. This is because when the noise power is lower, $R(N)$ is easier to reach R_{min} with the increasing of **N**. The smaller iterative starting point of **N** means that the iterative range of **N** increases, which may add some computational complexity for finding N_{opt}. Table 4 gives N_{opt} and iterations with different σ^2.

Table 4. N_{opt} and iterations with different σ^2.

σ^2 (dBm)	N_{opt}	EE_{max}(N) (Mbits/J)	Iterations of Bubble Sort Method	Iterations of BSCI Algorithm
−90	235	1.2116	6189921	12
−94	197	1.4504	6295926	11
−98	168	1.7049	6363528	12
−102	145	1.9732	6406410	12
−106	128	2.2531	6435078	12

According to Table 4, as the noise power decreases, the number of iterations requirement using the Bubble Sort method increases. Even if the iterative range of **N** changes, the computational complexity using BSCI algorithm changes little, N_{opt} can still be found quickly without bringing unexpected complexity to the system. Therefore, the BSCI algorithm has better performance.

As can been from Table 4, it is easier to reach the $EE_{max}(N)$ with lower noise power, and fewer mirrors are required. This is because the growth rate of the numerator in Formula (27) becomes faster when the SNR is smaller. Therefore, reducing noise power is an effective way to obtain higher EE using fewer mirrors. In VVLC, the background light is the main source of the noise. Although the background light noise cannot be eliminated, the optical filters can be considered to reduce the interference of background light and noise power which can improve EE and resource efficiency.

4.2.3. EE Performance with Different y_s

y_s represents the distance between the vehicle and the IRS in the direction of the road. Since the vehicle is moving, y_s is dynamically changing. Figure 6 illustrates the EE versus **N** with different y_s when $R_{min} = 40$ Mbps and $\sigma^2 = -98$ dBm.

As can be seen from Figure 6, EE increases first and then decreases when $y_s = 10, 20, 30, 40, 50$ m, which is a unimodal function. When the y_s is smaller, the larger $EE(N)$ is obtained using the same number of mirrors. According to Formula (6), the increase of y_s means that $R_{i,j}D$ and $R_{i,j}S$ are increase, resulting that the channel gain corresponding to each mirror decreases. The reduction of channel gain makes the received power and the SNR smaller. In this way, $R(N)$ and EE will also be reduced.

Like the analysis in Section 4.2.3, with the smaller y_s, the $EE(N)$ can reach the maximum value using fewer mirrors and $EE_{max}(N)$ is also larger at the same time. Taking $y_s = 10$ m as an example, the EE can reach the maximum value using 168 mirrors. The N_{opt} is reduced 74.6% and $EE_{max}(N)$ is increased by 4.15 times compared to $y_s = 50$ m.

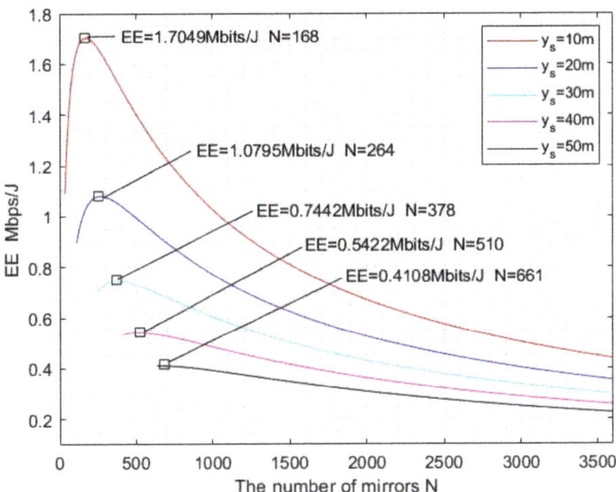

Figure 6. EE versus N with different y_s.

The smaller y_s causes the smaller the iterative starting point of N, so that the iterative range of N used to solve N_{opt} becomes larger. If the iterative range of N is larger, it is easy to cause the amount of computation becomes larger. Table 5 gives N_{opt} and iterations with different y_s.

Table 5. N_{opt} and iterations with different y_s.

y_s (m)	N_{opt}	$EE_{max}(N)$ (Mbits/J)	Iterations of Bubble Sort Method	Iterations of BSCI Algorithm
10	168	1.7049	6363528	12
20	264	1.0795	6098778	12
30	378	0.7442	5666661	12
40	510	0.5422	5089645	11
50	661	0.4108	4394130	11

As can be seen from Table 5, with the increasing of y_s, the iterative number of the Bubble Sort method decreases. This is because the increase of y_s causes the starting iterative point of N to become large, thus the range of iteration reduces. Compared with the bubbling method, the BSCI algorithm has fewer iterations and computations.

When the vehicle is moving, the distance between the vehicle and the IRS is constantly changing. When the distance is closer, N_{opt} is smaller and $EE(N)$ is higher. To improve the performance of the N_{opt} and $EE(N)$, the distance between vehicle and IRS should be optimized. The multiple IRSs can be installed using the existing traffic infrastructures, and the distance between adjacent IRS is not too large, so that the distance is controlled within an appropriate range, which can solve this problem.

5. Conclusions

Energy efficiency is an important indicator to measure the energy consumption of communication systems. In this paper, the VLC system via mirror array-based IRS for parallel vehicles is designed first, and the calculation formula of channel gain is given. Then, the achievable rate and power consumption of the system are analyzed, and the calculation method of EE is given. On this basis, considering the non-negative and real of the transmitted signal, the maximum power consumption satisfied luminous ability and eye safety, the minimum achievable rate, and the required BER, the optimization problem of the number of mirrors under EE maximization is proposed. Under the existing constraints, it is proved that $EE(N)$ is a unimodal function. To quickly find the optimal value of the number of mirrors, the BSCI algorithm is proposed. By comparing the optimal number of mirrors corresponding to different R_{min}, σ^2, y_s, we can know that different parameter changes will bring the different iterative range of the number of mirrors, and effect the optimal number of mirrors. Compared with the Bubble Sort method, the BSCI algorithm reduces the amount of computation by 10^5 orders of magnitude, and can quickly find the optimal number of mirrors and the maximum value of EE, which is an effective algorithm.

The numerical results show that when the EE corresponding to R_{min} is less than the $EE_{max}(N)$, $EE(N)$ increases first and then decreases. Otherwise, $EE(N)$ decreases monotonically, and the $EE_{max}(N)$ obtained at this time is smaller than that in the previous case. Therefore, it is necessary to select an appropriate R_{min} according to the actual communication needs of the vehicle. This requires consideration of the tradeoff between the EE and achievable rate. When noise power increases, $EE(N)$ becomes smaller with the same number of mirrors. Therefore, it is necessary to reduce noise power to obtain a smaller number of optimized mirrors and higher EE, especially background light noise. The use of optical elements, such as optical filters, can be considered. As y_s increases, the distance between the vehicle and IRS is longer, resulting in the optimal number of mirrors increasing and $EE(N)$ decreasing. To solve this problem, it can be considered to install multiple IRSs that the distance between the vehicle and the IRS is within a controllable range, which can improve the efficiency of the mirrors in the IRS and the performance of EE.

Author Contributions: Conceptualization, L.Z. and H.Z.; methodology, L.Z.; validation, L.Z., H.Z. and J.L.; formal analysis, W.Z.; investigation, L.Z.; data curation, H.Z.; writing—original draft preparation, L.Z.; writing—review and editing, J.L.; visualization, L.Z.; supervision, J.L.; project administration, J.L.; funding acquisition, J.L. All authors have read and agreed to the published version of the manuscript.

Funding: This research was funded by the National Natural Science Foundation of China under Grant 61966007 and 61961007, the Basic Ability Improvement Project of Young and Middle-Aged Teachers in Guangxi Universities, grant number 2021KY0217; Key Laboratory of Cognitive Radio and Information Processing, Ministry of Education under Grants CRKL170110, CRKL180201, and CRKL180106, and the Guangxi Key Laboratory of Wireless Wideband Communication and Signal Processing, Guilin University of Electronic Technology (GXKL0619204, GXKL06200116).

Institutional Review Board Statement: Not applicable.

Informed Consent Statement: Not applicable.

Data Availability Statement: Not applicable.

Acknowledgments: The authors wish to thank the anonymous reviewers for their valuable suggestions.

Conflicts of Interest: The authors declare no conflict of interest.

References

1. Reza, S.; Oliveira, H.S.; Machado, J.J.M.; Tavares, J. Urban Safety: An Image-Processing and Deep-Learning-Based Intelligent Traffic Management and Control System. *Sensors* **2021**, *21*, 7705. [CrossRef] [PubMed]
2. Lv, Z.; Lou, R.; Singh, A.K. AI Empowered Communication Systems for Intelligent Transportation Systems. *IEEE Trans. Intell. Transp. Syst.* **2021**, *22*, 4579–4587. [CrossRef]
3. Zadobrischi, E.; Cosovanu, L.-M.; Dimian, M. Traffic Flow Density Model and Dynamic Traffic Congestion Model Simulation Based on Practice Case with Vehicle Network and System Traffic Intelligent Communication. *Symmetry* **2020**, *12*, 1172. [CrossRef]
4. Yang, Y.; Hua, K. Emerging Technologies for 5G-Enabled Vehicular Networks. *IEEE Access* **2019**, *7*, 181117–181141. [CrossRef]
5. He, R.; Schneider, C.; Ai, B.; Wang, G.; Zhong, Z.; Dupleich, D.A.; Thomae, R.S.; Boban, M.; Luo, J.; Zhang, Y. Propagation Channels of 5G Millimeter-Wave Vehicle-to-Vehicle Communications: Recent Advances and Future Challenges. *IEEE Veh. Technol. Mag.* **2020**, *15*, 16–26. [CrossRef]
6. do Vale Saraiva, T.; Campos, C.A.V.; Fontes, R.d.R.; Rothenberg, C.E.; Sorour, S.; Valaee, S. An Application-Driven Framework for Intelligent Transportation Systems Using 5G Network Slicing. *IEEE Trans. Intell. Transp. Syst.* **2021**, *22*, 5247–5260. [CrossRef]
7. Ning, Z.; Zhang, K.; Wang, X.; Obaidat, M.S.; Guo, L.; Hu, X.; Hu, B.; Guo, Y.; Sadoun, B.; Kwok, R.Y.K. Joint Computing and Caching in 5G-Envisioned Internet of Vehicles: A Deep Reinforcement Learning-Based Traffic Control System. *IEEE Trans. Intell. Transp. Syst.* **2021**, *22*, 5201–5212. [CrossRef]
8. Chen, C.; Yuru, T.; Yueping, C.; Min, L. Fairness-aware hybrid NOMA/OFDMA for bandlimited multi-user VLC systems. *Opt. Express* **2021**, *29*, 42265–42275. [CrossRef]

9. Chen, C.; Zhong, X.; Fu, S.; Jian, X.; Liu, M.; Yang, H.; Alphones, A.; Fu, H.Y. OFDM-Based Generalized Optical MIMO. *J. Lightw. Technol.* **2021**, *39*, 6063–6075. [CrossRef]
10. Miao, P.; Yin, W.; Peng, H.; Yao, Y. Study of the Performance of Deep Learning-Based Channel Equalization for Indoor Visible Light Communication Systems. *Photonics* **2021**, *8*, 453. [CrossRef]
11. Chen, C.; Fu, S.; Jian, X.; Liu, M.; Deng, X.; Ding, Z. NOMA for Energy-Efficient LiFi-Enabled Bidirectional IoT Communication. *IEEE Trans. Commun.* **2021**, *69*, 1693–1706. [CrossRef]
12. Arai, S.; Tang, Z.; Nakayama, A.; Takata, H.; Yendo, T. Rotary LED Transmitter for Improving Data Transmission Rate of Image Sensor Communication. *IEEE Photonics J.* **2021**, *13*, 1–11. [CrossRef]
13. Takahashi, K.; Kamakura, K.; Kinoshita, M.; Yamazato, T. Luminance Inversion for Parallel Transmission Visible Light Communication Between LCD and Image Sensor Camera. *J. Lightw. Technol.* **2021**, *39*, 6759–6767. [CrossRef]
14. Sun, X.; Shi, W.; Cheng, Q.; Liu, W.; Wang, Z.; Zhang, J. An LED Detection and Recognition Method Based on Deep Learning in Vehicle Optical Camera Communication. *IEEE Access* **2021**, *9*, 80897–80905. [CrossRef]
15. Zhu, Z.; Wei, S.; Liu, R.; Hong, Z.; Zheng, Z.; Fan, Z.; Ma, D. Freeform surface design for high-efficient LED low-beam headlamp lens. *Opt. Commun.* **2020**, *477*, 126269. [CrossRef]
16. Singh, R.; Mochizuki, M.; Yamada, T.; Nguyen, T. Cooling of LED headlamp in automotive by heat pipes. *Appl. Therm. Eng.* **2020**, *166*, 114733. [CrossRef]
17. Memedi, A.; Dressler, F. Vehicular Visible Light Communications: A Survey. *IEEE Commun. Surv. Tutor.* **2021**, *23*, 161–181. [CrossRef]
18. Turan, B.; Coleri, S. Machine Learning Based Channel Modeling for Vehicular Visible Light Communication. *IEEE Trans. Veh. Technol.* **2021**, *70*, 9659–9672. [CrossRef]
19. Alsalami, F.M.; Ahmad, Z.; Zvanovec, S.; Haigh, P.A.; Haas, O.C.L.; Rajbhandari, S. Statistical channel modelling of dynamic vehicular visible light communication system. *Veh. Commun.* **2021**, *29*, 100339. [CrossRef]
20. Karbalayghareh, M.; Miramirkhani, F.; Eldeeb, H.B.; Kizilirmak, R.C.; Sait, S.M.; Uysal, M. Channel Modelling and Performance Limits of Vehicular Visible Light Communication Systems. *IEEE Trans. Veh. Technol.* **2020**, *69*, 6891–6901. [CrossRef]
21. Luo, P.; Ghassemlooy, Z.; Minh, H.L.; Bentley, E.; Burton, A.; Tang, X. Performance analysis of a car-to-car visible light communication system. *Appl. Opt.* **2015**, *54*, 1696–1706. [CrossRef]
22. Huang, C.; Zappone, A.; Alexandropoulos, G.C.; Debbah, M.; Yuen, C. Reconfigurable Intelligent Surfaces for Energy Efficiency in Wireless Communication. *IEEE Trans. Wirel. Commun.* **2019**, *18*, 4157–4170. [CrossRef]
23. Pan, C.; Ren, H.; Wang, K.; Elkashlan, M.; Nallanathan, A.; Wang, J.; Hanzo, L. Intelligent Reflecting Surface Aided MIMO Broadcasting for Simultaneous Wireless Information and Power Transfer. *IEEE J. Sel. Areas Commun.* **2020**, *38*, 1719–1734. [CrossRef]
24. Du, H.; Zhang, J.; Cheng, J.; Ai, B. Millimeter Wave Communications with Reconfigurable Intelligent Surfaces: Performance Analysis and Optimization. *IEEE Trans. Commun.* **2021**, *69*, 2752–2768. [CrossRef]
25. Sur, S.N.; Bera, R. Intelligent reflecting surface assisted MIMO communication system: A review. *Phys. Commun.* **2021**, *47*, 101386. [CrossRef]
26. Masini, B.M.; Silva, C.M.; Balador, A. The Use of Meta-Surfaces in Vehicular Networks. *J. Sens. Actuator Netw.* **2020**, *9*, 15. [CrossRef]
27. Makarfi, A.U.; Rabie, K.M.; Kaiwartya, O.; Adhikari, K.; Li, X.; Quiroz-Castellanos, M.; Kharel, R. Reconfigurable Intelligent Surfaces-Enabled Vehicular Networks: A Physical Layer Security Perspective. *arXiv* **2020**, arXiv:2004.11288.
28. Dingel, B.B.; Tsukamoto, K.; Mikroulis, S.; Deng, P.; Kavehrad, M.; Lou, Y. MEMS-based beam-steerable free-space optical communication link for reconfigurable wireless data center. In Proceedings of the Broadband Access Communication Technologies XI, San Francisco, CA, USA, 28 January 2017; Volume 10128, p. 1012805.
29. Cao, Z.; Zhang, X.; Osnabrugge, G.; Li, J.; Vellekoop, I.M.; Koonen, A.M.J. Reconfigurable beam system for non-line-of-sight free-space optical communication. *Light Sci. Appl.* **2019**, *8*, 69. [CrossRef]
30. Jamali, V.; Ajam, H.; Najafi, M.; Schmauss, B.; Schober, R.; Poor, H.V. Intelligent Reflecting Surface-assisted Free-space Optical Communications. *IEEE Commun. Mag.* **2021**, *59*, 57–63. [CrossRef]
31. Abdelhady, A.M.; Salem, A.K.S.; Amin, O.; Shihada, B.; Alouini, M.-S. Visible Light Communications via Intelligent Reflecting Surfaces: Metasurfaces vs Mirror Arrays. *IEEE Open J. Commun. Soc.* **2021**, *2*, 1–20. [CrossRef]
32. Alizadeh Jarchlo, E.; Eso, E.; Doroud, H.; Siessegger, B.; Ghassemlooy, Z.; Caire, G.; Dressler, F. Li-Wi: An upper layer hybrid VLC-WiFi network handover solution. *Ad Hoc Netw.* **2022**, *124*, 102705. [CrossRef]
33. Basar, E. Reconfigurable Intelligent Surface-Based Index Modulation: A New Beyond MIMO Paradigm for 6G. *IEEE Trans. Commun.* **2020**, *68*, 3187–3196. [CrossRef]
34. Zhang, J.; Bjornson, E.; Matthaiou, M.; Ng, D.W.K.; Yang, H.; Love, D.J. Prospective Multiple Antenna Technologies for Beyond 5G. *IEEE J. Sel. Areas Commun.* **2020**, *38*, 1637–1660. [CrossRef]
35. Bjornson, E.; Ozdogan, O.; Larsson, E.G. Intelligent Reflecting Surface Versus Decode-and-Forward: How Large Surfaces are Needed to Beat Relaying? *IEEE Wirel. Commun. Lett.* **2020**, *9*, 244–248. [CrossRef]
36. Wang, J.; Hu, Q.; Wang, J.; Chen, M.; Wang, J. Tight bounds on channel capacity for dimmable visible light communications. *J. Lightw. Technol.* **2013**, *31*, 3771–3779. [CrossRef]

37. Ma, S.; Yang, R.; Li, H.; Dong, Z.-L.; Gu, H.; Li, S. Achievable rate with closed-form for SISO channel and broadcast channel in visible light communication networks. *J. Lightw. Technol.* **2017**, *35*, 2778–2787. [CrossRef]
38. Ma, S.; Yang, R.; He, Y.; Lu, S.; Zhou, F.; Al-Dhahir, N.; Li, S. Achieving Channel Capacity of Visible Light Communication. *IEEE Syst. J.* **2021**, *15*, 1652–1663. [CrossRef]
39. Sun, S.; Yang, F.; Song, J. Sum Rate Maximization for Intelligent Reflecting Surface-Aided Visible Light Communications. *IEEE Commun. Lett.* **2021**, *25*, 3619–3623. [CrossRef]
40. Ma, S.; Zhang, T.; Lu, S.; Li, H.; Wu, Z.; Li, S. Energy Efficiency of SISO and MISO in Visible Light Communication Systems. *J. Lightw. Technol.* **2018**, *36*, 2499–2509. [CrossRef]
41. Teixeira, L.; Loose, F.; Barriquello, C.H.; Reguera, V.A.; Costa, M.A.D.; Alonso, J.M. On Energy Efficiency of Visible Light Communication Systems. *IEEE J. Emerg. Sel. Top. Power Electron.* **2021**, *9*, 6396–6407. [CrossRef]
42. Cao, B.; Chen, M.; Yang, Z.; Zhang, M.; Zhao, J.; Chen, M. Reflecting the Light: Energy Efficient Visible Light Communication with Reconfigurable Intelligent Surface. In Proceedings of the 2020 IEEE 92nd Vehicular Technology Conference, Victoria, BC, Canada, 18 November–16 December 2020.
43. Akanegawa, M.; Tanaka, Y.; Nakagawa, M. Basic study on traffic information system using LED traffic lights. *IEEE Trans. Intell. Transp. Syst.* **2001**, *2*, 197–203. [CrossRef]
44. Komine, T.; Nakagawa, M. Fundamental analysis for visible-light communication system using LED lights. *IEEE Trans. Consum. Electron.* **2004**, *50*, 100–107. [CrossRef]

 *photonics*

Article

Propagation Characteristics Comparisons between mmWave and Visible Light Bands in the Conference Scenario

Baobao Liu [1], Pan Tang [1,*], Jianhua Zhang [1], Yue Yin [1], Guangyi Liu [2] and Liang Xia [2]

[1] State Key Laboratory of Networking and Switching Technology, Beijing University of Posts and Telecommunications, Beijing 100876, China; lbb@bupt.edu.cn (B.L.); jhzhang@bupt.edu.cn (J.Z.); yinyue18@bupt.edu.cn (Y.Y.)

[2] China Mobile Research Institute, Beijing 100053, China; liuguangyi@chinamobile.com (G.L.); xialiang@chinamobile.com (L.X.)

* Correspondence: tangpan27@bupt.edu.cn

Abstract: Millimeter-wave (mmWave) communications and visible light communications (VLC) are proposed to form hybrid mmWave/VLC systems. Furthermore, channel modeling is the foundation of system design and optimization. In this paper, we compare the propagation characteristics, including path loss, root mean square (RMS) delay spread (DS), K-factor, and cluster characteristics, between mmWave and VLC bands based on a measurement campaign and ray tracing simulation in a conference room. We find that the optical path loss (OPL) of VLC channels is highly dependent on the physical size of the photodetectors (PDs). Therefore, an OPL model is further proposed as a function of the distance and size of PDs. We also find that VLC channels suffer faster decay than mmWave channels. Moreover, the smaller RMS DS in VLC bands shows a weaker delay dispersion than mmWave channels. The results of K-factor indicate that line-of-sight (LOS) components mainly account for more power for mmWave in LOS scenarios. However, non-LOS (NLOS) components can be stronger for VLC at a large distance. Furthermore, the K-Power-Means algorithm is used to perform clustering. The fitting cluster number is 5 and 6 for mmWave and VLC channels, respectively. The clustering results reveal the temporal sparsity in mmWave bands and show that VLC channels have a large angular spread.

Keywords: visible light communications (VLC); mmWave communications; channel modeling; channel propagation characteristics; path loss; delay spread (DS); Ricean K-factor; cluster characteristics

Citation: Liu, B.; Tang, P.; Zhang, J.; Yin, Y.; Liu, G.; Xia, L. Propagation Characteristics Comparisons between mmWave and Visible Light Bands in the Conference Scenario. *Photonics* **2022**, *9*, 228. https://doi.org/10.3390/photonics9040228

Received: 26 February 2022
Accepted: 28 March 2022
Published: 1 April 2022

1. Introduction

Motivated by the rising spectrum needs, millimeter-wave (mmWave) communications located in 30–300 GHz (Figure 1) have received great attention in recent years [1–4], and the channel models have been standardized by groups around the world, such as the 3rd Generation Partnership Project (3GPP) (for 0.5–100 GHz) [5]. Furthermore, the sixth-generation (6G) wireless communication systems are desired for high speed, low latency, and high reliability, which the radio frequency (RF) technologies cannot support. To address such issues, visible light communication (VLC) has emerged to be a key technology in 6G [6–8]. VLC employs the unlicensed frequency spectrum resources at 400–800 THz (Figure 1), which can provide a high data transmission rate and a strong resistance to electromagnetic interferences [9,10]. However, the complex applications in 6G have different requirements, which may not be satisfied by a single technology, whether RF or VLC technologies. This requires the combination of multiple communication technologies, e.g., hybrid mmWave/VLC systems [11–13]. These two technologies are expected to complement each other to form a reliable communication system in the future.

mmWave and VLC channel modeling is the foundation of designing and optimizing hybrid mmWave/VLC communication systems, which can be used to evaluate the performance and determine the performance limit of wireless communication systems [14].

Moreover, it is also important to understand the differences in the propagation characteristics of VLC and mmWave channels, based on which can exert the respective advantages of mmWave and VLC to design the hybrid mmWave/VLC systems. In the following, we give a review of related works.

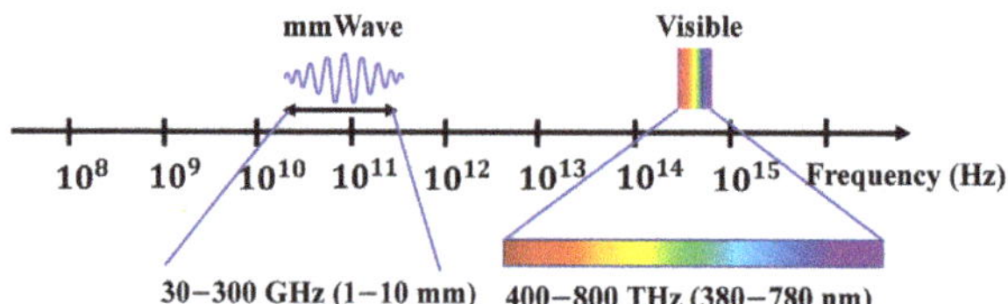

Figure 1. The comparison of frequency spectrum resources in mmWave and VLC bands. Generally, mmWave is located in 30–300 GHz. Moreover, some frequencies, e.g., 24 and 28 GHz, are also included in mmWave bands due to their similar characteristics to mmWave.

1.1. Literature Review

In [13], the authors provided an overview of RF and VLC systems, and they compared the basic system components and modulation techniques of RF and VLC technologies. In [15], the channel frequency response (CFR) of mmWave and VLC channels was investigated. The results show that the channel for VLC is frequency-selective and lowpass, while the channel for mmWave is almost flat. The same results can be found in [16], where the path loss (mmWave) and received power (vehicular VLC) are also given. In [17], the authors presented the results of path loss and time dispersion characteristics for mmWave (measurement) and VLC (simulation), respectively, under different configurations, e.g., room size, transceiver deployment, and operating frequency. They find that mmWave and VLC channels share some common characteristics but also some differences in path loss and time dispersion.

Generally, the existing research of propagation characteristics comparisons between mmWave and VLC bands under the same conditions (conference scenarios and parameters setting) is rare, but it is worth greater attention.

1.2. Contributions of This Paper

The key to propagation characteristics comparisons between mmWave and VLC bands requires the same conditions. In our work, the comparisons are made based on the same conference scenario. The contributions of this paper are as follows:

- Channel characteristics comparisons between mmWave and VLC bands are performed based on the same conference scenario and parameter settings.
- A unique optical path loss (OPL) model dependent on the physical size of photodetectors (PDs) is first proposed for VLC based on the widely used floating-intercept (FI) model in mmWave. The size of PDs can be estimated to meet the required coverage range by using this model while designing systems.
- The large-scale fading characteristics and multipath-related characteristics, including root mean square (RMS) delay spread (DS), K-factor, and cluster characteristics, between mmWave and VLC bands are compared fairly based on the same scenario.

The rest of this paper is organized as follows. Section 2 introduces the scenarios and parameter settings for mmWave and VLC, respectively. Propagation characteristics comparisons are presented in Section 3. Finally, Section 4 concludes this paper.

2. Scenarios and Setup

2.1. Measurement Scenario and Setup in mmWave Bands

A mmWave channel measurement campaign at 28 GHz was performed with a wideband correlation sounder in a conference room (No. 510) of the Scientific Technology Building at the Beijing University of Posts and Telecommunications (BUPT), as shown in

Figure 2a. The map of this room and the 16 measured positions are marked in Figure 2b. The transmitter (TX) is placed at the corner of this conference room while the receiver (RX) is placed at the 16 measured positions successively. The measured distance is within the range of 3 to 10.51 m and all of these measurements are in line-of-sight (LOS) scenarios. In order to obtain the large-scale parameters and small-scale parameters of channels, we used the measurement platform to conduct two kinds of measurements in this conference room.

- Setup one: Two omnidirectional biconical antennas (360° and 40° half power beam width (HPBW) in azimuth and elevation, respectively) were used at TX and RX sides to collect all multipath components (MPCs), and then 1000 channel impulse responses (CIRs) samples were measured in each position.
- Setup two: An omnidirectional biconical antenna was used at TX side while a directional horn antenna (10° and 11° HPBW in azimuth and elevation, respectively) was mounted in an electrical positioner at RX side. In these virtual measurements, the TX antenna was fixed and the RX antenna was rotated in steps of 5° in azimuth from 0° to 360°, and there were three different elevations, $-10°, 0°$, and $10°$. Channel characteristics in the spatial domain can be obtained through these virtual measurements. In each horn antenna pointing direction, we measured 1000 CIR samples.

(**a**) Measurement scenario in mmWave bands

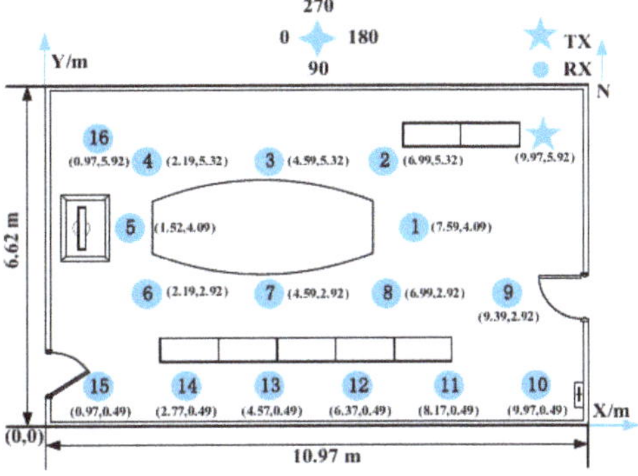

(**b**) Map for measurement scenario

Figure 2. Measurement scenario and map for this scenario, in which there are 16 measurement positions in total.

Based on these two measurements, we can obtain the path loss, RMS DS, K-factor, and spatial characteristics of mmWave channels. Furthermore, the parameters of the measurement setup are listed in Table 1.

Table 1. Parameters of measurement setup.

Parameter	Value
Central frequency	28 GHz
RF bandwidth	600 MHz
Chip sequence length	511
Chip rate	400 MHz
Delay resolution	2.5 ns
Pulse repetition interval	1277.5 ns
Biconical antenna/Horn antenna gain	2.93 dBi/25 dBi
Biconical antenna/Horn antenna polorization	Vertical/Vertical
Biconical antenna/Horn antenna azimuth HPBW	360°/10°
Biconical antenna/Horn antenna elevation HPBW	40°/11°
TX antenna/RX antenna height	1.68 m/1.68 m

2.2. Simulation Scenario and Setup in VLC Bands

The VLC channel characteristics are investigated based on the ray tracing features of Zemax® [18] due to the lack of a channel sounding platform for VLC, which is limited by the narrow −3 dB bandwidth (typically around a few MHz) of optical transmitters (LEDs) [19]. VLC channel models developed by this ray tracing method were accepted as reference channel models in IEEE 802.15.13 and IEEE 802.11bb [20]. This channel modeling approach has been demonstrated by experimental results in [21].

In this work, we adopt this well-established realistic channel modeling approach [22]. First, a 3D simulation environment is created, as shown in Figure 3a, where the geometry of the simulation environment can be specified accurately based on the conference room in Figure 2a. Moreover, the CAD models, e.g., tables, desks, and display screen, designed according to the actual measured size in the conference room (No. 510), are imported into the software. The map of this simulation environment and the 16 detector positions are shown in Figure 3b. The wavelength-dependent spectral reflection characteristics of the surface materials (i.e., ceiling, floor, walls, furniture, etc.) are given in Figure 4, which are from [23]. Moreover, we consider the mixed reflections including diffuse and specular reflections by adjusting the scatter fraction (SF) parameter. This parameter changes between 0 and 1 such that zero indicates the purely specular reflections and unity indicates the purely diffuse case. The coating materials of objects in this environment and SF setting are given in Table 2. The specifications of the light sources are shown in Figure 5. These are commercially available light sources (PAR20) with 20° half viewing angle [24], i.e., 40° viewing angle. Considering the 360° and 40° HPBW in azimuth and elevation of mmWave antennas, we use nine LEDs (PAR20) with different pointing from 0° to 360° (interval of 40°) at TX. Therefore, the same coverage as mmWave antennas can be achieved in azimuth (360°) and elevation (40°), as illustrated in Figure 5a. Figure 5b,c illustrate the intensity profiles and the normalized optical spectrum of LED, respectively. Moreover, a spherical PD, as in Figure 3a, called detector polar in Zemax®, is adopted as the RX antenna. This spherical PD is used in order to collect all the MPCs arriving at the PD, as mmWave antenna does. The photosensitive area of PD can be changed by the radial size (RS), i.e., the radius of detector polar in our simulation. Note that the LEDs and detector are positioned at the same height of 1.68 m as the mmWave antennas at RX.

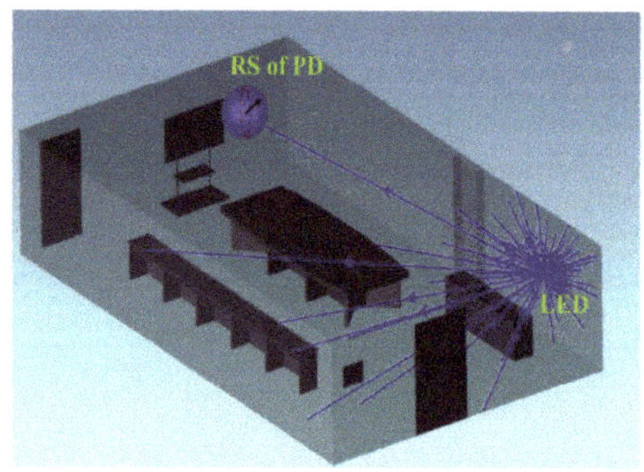

(a) Simulation scenario in VLC bands

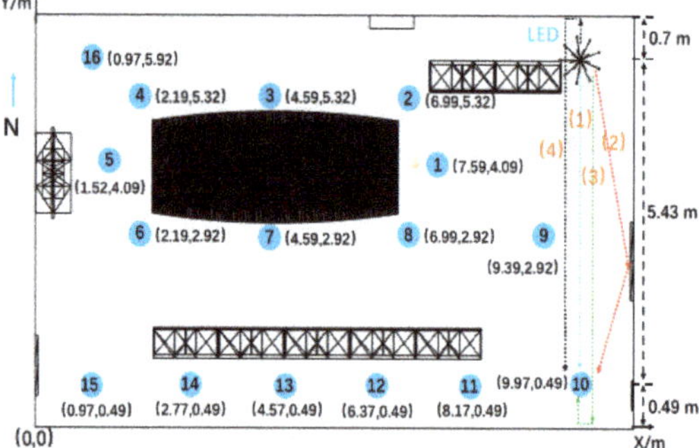

(b) Map for simulation scenario

Figure 3. Simulation scenario and map for this scenario, in which there are 16 detector positions in total.

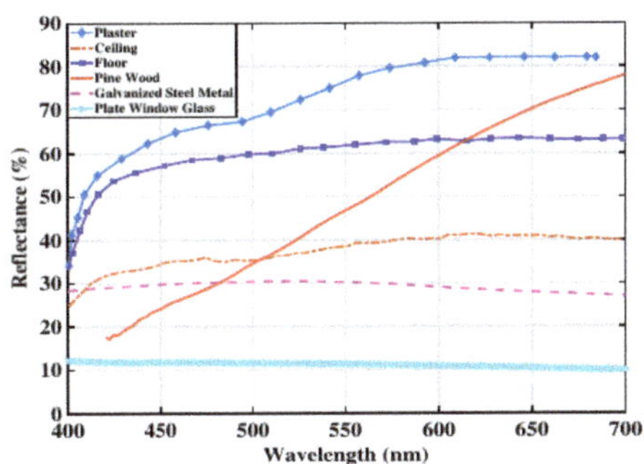

Figure 4. Spectral reflectance of various materials.

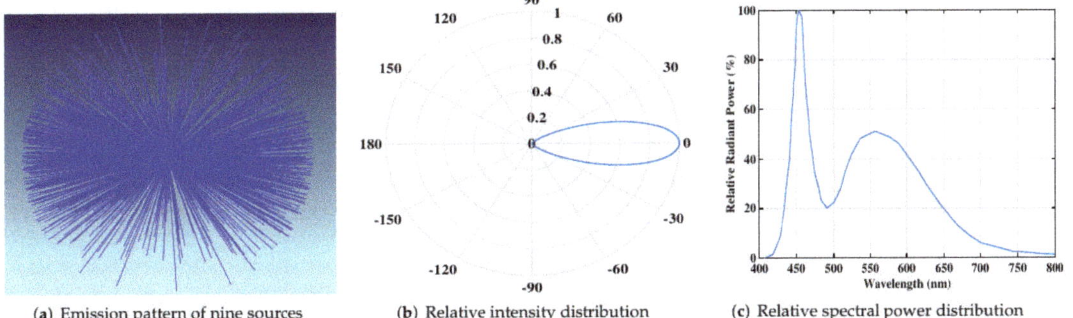

(a) Emission pattern of nine sources (b) Relative intensity distribution (c) Relative spectral power distribution

Figure 5. The specifications of the light sources, including the emission pattern of LEDs, relative intensity distribution, and relative spectral power distribution.

Table 2. Parameters of simulation setup.

Item	Parameter	Value
Room	Size	$10.97 \times 6.62 \times 2.40\ \text{m}^3$
Reflections specifications	Number of reflections Material reflectance Type of reflections	4 Wavelength-dependent Specular and diffuse reflections
Coating material	Walls Ceiling Floor Table, desks, and doors TV screen TV shelf and electric box	Plaster Ceiling Floor Pine Wood Plate Window Glass Galvanized Steel Metal
Scatter fraction	TV screen Desks and doors Other objects	0.2 0.5 1
TX	Model of LED Number of LED Optical power of each LED Analysis rays Minimum relative ray intensity	PAR20 9 2 W 10^7 10^{-3}
Channel	Length d_{LOS} Delay resolution	3–10.51 m 0.2 ns
RX	Type of receiver Radial size	Detector Polar 10 mm

The non-sequential ray tracing features of Zemax® can be used to calculate the received power and path lengths from the light source to detector for each ray, which can be further used to obtain the CIRs between the light source and the detector point. In our simulation, the detector is placed at the 16 positions, respectively. Here, Monte Carlo analysis and Sobol sampling are adopted as the random ray tracing methods. The number of reflections was determined based on the simulation of particular ray propagation, where the normalized intensity dropped to 10^{-3}, i.e., the value of parameter "minimum relative ray intensity" in Zemax®. All other key parameters can be found in Table 2.

3. Channel Characteristics Comparisons

3.1. Channel Impulse Responses

For a wideband mmWave channel, the receiver can resolve multiple paths according to their delay and the CIRs are commonly represented by the superposition of many plane waves. Furthermore, in our measurements, only two students who operated the measurement platform stayed in the conference room, and they tried to keep stationary during the measurement. Thus, time variance can be ignored here. The CIRs are given as follows:

$$h(\tau) = V\delta(\tau - \tau_0) + \sum_{n=1}^{N} a_n e^{j\phi_n}\delta(\tau - \tau_n), \tag{1}$$

where V presents the deterministic (typically LOS) component with the excess delay τ_0. The parameters a_n, τ_n, and ϕ_n are the amplitude, excess delay, and initial phase of the nth stochastic component, i.e., typically non-LOS (NLOS) component, respectively. N denotes the number of multiple paths.

For VLC channels, the ray tracing process will end with an output file, including the detected optical power and path lengths from sources to the detector for each ray. Then, we can obtain the CIRs based on these data, as in [18]:

$$h(\tau) = \sum_{i=1}^{N_r} P_i\delta(\tau - \tau_i), \tag{2}$$

where the P_i and τ_i are the received optical power and propagation delay of ith ray, and N_r denotes the number of received rays at the detector.

The CIRs at the 10th position are given in Figure 6a,b for mmWave and VLC, respectively. It is observed from Figure 6b that the delay of the first peak (LOS) is 18 ns, with which multiplied by the speed of light (3×10^8 m/s) we can obtain the LOS distance (5.4 m). It can be verified that the LOS distance is equal to the distance from TX to the 10th position (5.43 m) in Figure 3b, as the blue line shows. The second peak in Figure 6b is close to the LOS, so we can infer that it mainly comes from the reflections of the east door (red lines in Figure 3b). Moreover, the delay of the third and fourth peaks is 21.4 ns and 22.8 ns, i.e., the propagation distances are 6.42 m and 6.84 m, respectively. The green route in Figure 3b means $T_X \rightarrow$ the south wall $\rightarrow R_X$ and the path distance is 6.41 m ($0.49 \times 2 + 5.43 \approx 6.42$ m). Therefore, the third peak mainly comes from the reflections from the south wall. Furthermore, the fourth peak is mainly from the north wall, as the black route presents with a path distance of 6.83 m ($0.7 \times 2 + 5.43 \approx 6.84$ m) in Figure 3b.

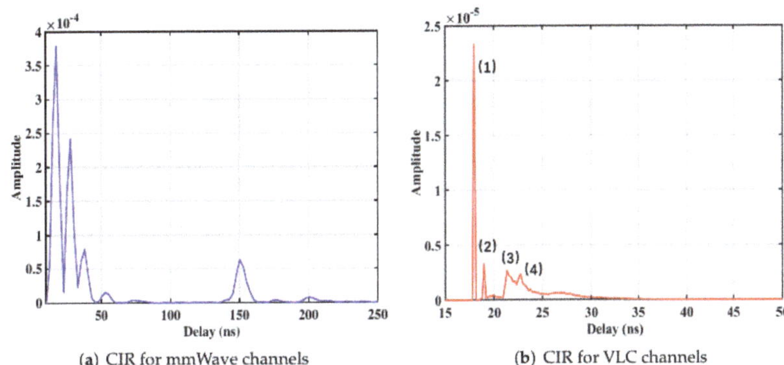

Figure 6. CIR at the 10th position.

We find that mmWave channels have more peaks than those in VLC channels, which can also be found at other points. This result indicates that the mmWave signals can be received at RX through the reflections of more scatters but a few reflections by specific scatters can reach the detector for VLC. Moreover, mmWave channels present a larger delay fluctuation (130 ns) than that of VLC channels (20 ns), which can be explained by the diffuse reflections of VLC, i.e., the light rays split into many random rays and they are reflected in all directions when interacting with scatters. Finally, only small rays can reach the detector.

3.2. Path Loss

Path loss is the reduction in power density of a radio wave as it propagates [25]. In wireless communication systems, the path loss is of high importance for coverage prediction and interference analysis [26,27]. For mmWave channels, the path loss is a function of distance and can be written by [14]:

$$\mathrm{PL}(d) = \mathrm{P_T} + \mathrm{G_T} + \mathrm{G_R} + \mathrm{G_S} + 20\log_{10}(\frac{\lambda}{4\pi d}) - \mathrm{P_R}, \tag{3}$$

where $\mathrm{G_T}, \mathrm{G_R}$, and $\mathrm{G_S}$ are the antenna gain of TX, RX, and the system gain, respectively; λ is the wavelength; d is the spatial distance between TX and RX; $\mathrm{P_T}$ is the transmitter power, and $\mathrm{P_R} = 10\log_{10}(\sum_{n=1}^{N}(V^2 + a_n^2))$ denotes the received power in our measurements.

In each measured position, we can collect 1000 samples of path loss from the first kind of measurements. We can model path loss in this conference room based on these samples with the widely used floating-intercept (FI) [5] and close-in (CI) [28] model. The FI and CI model can be expressed as

$$\mathrm{PL^{FI}}(d) = \beta + 10\alpha\log_{10}(d) + X_\sigma{}^{\mathrm{FI}}, \tag{4}$$

$$\mathrm{PL^{CI}}(f,d) = 20\log_{10}(\frac{4\pi d_0 f}{c}) + 10n_{CI}\log_{10}(\frac{d}{d_0}) + X_\sigma{}^{\mathrm{CI}} \quad \text{for } d \geq d_0,\ d_0 = 1\,\mathrm{m}, \tag{5}$$

where β is the intercept, α and n_{CI} are the parameters of fit, i.e., path loss exponent (PLE), X_σ is a zero-mean Gaussian variable representing the shadowing, and d_0 is the reference distance.

For VLC channels, the impulse responses of LEDs and PDs have not been standardized as with the antennas in mmWave bands. Therefore, the characteristics of LEDs and PDs are considered in the existing channel models [29–31] for VLC. The channel path loss coupled with transceivers is defined as OPL, which can be expressed as [18,32]:

$$\mathrm{OPL}(d) = -10\log_{10}\left(\int_0^\infty h_d(\tau)d\tau\right), \tag{6}$$

where the $h_d(\tau)$ denotes the CIR for the distance d. The CI model cannot be applied to model the OPL in VLC bands, because this model requires a specific frequency (i.e., f in (5)), while the LEDs work with mixed wavelengths, as shown in Figure 5c. Therefore, we choose the FI Model (4) to fit the OPL for VLC channels.

Figure 7 shows the path loss fitting results in mmWave bands. The PLE in the FI model is 1.31, which is slightly smaller than the PLE (1.57) in the CI model. This difference can also be found by investigating the intercept of these two slopes. The intercept β in the FI model is 63.5 dB, while it is 61.34 dB in the CI model. On the other hand, the standard derivation σ in the CI model is 0.1 dB larger than that in the FI model. Generally, these two models can give good insights into the path loss here, and the FI model appears to fit better with the path loss samples, i.e., smaller standard derivation. We also plot the path loss model defined in the 3GPP technical report (TR) 38.901 for the indoor office LOS scenario [5]. We can find that the PLE and σ parameters in this model are both larger than the fitted parameters here. This can be explained by the fact that the typical office room in 3GPP TR 38.901 has more complicated structures than the conference room in Figure 2a, so that the signals are more likely to be obstructed by objects, e.g., the tables, chairs, and plants. Moreover, the path loss model in free space is also presented in Figure 7, and it is apparent that the path loss in free space is larger than the path loss in our measured conference room and the office room. This is within expectation because more reflected signals from walls and ceilings in indoor rooms can be received by the RX.

Figure 8a shows the photosensitive area of commercial PDs [33] and r denotes the RS of PDs. It is clear that the OPL is related to the size of PDs. Figure 8b presents the OPL fitting results in VLC bands with RS set to be 0.2 and 10 mm, respectively. It is noteworthy that the OPL of VLC is highly dependent on the size of PDs, which is set by RS in our simulation. The OPL varies between 38.7 and 47.7 dB when RS is 10 mm. However, it varies from 73.6 to 87.2 dB with RS set to be 0.2 mm. Note that the OPL increases by more than 35 dB at the same position when we change the size of PDs in VLC bands. Consistent with our expected results, the OPL will increase with smaller detectors because of the decrease in received rays.

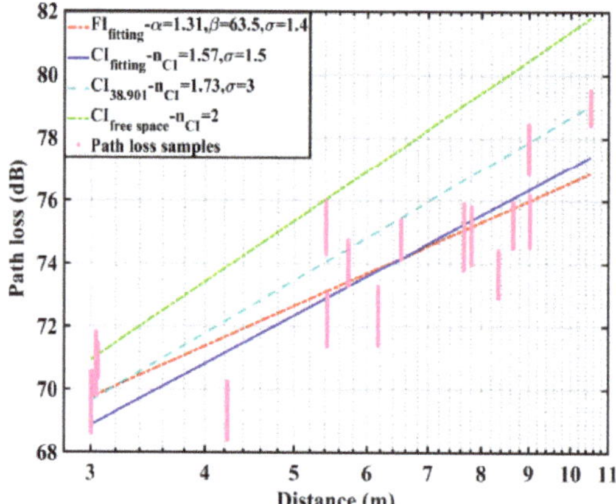

Figure 7. Path loss fitting results for mmWave channels.

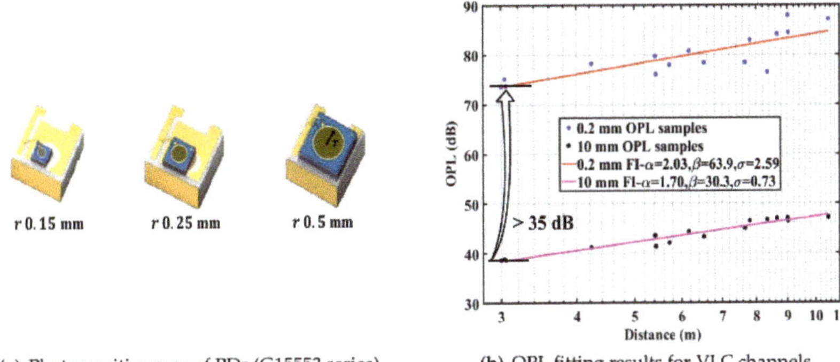

(a) Photosensitive area of PDs (G15553 series) (b) OPL fitting results for VLC channels

Figure 8. Different sizes of commercial PDs and OPL fitting results for VLC channels.

Furthermore, the RS is set in steps of 0.5 mm from 10 to 0.5 mm in our simulations, so that we can obtain the OPL samples with different sizes of PDs in Figure 9. We can model the OPL based on these samples with the FI model. In addition, we propose an OPL model for VLC channels as a function of the spatial distance and size of PDs based on the FI Model (4) through curve fitting techniques. This proposed OPL model can be used to estimate the physical size of PDs to meet the required coverage range while designing systems, which can be expressed as:

$$\text{OPL}(d,r) = \beta + 10\alpha\log_{10}(d) + \gamma\log_{10}(\frac{1}{r^3}) + X_\sigma, \quad (7)$$

where r denotes the RS of PDs. The parameter α, i.e., PLE, is 1.73, which is larger than the PLE in mmWave (1.31). This result reveals that VLC channels suffer faster decay than mmWave channels with increasing distance in the conference room. The parameter β (50.84) is smaller than that in mmWave bands (63.5). This is mainly because the path loss is always lower than that in mmWave bands when RS is set to be greater than 1 mm, i.e., most of the RS sets (0.2, 0.5–10 mm) in our simulation present the lower OPL in VLC bands. The parameter γ (6.98) is defined as a coefficient related to the size of PDs. Moreover, the standard derivation σ representing shadowing fading is 0.99.

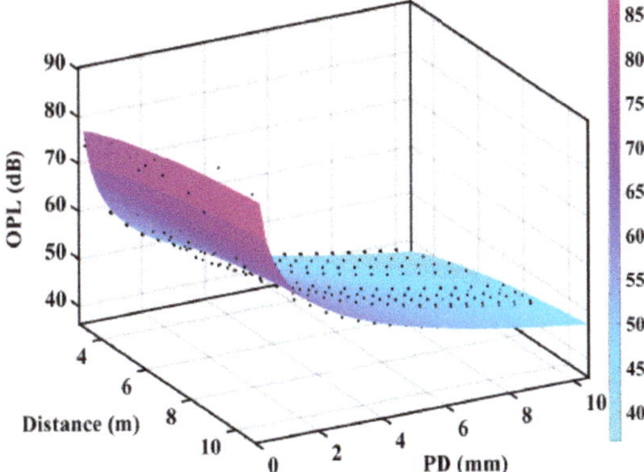

Figure 9. OPL fitting results for VLC channels considering the physical size of PDs.

3.3. RMS DS

RMS DS describes the time dispersion of MPCs, and it can be used to calculate the coherence bandwidth [34]. Furthermore, large DS may cause inter-symbol interference. The RMS DS can be calculated as follows [35]:

$$\tau_{\text{RMS}} = \sqrt{\frac{\sum_{n=1}^{N}(\tau_n - \tau_{\text{mean}})^2 P(\tau_n)}{\sum_{n=1}^{N} P(\tau_n)}}, \tag{8}$$

$$\tau_{\text{mean}} = \frac{\sum_{n=1}^{N} \tau_n P(\tau_n)}{\sum_{n=1}^{N} P(\tau_n)}, \tag{9}$$

where n is the index of paths, τ_n and $P(\tau_n)$ are the delay and power of the nth path, and τ_{mean} is the mean excess delay given by (9). Figure 10 presents the RMS DS for mmWave and VLC channels calculated by (8) and (9). To model the RMS DS, the distance-dependent model is selected as a candidate model in 3GPP [5,36]. We can model the RMS DS with the distance-dependent model as shown in [37],

$$\tau_{\text{rms}} \propto d^c, \tag{10}$$

where d is the distance between TX and RX, and c is a constant, which in this case is 0.15 and −0.04 for mmWave and VLC channels, respectively. Moreover, the confidence interval for c with the 95% confidence level is [0.141–0.159] and [−0.256–0.172] for mmWave and VLC channels, respectively. Note that the values of c in the confidence interval for mmWave are always positive, which indicates that the RMS DS will increase with distance for mmWave channels in this conference room. Moreover, this paper shows smaller RMS DS than that in [37] (28–38 ns). This is mainly because the environment in [37] is larger and more complicated, resulting in greater RMS DS. On the contrary, both positive and negative values appear in the confidence interval for VLC channels. This result indicates that RMS DS is little dependent on the distance for VLC channels in this conference room. This can be explained by the fact that the light rays experiencing complicated reflections are attenuated greatly and reflected in all directions, so that they can barely reach the RX, resulting in the small RMS DS in VLC bands.

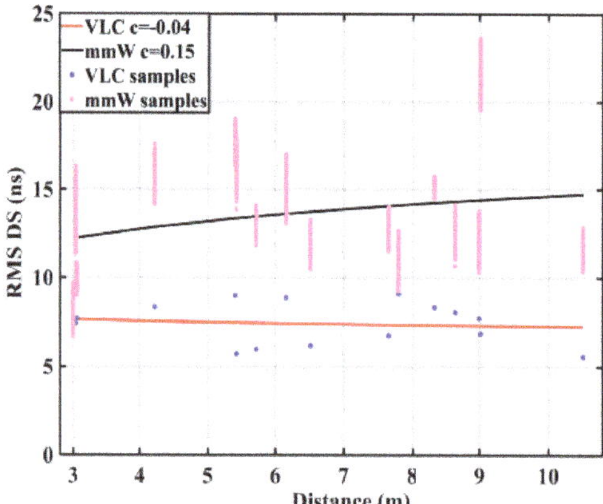

Figure 10. RMS DS fitting results for mmWave and VLC channels.

In Figure 11, we plot the cumulative distribution function (CDF) of RMS DS for mmWave and VLC channels, respectively. The μ and σ parameters in normal fitting are −7.88 ($10^{-7.88} \approx 13.18$ ns) and 0.1 for mmWave channels, while they are −8.13 ($10^{-8.13} \approx 7.41$ ns) and 0.07, respectively, for VLC channels. This indicates that the delay dispersion is weaker in VLC bands than that in mmWave bands, i.e., propagation distances of multipath rays are mostly close due to the few received multipath rays experiencing complicated reflections with a large excess delay in VLC bands.

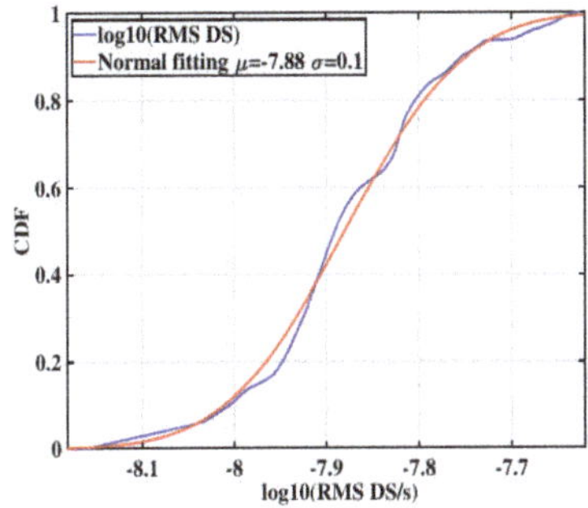

(a) The CDF of RMS DS for mmWave channels

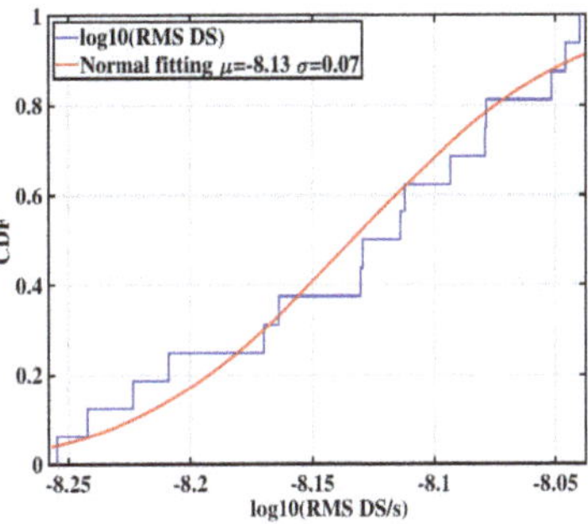

(b) The CDF of RMS DS for VLC channels

Figure 11. The CDF results of RMS DS.

3.4. K-Factor

K-factor is a measure of the severity for the small-scale fading. Knowledge of K-factor statistics is essential for the design and performance analysis of wireless systems. Furthermore, K-factor is defined as the ratio of the power of the deterministic MPC (typically LOS) and the power of all other stochastic MPCs (typically NLOS). Thus, we can obtain the K-factor according to (1), as shown in (11). However, it is difficult to distinguish which one is a LOS component and which one is an NLOS component from the measured CIRs for mmWave channels. Sometimes, one delay bin may contain both the LOS and NLOS components. Here, we applied the moment method to estimate the K-factor based upon the analysis of frequency selectivity for mmWave channels [38,39].

$$K = \frac{|V|^2}{\sum_{n=1}^{N} |a_n|^2}. \tag{11}$$

For VLC channels, we obtain the channel characteristics by the ray tracing method. Therefore, we can distinguish the LOS and NLOS components clearly based on the source output file. The K-factor for VLC channels can be expressed according to its definition as:

$$K = \frac{\sum_{n_{LOS}=1}^{N_{LOS}} P(n_{LOS})}{\sum_{n_{NLOS}=1}^{N_{NLOS}} P(n_{NLOS})}, \tag{12}$$

where the N_{LOS} and N_{NLOS} are the numbers of LOS and NLOS rays, respectively.

The CDF of the estimated K-factor and its normal fitting for mmWave channels is shown in Figure 12a. It shows that 20% of the K-factor values are less than 0 dB and these values are from the 10th (5.43 m) and 16th (9.0 m) measured points. It appears that NLOS components account for more power in these two points, especially in the 16th point. This is mainly because these two points are close to the walls, which means more reflection components, resulting in the small K-factor. However, 74% of the K-factor values for VLC channels are less than 0 dB in Figure 12b, and this happens when the distance between TX and RX is larger than 5.43 m (10th point). This indicates that NLOS components are stronger at a large distance in this conference room. Note that the K-factor in VLC bands is 9 dB lower than that in mmWave bands when the distance is 3 m. This can be explained by the definition of K-factor in (12). It is clear that whether there are one or nine LEDs placed at the TX side, the number of LOS components (N_{LOS}) is fixed. In our simulation, we place nine LEDs successively at TX to obtain the same coverage as mmWave antennas in azimuth (360°), which is required for a fair comparison. This brings more MPCs received by the detector, i.e., more NLOS rays (N_{NLOS}), resulting in the small K-factor. In addition, the μ parameter in the normal fitting is 2.88 dB for mmWave channels, while it is −1.03 dB for VLC channels. These results suggest that it is better to select mmWave technologies when large coverage is required, i.e., VLC technologies are expected for short-range communication.

In Figure 12c,d, we plot both the RMS DS and K-factor with the distance. The distance dependence of these two parameters seems not apparent. Furthermore, we can see that these two parameters appear to present a similar trend, varying with the distance, i.e., the K-factor is proportional to the RMS DS in the conference room. Moreover, the correlation coefficients calculated by (13) are −0.22 and 0.77 for mmWave and VLC channels, respectively. The coefficient −0.22 shows a negative correlation between RMS DS and K-factor, which is demonstrated by the fact that the valley value corresponds to the peak value when the distance is 8 m in Figure 12c. It can also be verified that RMS DS and K-factor in the VLC bands have a higher correlation (0.77 > 0.22), as seen in Figure 12d.

$$\rho_{X,Y} = \frac{cov(X,Y)}{\sqrt{var(X)var(Y)}}, \tag{13}$$

where $cov(\cdot)$ and $var(\cdot)$ represent the covariance and variance, respectively.

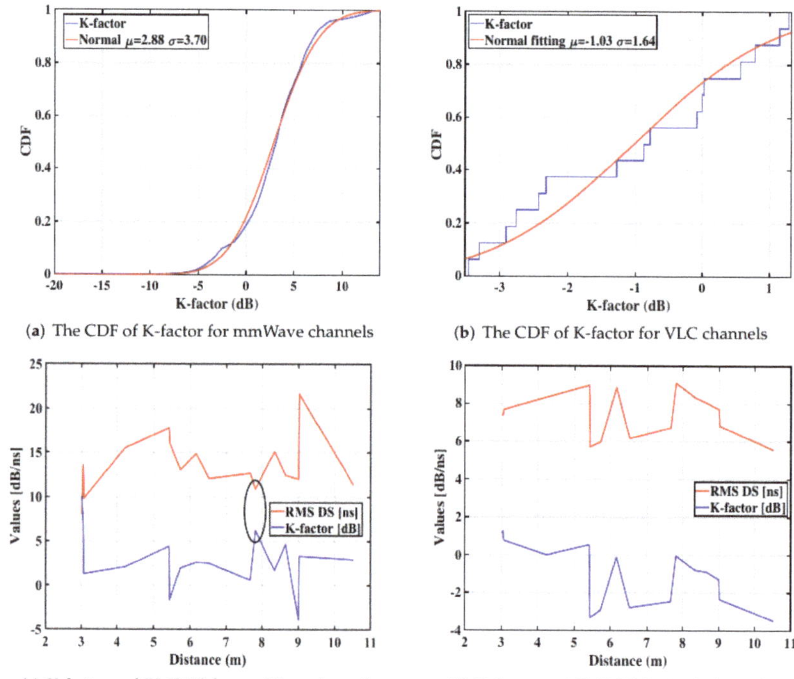

(a) The CDF of K-factor for mmWave channels

(b) The CDF of K-factor for VLC channels

(c) K-factor and RMS DS for mmWave channels

(d) K-factor and RMS DS for VLC channels

Figure 12. The results of K-factor and RMS DS.

3.5. Cluster Characteristics

The clustered delay line (CDL) model has been used widely in wireless mobile channel standardization, e.g., 3GPP TR 38.901 [5]. In this model, a cluster means a group of similar MPCs, and MPCs can be clustered from delay and angle dimensions.

Here, we use the K-Power-Means algorithm [40] with a novel initial step [41] to perform clustering. For mmWave channels, multipath information, including amplitude, delay, azimuth angle of arrival (AOA), and elevation angle of arrival (EOA), is estimated by the space-alternating generalized expectation-maximization (SAGE) algorithm [42].

For VLC channels, the ray tracing method is adopted and the parameters such as AOA and EOA can be obtained by geometric calculation.

In Figure 13a,b, the clustering results at the 10th position are presented. Clusters for mmWave and VLC channels are marked with M and R, respectively. Moreover, the points with the same color belong to one cluster. It is clear that there are six clusters divided for mmWave while there are five clusters for VLC at the same position. Obviously, these two clustering results are different despite the similar cluster numbers that they have. It is apparent that the MPCs for mmWave can be distinguished more easily in the delay dimension for this measured scenario because the clusters seem to be located in different delay bins, with small fluctuations in AOA and EOA within one cluster. This shows the temporal sparsity in mmWave bands in this conference scenario. However, the difference can be found by investigating the clusters in Figure 13b, which shows that the delay of different clusters is close while the angle dimension (AOA and EOA) varies dramatically, i.e., large angular spread in VLC bands, and we can obtain the same results at other points in this room. This can be explained by the radio wave propagation mechanisms. The light rays have a much smaller wavelength than mmWave signals. The reflection case occurs with similar directions when mmWave signals arrive the surfaces of materials, while the light rays are scattered randomly in all directions, resulting in a large angular spread, in VLC bands.

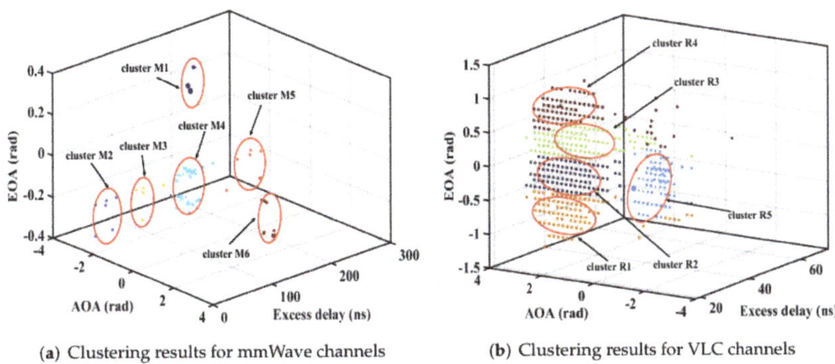

(**a**) Clustering results for mmWave channels

(**b**) Clustering results for VLC channels

Figure 13. Clustering results in the 10th position.

In the standardization channel model, the number of clusters is commonly fixed and made equal to what exists in bands below 6 GHz. For example, the number of clusters is 15 for indoor LOS scenarios in [5]. However, sparsity in the mmWave bands makes this assumption unreasonable [43,44]. The typical cluster numbers reported in [45,46] are small and random with a Poisson distribution. In our work, Figure 14a,b present the statistics of cluster numbers. The Poisson distribution is used to fit the cluster numbers. The fitting parameter λ denotes the average value of cluster numbers, and they are 5.71 and 6.38 for mmWave and VLC, respectively. Here, we consider that the λ is 5 in mmWave bands, and it is 6 in VLC bands because the cluster numbers are expected to be integers.

To confirm our results in the conference scenario, the statistical test method in Statistics, e.g., Kolmogorov–Smirnov test (KS-test) [47], is used to decide if the cluster numbers follow the Poisson distribution. The confidence interval and significance level are 95 % and 0.05, respectively. The test result, i.e., the significance probability p-value for mmWave and VLC, is 0 (<0.05) and 0.547 (>0.05), respectively. The results indicate that VLC cluster numbers follow the Poisson distribution, while the mmWave cluster numbers may not follow the Poisson distribution. We can infer that mmWave cluster numbers may follow the normal distribution from its histogram in Figure 14a, and correspondingly we plot the normal curve to fit the cluster numbers. The fitting result shows that the histogram and normal curve of mmWave cluster numbers can fit well together, which indicates that mmWave cluster numbers follow the normal distribution [48]. In general, the results indicate that the cluster numbers are overestimated in the standardization channel model, and this work can be supplemental in the 3GPP standardization works.

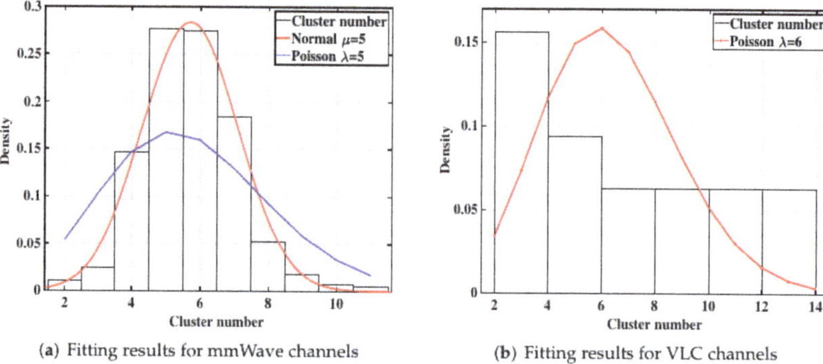

(a) Fitting results for mmWave channels (b) Fitting results for VLC channels

Figure 14. Fitting results of cluster number.

4. Conclusions and Future Work

This paper focuses on the propagation characteristics comparisons between mmWave and VLC bands in indoor scenarios based on the measurement campaign and ray tracing simulation. The results are analyzed based on the conference room. We find that the OPL of VLC channels is highly dependent on the physical size of PDs and an OPL model considering the size of PDs is further proposed. The size of PDs can be estimated to meet the required coverage range while designing systems by using this proposed model. In addition, the PLE is 1.73 for VLC channels, which is larger than that for mmWave channels (1.31). This result reveals that the VLC channels suffer faster decay than mmWave channels with increasing distance in this room. Moreover, mmWave signals can travel farther with a larger RMS DS (13.18 ns) than that of VLC channels (7.41 ns), and the smaller RMS DS in VLC bands shows the weaker delay dispersion compared to mmWave channels. Moreover, the RMS DS of mmWave increases with distance, while it is little dependent on the distance for VLC channels. Furthermore, the mean K-factor for mmWave is 2.88 dB with −1.03 dB for VLC, and LOS components mainly account for more power (80% of K-factor > 0 dB) for mmWave in LOS scenarios, while NLOS components can be stronger (74% of K-factor < 0 dB) at a large distance for VLC channels. The results suggest that it is better to select mmWave technologies when large coverage is required, i.e., VLC technologies are expected for short-range communication. On the other hand, RMS DS and K-factor appear to present a similar trend, varying with distance, i.e., RMS DS is proportional to K-factor for both mmWave and VLC channels. However, this requires further verification. Furthermore, the K-Power-Means algorithm is used to perform clustering from the dimensions of delay and angle. The fitting cluster number is 5 and 6 for mmWave and VLC, respectively. The clustering results reveal the temporal sparsity in mmWave bands and show that VLC channels have a large angular spread.

The propagation characteristics for VLC channels are investigated based on ray tracing methods. Although this channel modeling approach has been demonstrated by experimental results, there are differences between simulation and experimental measurements considering the fact that the reflection coefficients of walls and objects inside the conference room could not be measured and representative values were assumed in the simulations. A channel sounding platform for VLC channels is still needed, which will be our further focus.

Author Contributions: Conceptualization, B.L.; methodology, B.L.; software, B.L. and P.T.; validation, B.L., P.T. and Y.Y.; investigation, B.L. and P.T.; resources, B.L. and P.T.; data curation, B.L. and P.T.; writing—original draft preparation, B.L.; writing—review and editing, P.T. and J.Z.; visualization,

B.L.; supervision, J.Z., G.L. and L.X. All authors have read and agreed to the published version of the manuscript.

Funding: This work is supported by the National Science Fund for Distinguished Young Scholars (No. 61925102), the National Key R&D Program of China (No. 2020YFB1805002), the National Natural Science Foundation of China (No. 62031019 and No. 62101069), and BUPT-CMCC Joint Innovation Center (CMYJY-202000276).

Institutional Review Board Statement: Not applicable.

Informed Consent Statement: Not applicable.

Data Availability Statement: Not applicable.

Conflicts of Interest: The authors declare no conflict of interest.

References

1. Sun, S.; Rappaport, T.S.; Shafi, M.; Tang, P.; Zhang, J.; Smith, P.J. Propagation models and performance evaluation for 5G millimeter-wave bands. *IEEE Trans. Veh. Technol.* **2018**, *67*, 8422–8439. [CrossRef]
2. Tang, P.; Zhang, J.; Molisch, A.F.; Smith, P.J.; Shafi, M.; Tian, L. Estimation of the K-factor for temporal fading from single-snapshot wideband measurements. *IEEE Trans. Veh. Technol.* **2018**, *68*, 49–63. [CrossRef]
3. Zhang, J.; Shafi, M.; Molisch, A.F.; Tufvesson, F.; Wu, S.; Kitao, K. Channel models and measurements for 5G. *IEEE Commun. Mag.* **2018**, *56*, 12–13. [CrossRef]
4. Zhang, J.; Kang, K.; Huang, Y.; Shafi, M.; Molisch, A.F. Millimeter and THz wave for 5G and beyond. *China Commun.* **2019**, *16*, iii–vi.
5. 3GPP. Study on Channel Model for Frequencies from 0.5 to 100 GHz. 3rd Generation Partnership Project, Technical Report TR 38.901 V16.1.0. 2019. Available online: https://portal.3gpp.org/desktopmodules/Specifications/SpecificationDetails.aspx?specificationId=3173 (accessed on 12 December 2021).
6. Chi, N.; Zhou, Y.; Wei, Y.; Hu, F. Visible light communication in 6G: Advances, challenges, and prospects. *IEEE Veh. Technol. Mag.* **2020**, *15*, 93–102. [CrossRef]
7. Zhang, J.H.; Tang, P.; Yu, L.; Jiang, T.; Tian, L. Channel measurements and models for 6G: Current status and future outlook. *Front. Inf. Technol. Electron. Eng.* **2020**, *21*, 39–61. [CrossRef]
8. Chen, C.; Yang, H.; Du, P.; Zhong, W.D.; Alphones, A.; Yang, Y.; Deng, X. User-centric MIMO techniques for indoor visible light communication systems. *IEEE Syst. J.* **2020**, *14*, 3202–3213. [CrossRef]
9. Gunawan, W.H.; Liu, Y.; Chow, C.W.; Chang, Y.H.; Yeh, C.H. High speed visible light communication using digital power domain multiplexing of orthogonal frequency division multiplexed (OFDM) signals. *Photonics* **2021**, *8*, 500. [CrossRef]
10. Ji, R.; Wang, S.; Liu, Q.; Lu, W. High-speed visible light communications: Enabling technologies and state of the art. *Appl. Sci.* **2018**, *8*, 589. [CrossRef]
11. Delgado-Rajo, F.; Melian-Segura, A.; Guerra, V.; Perez-Jimenez, R.; Sanchez-Rodriguez, D. Hybrid RF/VLC network architecture for the internet of things. *Sensors* **2020**, *20*, 478. [CrossRef]
12. Qin, D.; Wang, Y.; Zhou, T. Performance analysis of hybrid radio frequency and free space optical communication networks with cooperative spectrum sharing. *Photonics* **2021**, *8*, 108. [CrossRef]
13. Abuella, H.; Elamassie, M.; Uysal, M.; Xu, Z.; Serpedin, E.; Qaraqe, K.A.; Ekin, S. Hybrid RF/VLC systems: A comprehensive survey on network topologies, performance analyses, applications, and future directions. *IEEE Access* **2021**, 9, 160402–160436. [CrossRef]
14. Molisch, A.F. *Wireless Communications*, 2nd ed.; John Wiley & Sons: New York, NY, USA, 2012.
15. Tavakkolnia, I.; Cheadle, D.; Bian, R.; Loh, T.H.; Haas, H. High speed millimeter-wave and visible light communication with off-the-shelf components. In Proceedings of the 2020 IEEE Globecom Workshops (GC Wkshps), Taiwan, China, 7–11 December 2020; pp. 1–6.
16. Uyrus, A.; Turan, B.; Basar, E.; Coleri, S. Visible light and mmWave propagation channel comparison for vehicular communications. In Proceedings of the 2019 IEEE Vehicular Networking Conference (VNC), Los Angeles, CA, USA, 4–6 December 2019; pp. 1–7.
17. Feng, L.; Yang, H.; Hu, R.Q.; Wang, J. MmWave and VLC-based indoor channel models in 5G wireless networks. *IEEE Wirel. Commun.* **2018**, *25*, 70–77. [CrossRef]
18. Miramirkhani, F.; Uysal, M. Channel modeling and characterization for visible light communications. *IEEE Photonics J.* **2015**, *7*, 1–16. [CrossRef]
19. Rajagopal, S.; Roberts, R.D.; Lim, S.K. IEEE 802.15. 7 visible light communication: Modulation schemes and dimming support. *IEEE Commun. Mag.* **2012**, *50*, 72–82. [CrossRef]
20. Uysal, M.; Miramirkhani, F.; Narmanlioglu, O.; Baykas, T.; Panayirci, E. IEEE 802.15. 7r1 reference channel models for visible light communications. *IEEE Commun. Mag.* **2017**, *55*, 212–217. [CrossRef]

21. Eldeeb, H.B.; Uysal, M.; Mana, S.M.; Hellwig, P.; Hilt, J.; Jungnickel, V. Channel modelling for light communications: Validation of ray tracing by measurements. In Proceedings of the 2020 12th International Symposium on Communication Systems, Networks and Digital Signal Processing (CSNDSP), Porto, Portugal, 20–22 July 2020; pp. 1–6.
22. Eldeeb, H.B.; Elamassie, M.; Sait, S.M.; Uysal, M. Infrastructure-to-Vehicle Visible Light Communications: Channel Modelling and Performance Analysis. *IEEE Trans. Veh. Technol.* **2022**, 71, 2240–2250. [CrossRef]
23. ASTER Spectral Library-Version 2.0. Available online: http://speclib.jpl.nasa.gov (accessed on 12 December 2021).
24. CREE LEDs. Available online: https://www.creelighting.com/products/indoor/lamps/ (accessed on 12 December 2021).
25. Tang, P.; Zhang, J.; Tian, H.; Chang, Z.; Men, J.; Zhang, Y.; Tian, L.; Xia, L.; Wang, Q.; He, J. Channel measurement and path loss modeling from 220 GHz to 330 GHz for 6G wireless communications. *China Commun.* **2021**, *18*, 19–32. [CrossRef]
26. Yu, X.; Zhang, J.; Haenggi, M.; Letaief, K.B. Coverage analysis for millimeter wave networks: The impact of directional antenna arrays. *IEEE J. Sel. Areas Commun.* **2017**, *35*, 1498–1512. [CrossRef]
27. Ashikhmin, A.; Li, L.; Marzetta, T.L. Interference reduction in multi-cell massive MIMO systems with large-scale fading precoding. *IEEE Trans. Inf. Theory* **2018**, *64*, 6340–6361. [CrossRef]
28. Maccartney, G.R.; Rappaport, T.S.; Sun, S.; Deng, S. Indoor office wideband millimeter-wave propagation measurements and channel models at 28 and 73 GHz for ultra-dense 5G wireless networks. *IEEE Access* **2015**, 3, 2388–2424. [CrossRef]
29. Miramirkhani, F. A path loss model for link budget analysis of indoor visible light communications. *Electrica* **2021**, *21*, 242–249. [CrossRef]
30. Lee, K.; Park, H.; Barry, J.R. Indoor channel characteristics for visible light communications. *IEEE Commun. Lett.* **2011**, *15*, 217–219. [CrossRef]
31. Qiu, Y.; Chen, H.H.; Meng, W.X. Channel modeling for visible light communications—A survey. *Wirel. Commun. Mob. Comput.* **2016**, *16*, 2016–2034. [CrossRef]
32. Elamassie, M.; Karbalayghareh, M.; Miramirkhani, F.; Kizilirmak, R.C.; Uysal, M. Effect of fog and rain on the performance of vehicular visible light communications. In Proceedings of the 2018 IEEE 87th Vehicular Technology Conference (VTC-Spring), Porto, Portugal, 3–6 June 2018; pp. 1–6.
33. Hamamatsu Products. Available online: http://www.hamamatsu.com.cn/product/23464.html (accessed on 12 December 2021).
34. Janssen, G.J.; Stigter, P.A.; Prasad, R. Wideband indoor channel measurements and BER analysis of frequency selective multipath channels at 2.4, 4.75, and 11.5 GHz. *IEEE Trans. Commun.* **1996**, *44*, 1272–1288. [CrossRef]
35. Hashemi, H.; Tholl, D. Statistical modeling and simulation of the RMS delay spread of indoor radio propagation channels. *IEEE Trans. Veh. Technol.* **1994**, *43*, 110–120. [CrossRef]
36. Jiang, T.; Zhang, J.; Tang, P.; Tian, L.; Zheng, Y.; Dou, J.; Asplund, H.; Raschkowski, L.; D'Errico, R.; Jamsa, T. 3GPP standardized 5G channel model for IIoT scenarios: A survey. *IEEE Internet Things J.* **2021**, *8*, 8799–8815. [CrossRef]
37. Karedal, J.; Wyne, S.; Almers, P.; Tufvesson, F.; Molisch, A.F. A measurement-based statistical model for industrial ultra-wideband channels. *IEEE Trans. Wirel. Commun.* **2007**, *6*, 3028–3037. [CrossRef]
38. Greenstein, L.J.; Michelson, D.G.; Erceg, V. Moment-method estimation of the Ricean K-factor. *IEEE Commun. Lett.* **1999**, 3, 175–176. [CrossRef]
39. Bernadó, L.; Zemen, T.; Karedal, J.; Paier, A.; Thiel, A.; Klemp, O.; Czink, N.; Tufvesson, F.; Molisch, A.F.; Mecklenbrauker, C.F. Multi-dimensional K-factor analysis for V2V radio channels in open sub-urban street crossings. In Proceedings of the 21st Annual IEEE International Symposium on Personal, Indoor and Mobile Radio Communications, Istanbul, Turkey, 26–30 September 2010; pp. 58–63.
40. Czink, N.; Cera, P.; Salo, J.; Bonek, E.; Nuutinen, J.P.; Ylitalo, J. Improving clustering performance using multipath component distance. *Electron. Lett.* **2006**, *42*, 33–35. [CrossRef]
41. Huang, C.; Zhang, J.; Nie, X.; Zhang, Y. Cluster characteristics of wideband MIMO channel in indoor hotspot scenario at 2.35 GHz. In Proceedings of the 2009 IEEE 70th Vehicular Technology Conference Fall (VTC-Fall), Anchorage, AK, USA, 20–23 September 2009; pp. 1–5.
42. Fleury, B.H.; Tschudin, M.; Heddergott, R.; Dahlhaus, D.; Pedersen, K.I. Channel parameter estimation in mobile radio environments using the SAGE algorithm. *IEEE J. Sel. Areas. Commun.* **1999**, *17*, 434–450. [CrossRef]
43. Jiang, T.; Zhang, J.; Shafi, M.; Tian, L.; Tang, P. The comparative study of SV model between 3.5 and 28 GHz in indoor and outdoor scenarios. *IEEE Trans. Veh. Technol.* **2019**, *69*, 2351–2364. [CrossRef]
44. Jameel, F.; Wyne, S.; Nawaz, S.J.; Chang, Z. Propagation channels for mmWave vehicular communications: State-of-the-art and future research directions. *IEEE Wirel. Commun.* **2018**, *26*, 144–150. [CrossRef]
45. Ju, S.; Xing, Y.; Kanhere, O.; Rappaport, T.S. Millimeter wave and sub-terahertz spatial statistical channel model for an indoor office building. *IEEE J. Sel. Areas Commun.* **2021**, *39*, 1561–1575. [CrossRef]
46. Samimi, M.K.; Rappaport, T.S. 3-D millimeter-wave statistical channel model for 5G wireless system design. *IEEE Trans. Microw. Theory Tech.* **2016**, *64*, 2207–2225. [CrossRef]

47. Frey, J. An exact Kolmogorov–Smirnov test for the Poisson distribution with unknown mean. *J. Stat. Comput. Simul.* **2012**, *82*, 1023–1033. [CrossRef]
48. Arrenberg, J. *Analysis of Multivariate Data with SPSS: Workbook with Detailed Examples;* Books on Demand Gmbh: Norderstedt, Germany, 2020.

Article

Hybrid POF/VLC Links Based on a Single LED for Indoor Communications

Juan Andrés Apolo *, Beatriz Ortega and Vicenç Almenar

Instituto de Telecomunicaciones y Aplicaciones Multimedia, ITEAM, Universitat Politècnica de València, Camino de Vera, 46022 Valencia, Spain; bortega@dcom.upv.es (B.O.); valmenar@dcom.upv.es (V.A.)
* Correspondence: juaapgon@teleco.upv.es

Abstract: A hybrid fiber/wireless link based on a single visible LED and free of opto-electronic intermediate conversion stages has been demonstrated for indoor communications. This paper shows the main guidelines for proper coupling in fiber/air/detector interfaces. Experimental demonstration has validated the design results with very good agreement between geometrical optics simulation and received optical power measurements. Different signal bandwidths and modulation formats, i.e., QPSK, 16-QAM, and 64-QAM, have been transmitted over 1.5 m polymer optical fiber (POF) and 1.5 m free-space optics (FSO). Throughputs up to 294 Mb/s using a 64-QAM signal have been demonstrated using a commercial LED, which paves the way for massive deployment in industrial applications.

Keywords: POF; FSO; LiFi; LED

Citation: Apolo, J.A.; Ortega, B.; Almenar, V. Hybrid POF/VLC Links Based on a Single LED for Indoor Communications. *Photonics* **2021**, *8*, 254. https://doi.org/10.3390/photonics8070254

Received: 25 May 2021
Accepted: 29 June 2021
Published: 2 July 2021

1. Introduction

Future smart factories in Industry 4.0 will require fast and reliable wireless connectivity to provide automation and real-time control of equipment [1]. Optical wireless communication (OWC) is a promising technology for enabling such industrial transformation [2,3] since it offers a huge amount of unregulated bandwidth, high security, low latency, and immunity to electromagnetic interference. However, OWC links are based on line-of-sight (LOS) signal reception, and therefore the blockage of direct transmission by shadowing or obstacles is its main disadvantage. Current OWC solutions utilize multiple-input multiple-out (MIMO) schemes where several transmitters and receivers are employed [4,5].

Among the OWC approaches, visible light communication (VLC) is an attractive choice for indoor networks because of its capability to combine lighting and communications while using low-cost narrow bandwidth light-emitting diodes (LEDs), also known as light-fidelity (Li-Fi) [6]. Recent works have demonstrated the feasibility of OFDM transmission with data rates of 15.7 Gbps, employing off-the-shelf LEDs [7]; allowing multiuser access, interference mitigation, and mobility support [8,9]; and enabling 100 Gbps LiFi access network [10] when laser-based lighting or micro-LEDs are considered.

Furthermore, recent works demonstrated the interest in heterogeneous networks based on hybrid polymer optical fiber (POF)/VLC links as a converged solution to integrate fiber backbone and in-building networks [11,12] where large core diameter and small bending radius of POFs lead to low cost and easy installation. These systems include a photodiode (PD) to provide opto-electrical conversion of the signal transmitted over the POF before supplying the recovered electrical signal to the VLC optical source, i.e., an LED as an illuminating source or a laser diode (LDs) if larger bandwidths and collimated beams are required.

Among different types of POFs, the 1-mm core size poly-methyl methacrylate (PMMA) step-index (SI)-POF is the most popular for indoor communications, whose major drawback is a low transmission bandwidth due to intermodal dispersion. However, in [13] a transmission rate of 14.77 Gb/s was demonstrated by using wavelength division multiplexed (WDM) signals emitted by several LDs. Moreover, bandwidth efficient advanced modulation formats, i.e., multilevel pulsed amplitude modulation (M-PAM) signals modulating a micro-LED, with larger bandwidth than conventional LEDs, allowed 5 Gb/s transmission over 10 m [14].

As an attractive simple high capacity indoor communication system, Correa et al. have recently proposed using large core optical fibers acting as the light source in a luminaire-free scheme for VLC transmission; 2 Gb/s was achieved using two WDM LDs over 1.6 m [15].

In this paper, we propose, for the first time to the authors' knowledge, the use of low-cost centralized LEDs to feed the fiber network with no further opto-electrical conversions after the fiber section to wirelessly connect the user equipment. Hence, no light source is required for the VLC link; in this way, an optical network industrial infrastructure can be significantly simplified.

The paper is organized as follows: Section 2 describes the VLC solution over POF. Section 3 reports the experimental results of the single LED hybrid POF/VLC link. Finally, conclusions are presented in Section 4.

2. Description of a Single LED Hybrid POF/VLC Link

Figure 1 shows the proposal for Li-Fi wireless communication between a factory network and fixed arms of robots, i.e., users, where both uplink and downlink could employ VLC technology with POF cabling providing multiple signal access points. The network wireless infrastructure may consist of multiple distributed access points with no overlapping coverage despite the proximity between users. As a result, interference is avoided, and the spectrum can be reused efficiently. Moreover, MIMO techniques with diversity receiver designs [16] can be employed in advanced implementations in order to prevent link blockage.

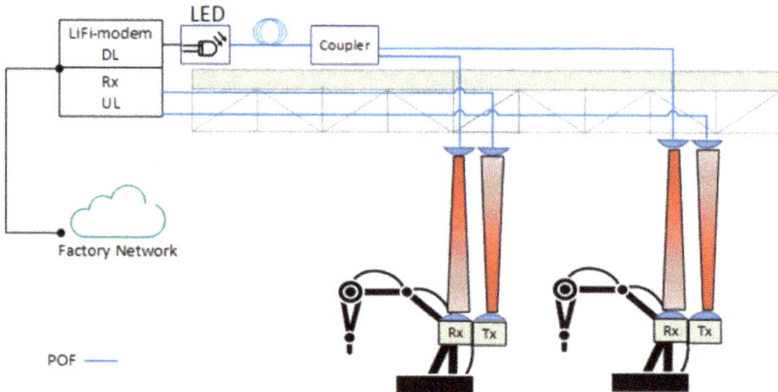

Figure 1. VLC-based infrastructure proposed for factory networks. DL: downlink, UL: uplink.

Therefore, a new hybrid POF/VLC link based on a single LED for indoor communications is proposed, as shown in Figure 2a. In this paper, we describe the implementation of a POF link as a front-haul downlink solution for feeding the LiFi system. The passive optical front-end at the ceiling only includes an optical lens with no optical-to-electric converter (O/E), i.e., optical source.

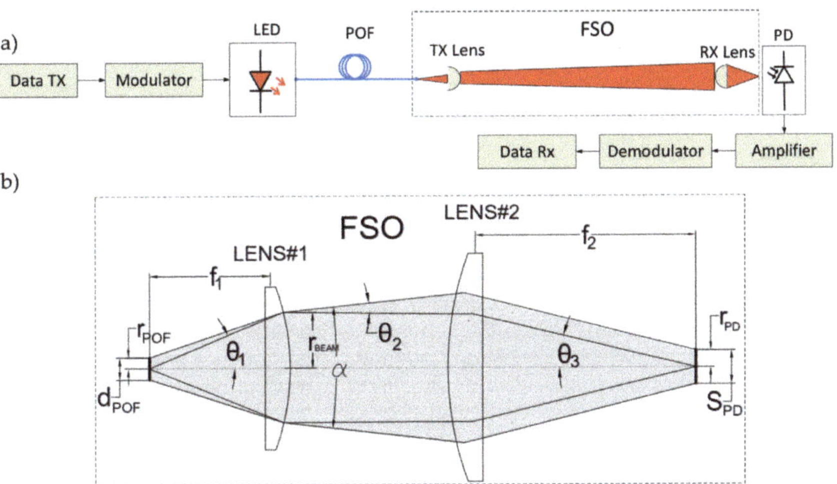

Figure 2. General scheme of the hybrid link: (**a**) block diagram of the experimental setup for the POF-VLC system, and (**b**) detail of FSO block: ray tracing diagram.

The transmitter consists of a low-cost commercial LED emitting light in the visible spectral range with directly modulated intensity. The output light is launched into the 1 mm diameter POF whose fiber-output is attached to a collimating lens to create the wireless interface. A lens is attached to the receiver input to focus the light onto a Ø1 mm photodetector. Then, a trans-impedance amplifier (TIA) is used as an amplification stage.

Collimators are often used to modify the divergent light emission from an optical fiber into a parallel light beam. Most commercial fiber optic collimators are designed for 125 μm diameter silica fibers with low numerical aperture (NA). However, an appropriate collimation system is essential to maximize the received optical power and ensure system operation. Moreover, a smaller collimator optic is preferred as it is cheaper to manufacture and requires less space for installation. Figure 2b shows the schematic of the free-space optics (FSO) link, including the geometrical optics and ray tracing, which are based on two lenses for proper light collimation as described below.

The purpose is to have efficient coupling (i.e., reduced insertion losses) and small collimator optics that minimizes cost. The numerical aperture provides the angle of acceptance of light from an optical system, i.e., the POF:

$$NA = n\,sin(\theta_1) \tag{1}$$

where n is the refractive index of the propagation medium and θ_1 is the angle of light emission from the optical element.

The fiber core size has a significant effect on the divergence angle of the collimated output beam, θ_2. Consider a nonpoint light source placed at the focal point of the lens, f_1, where the fiber diameter gives the size of the optical source, $d_{POF} = 2{\cdot}r_{POF}$. Using the paraxial approximation, the divergence angle of the output beam at lens #1 can be calculated as follows:

$$\theta_2(rad) = r_{POF}/f_1 \tag{2}$$

and the resulting beam radius at the exit of lens #1 corresponds to:

$$r_{BEAM} = \theta_1 * f_1 \tag{3}$$

In general, POFs have a relatively large core size to be considered as a point source. Therefore, the collimated beam will not be of constant diameter but will expand slightly as it propagates according to a beam divergence given by $\alpha = 2\theta_2$. To reduce this divergence, lenses with a longer focal length are required while maintaining the numerical aperture condition, $NA_{lens} > NA_{POF}$, since lenses with smaller numerical aperture than the fiber would imply coupling losses. No matter which lens is used, for a given optical source, the smaller beam that is desired, the greater divergence that is obtained.

We investigated the use of different lens combinations at both transmitter and receiver sides. Available commercial solutions, such as Prizmatix (Prizmatix Ltd. Azrieli Center. Building B, 7th Floor 26 Harokmim St. Holon 5885849, Israel) [17], offer 1/2′ and 1′ diameter POF collimators. Concretely, FCM1-05 offers the smallest divergence when using a 1 mm diameter and 0.5 NA POF, which has a full emission cone of 0.05 rad (2.86 deg).

We aim to reduce the beam divergence by optimizing the beam collimation in our system. According to Equation (2), a reduction down to $\alpha = 0.035$ rad (2 deg) leads to a focal length of the collimating lens f_1 larger than 28.5 mm for $r_{POF} = 0.5$ m. It is also necessary to satisfy the condition of $NA > 0.5$. Our analysis is limited to low-cost commercial lenses, aiming to find the smallest lens with the lowest divergence while maintaining the lens numerical aperture condition. Table 1 summarizes the optical characteristics of three commercial lenses and one Prizmatix solution that were selected to evaluate their performance.

Table 1. Optical characteristics of the employed lenses.

Commercial Lenses	Diameter (mm)	Focal Length (mm)	Back Focal Length (mm)	NA	Geometry	Divergence α (deg)
LA1951-A	25.4	25.3	17.6	0.5	Spherical	2.26
ACL25416U-A	25.4	16.0	7.3	0.79	Aspherical	3.58
ACL50832U	50.8	32	17	0.76	Aspherical	1.80
FCM05-05	12.7	~ 12	*	0.5	Spherical	5.73

* no info.

In an optical system only composed of lenses; the optical invariant, also called Lagrange invariant, states that the product of the object height and the marginal ray angle is constant [18]. From Figure 2b, the following expression can be obtained:

$$\theta_3 = \frac{r_{\text{BEAM}}}{f_2} \tag{4}$$

since the product $\theta_1 * r_{POF} = \theta_3 * r_{PD}$ is invariant, the beam radius at the PD input is given by:

$$r_{PD} = (\theta_1 * r_{POF})/\theta_3 \tag{5}$$

Replacing Equations (3) and (4) in (5), we obtain $r_{PD} = r_{POF} * f_2/f_1$ and the spot size $S_{PD} = 2 * r_{PD}$ is expressed as:

$$S_{PD} = (d_{POF} * f_2)/f_1 \tag{6}$$

where the focal distance of Lens #2 is f_2 and the PD is located at the focal point. Therefore, the minimum spot size is limited by the fiber diameter and the focal lengths of both lenses in the system. A shorter focal length in the receiver is required to reduce the spot size at photodetection. An aberration-free optical system is assumed. Aberrations introduced by the optical elements of the system may cause exceptions to these simple rules.

An optimized design of the FSO link requires the simulation of the system optical components. In this work, Zemax software (Zemax, LLC 10230 NE Points Drive, Suite 500 Kirkland, Washington 98033)

USA is employed to simulate nine different setups, which are summarized in Table 2. The simulator workspace is configured by placing the LED at the fiber input and the fiber-output at the focal distance of the collimating lens, ensuring proper face orientation, as depicted in Figure 2b. The distance between the collimating lens and the receiver lens is set from 0.25 m to 2 m, and the lenses are fully aligned. The optical power of the LED is set to –5.5 dBm according to the manufacturer parameters and experimental measurements.

Table 2. Lenses employed at both ends of the VLC link in different setups and corresponding obtained spot sizes at the PD input.

	Setup A	Setup B	Setup C	Setup D	Setup E	Setup F	Setup G	Setup H	Setup I
Fiber-output	ACL25416U-A	LA1951-A	FCM05-05	LA1951-A	LA1951-A	LA1951-A	FCM05-05	FCM05-05	FCM05-05
Receiver end	–	–	–	ACL25416U-A	LA1951-A	FCM05-05	ACL25416U-A	LA1951-A	FCM05-05
Spot size (mm)	–	–	–	0.62	1	0.47	1.33	2.1	1

According to the simulation, with specific lenses at the fiber-output and at the photodetector input, signal losses can be obtained as a distance function. Figure 3a compares the optical losses as a function of distance using different lenses in the transmitter, whereas no lens is employed at the detector (setups A, B, and C). As expected from the theoretical predictions, the lens with the longest focal length produces the least beam divergence and therefore the divergence losses over the distance will be lower (setup B). Figure 3b shows the simulated optical losses as a function of distance by using the optimal lens for collimation (LA1951-A) and different lenses at the detector input (setups D, E and F). The aspheric lens ACL25416U-A (setup D) better concentrates the input beam on the photodetector due to the size of the focusing point (0.62 mm), small enough for the photodetector to capture it without power reduction. Besides, an aspheric lens introduces less loss in the detection system than a spherical lens due to the correction of the aberrations inherent to the geometry of spherical lenses. Lastly, the Prizmatix FCM05-05 collimator is tested at the fiber-output and different lenses are used at the detector input (G, H, and I setups), as shown in Figure 3c. In all cases, lower optical power is collected than in setups D, E, and F. Note that G setup leads to higher received optical power, although setup I shows the smallest spot size (1 mm), but higher losses are introduced due to the reduced lens size (1/2′).

Experimental measurements were done by using Thorlabs (Thorlabs Inc. Newton, New Jersey, USA) S130C photodiode power sensor and the PM100D power meter console under several setups for the sake of validation of simulation results. As shown in Figure 3, there is good agreement between theoretical results and experimental measurements in spite of differences between simulated models and real components. Therefore, setup D has been proved to be as the best option for the transmission system employed in this work. Moreover, irradiance distribution is evaluated by nonsequential ray tracing in ZEMAX. The irradiance distribution variation can be shown as incoherent irradiance as a function of spatial position on the detector, as shown in Figure 4. The cross section corresponds to the photodetector (Ø1 mm). Three different cases are shown for the sake of comparison. First, no optics are used in the FSO link; second, only one collimating lens at the fiber-output is evaluated and no lens is employed at the receiver-end (setup B); finally, lens LA1951-A is employed at the fiber-output and lens ACL25416U-A at the receiver-end (setup D).

As shown in Figure 4a, the lack of an optical element at the fiber-output produces a highly divergent beam. The optical power incident on the photodiode over the distance is inadequate to establish communication. However, adding an optical element at the fiber-output to collimate the beam and homogenize the ray trajectories, as shown in Figure 4b, increases the incident power on the surface of the photodiode. Finally, by incorporating an optical element at the photodetector input, the collimated beam is concentrated (see Figure 4c) and thus the performance of the system is improved, as will be shown in the next section.

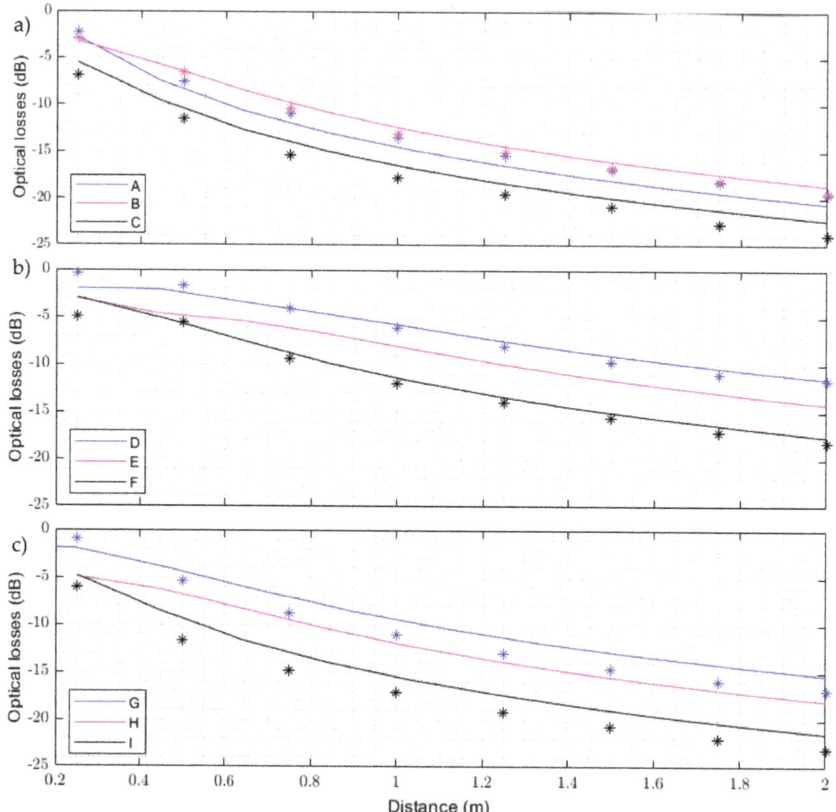

Figure 3. Experimental (symbols) and simulated (solid lines) optical losses vs. distance for different setups at the VLC link (defined in Table 2): (**a**) different lenses at the fiber-output and no lens at the receiver; (**b**) setups D, E, and F (Lens LA1951 is employed at the fiber-output in all of them); and (**c**) setups G, H, and I (FCM05-05 is employed at the fiber-output in all of them).

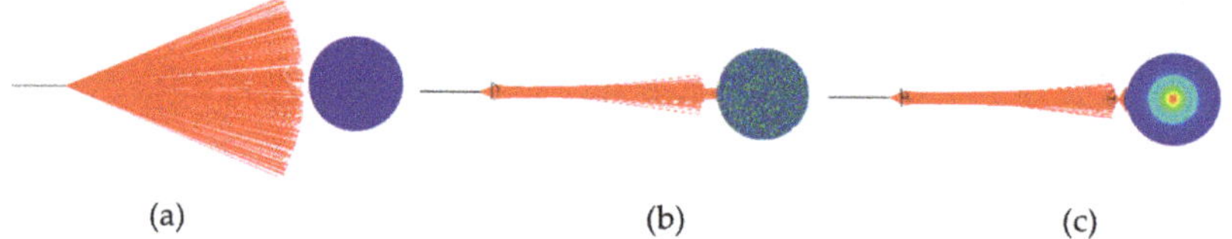

Figure 4. Rays trajectory and the corresponding cross section at the receiver: (**a**) no lenses are employed, (**b**) lens LA1951-A is employed at the fiber-output and no lens at the receiver-end (setup B), and (**c**) lens LA1951-A is employed at the fiber-output and ACL25416U-A at the receiver-end (setup D).

3. Transmission Measurements

3.1. Experimental Setup

In this section, we describe the experimental implementation of the single LED hybrid POF/VLC link. The basic blocks are shown in Figure 2a, where a digital QPSK or M-QAM signal is transmitted through a POF section followed by an OWC link. A low-cost LED (AVAGO SFH757) that emits light at a wavelength of 650 nm was employed as a transmitter.

The LED is directly modulated in intensity over the linear region, with information-carrying signals generated by an arbitrary waveform generator (AWG, Tektronix 7122C). A carrier frequency of 60 MHz was chosen for all the employed modulation formats, while the signal bandwidth was modified between 10 and 115 MHz. This AWG generates a fixed electrical power of –1.5 dBm, regardless of the modulation bandwidth. A Bias Tee (Mini-Circuit ZFBT-6GW+) is used to combine the DC bias current produced by a DC power supply (Keithley 2231A-30-3) and data signal from the AWG. The LED output is coupled into a 1.5 m long 1 mm step-index plastic fiber with 0.5 NA. The bare fiber is inserted directly into the LED housing. A collimating lens is placed at the fiber-output to create the wireless interface where an FC connector is employed at the fiber end. At the receiver-end, a lens is used to focus the light onto a Ø1 mm photodiode (Thorlabs, PDA10A) where the detected power is regulated by 1–12 dB ND absorption filters (Thorlabs, NEXXB). This photodetector incorporates a fixed gain TIA. Finally, the data is real-time demodulated on a digital phosphor oscilloscope (DPO, Tektronix 72004C).

3.2. System Performance for Different Bandwidths and Modulation Formats

This section reports the system performance obtained under two different lenses setup for the sake of comparison, i.e., setup H based on a commercial collimating solution and setup D proposed as the optimal solution in the previous section. Figure 5 shows the transmission results of three modulation formats, i.e., QPSK, 16-, and 64-QAM signals with different bandwidths under setup H (see Table 2) over 1 and 1.5 m of FSO link. The error vector magnitude (EVM) of the recovered signals was measured in order to test the viability of the setups and evaluate the maximum achievable capacity. The RoP at the detector over 1 m distance was –22.65 dBm, and the measured EVM results for the transmitted QPSK, 16-, and 64-QAM signals were below their valid maximum levels, which are 17.5%, 12.5%, and 8% [19], respectively, as shown in Figure 5a. Higher modulation orders have not been evaluated since they require lower EVM thresholds (i.e., 3.5% for 256-QAM) that can only be attained for shorter distances.

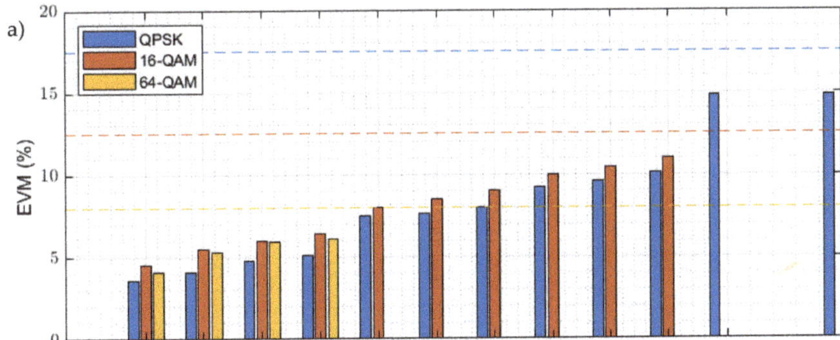

Figure 5. *Cont.*

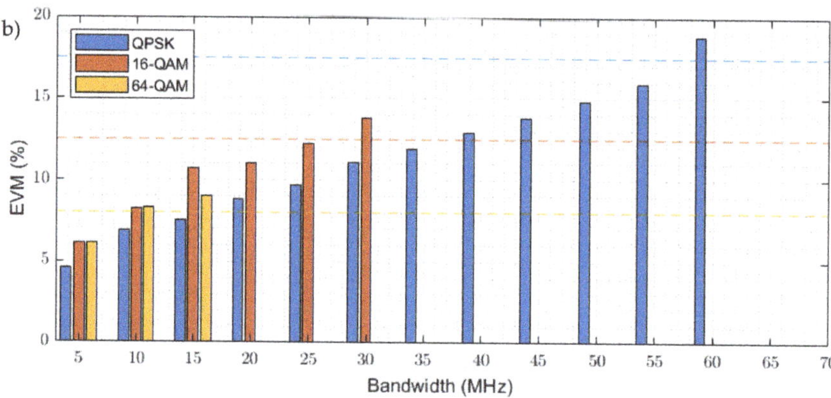

Figure 5. EVM measurements for setup H under different bandwidths and modulation formats at different distances: (**a**) 1 m, (**b**) 1.5 m.

Figure 5b shows the obtained EVM measurements over 1.5 m FSO link using the same setup, in this case the RoP was –26.23 dBm, and the estimated bandwidths satisfying the corresponding EVM levels were estimated to be around 55 MHz, 25 MHz, and 5 MHz for QPSK, 16-, and 64-QAM, respectively. Finally, some tests were done over 2 m (RoP of –28.58 dBm), but only 20 MHz QPSK and 5 MHz 16-QAM signals gave EVM measurements under their reference levels.

By using the same setup (H), the EVM performance of the system was characterized in terms of RoP. As depicted in Figure 6, the EVM increases significantly with the signal bandwidth and decreases for larger RoP values. Figure 6a shows EVM measurements for different bandwidths using QPSK modulation and demonstrates that the transmission of a 60 MHz QPSK signal requires a received optical power of –22.8 dBm, while –21.8 dBm and –21.3 dBm RoP are required for 50 MHz 16-QAM and 28 MHz 64-QAM signals, respectively. Signal constellations at the highest measured bandwidths are shown as insets for –20.76 dBm RoP. If we compare the curves obtained with 10 MHz and 20 MHz bandwidths (or 20 and 40 MHz) for a fixed EVM (equivalent to a fixed SNR), there is a 1.5 dB RoP penalty that corresponds to a 3 dB difference in electrical power that comes from doubling the noise when the bandwidth is doubled. However, this relationship does not hold for larger bandwidths where other channel distortions apart from noise are present; then, the RoP penalty increases.

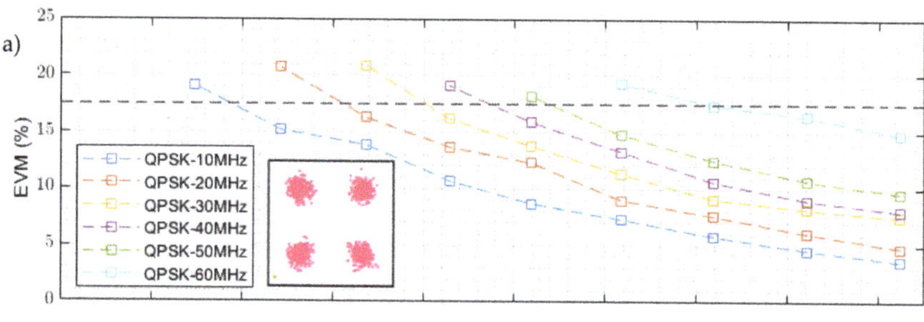

Figure 6. *Cont.*

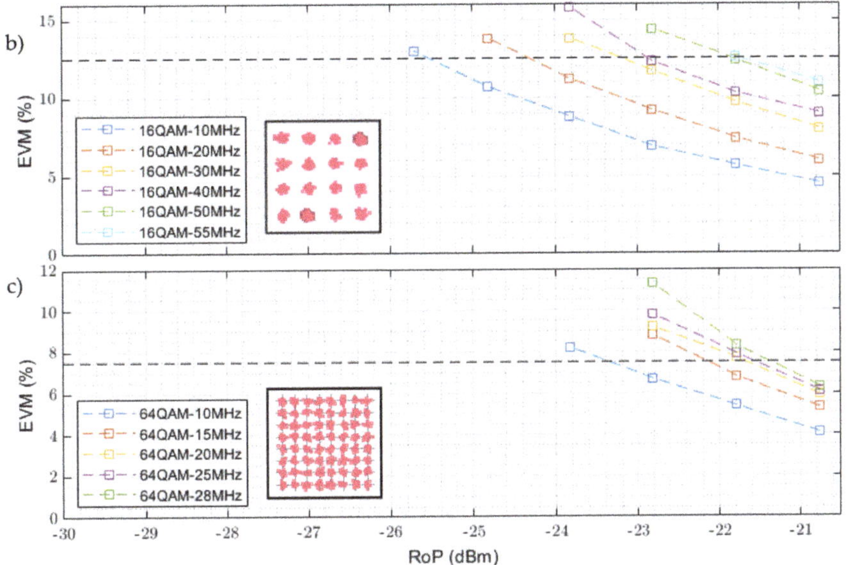

Figure 6. EVM performance vs RoP for different bandwidth signals (Setup H): (**a**) QPSK modulation, (**b**) 16-QAM modulation, and (**c**) 64-QAM modulation.

The following experiments employed setup D, which leads to maximum RoP according to Figure 3, which increases the capacity of the system. Figure 6 shows the EVM performance for different modulation formats vs. RoP and, in this case, transmission of 115 MHz QPSK signals was demonstrated with –16 dBm RoP, 65 MHz 16-QAM signal with –16.6 dBm and 45 MHz 64-QAM with –16.2 dBm. According to Figure 3b where optical losses was experimentally and theoretically characterized for different setups in terms of distance, this optical power level is achieved in setup D after 1.70 m free space optics propagation that is suitable for industrial applications, as described above. Constellations of the highest bandwidth signals are included as insets in Figure 7 for −15.7 dBm RoP.

In accordance with results depicted in Figure 3, setup D gives higher received optical power values, and, therefore, higher throughputs can be achieved using larger bandwidths and higher order M-QAM signals. To give a better insight, Figure 8 compares maximum achievable throughputs in terms of distance for both H and D setups. With setup D, a maximum throughput of 294 Mb/s can be achieved at 1.5 m, but as distance is increased, the RoP worsens and, since the noise level is fixed, the modulation bandwidth must be reduced, accordingly, giving a reduced bit rate. It can be seen that up to 2 m distance, 64-QAM is the best option giving throughputs higher than 200 Mb/s. For longer distances, the SNR reduction requires the use of QPSK with lower bandwidths as the distance increases, giving at 2.8 m a bit rate of 80 Mb/s, which is also suitable for industrial connectivity. At lower distances the QPSK throughput is limited by the 110 MHz maximum bandwidth of the employed components. On the other hand, when the H setup is chosen, the lower RoP makes more interesting the use of 16-QAM instead of 64-QAM, and again, an increasing distance requires a reduction of bandwidth until 1.3 m; from this point, a QPSK modulation should be used.

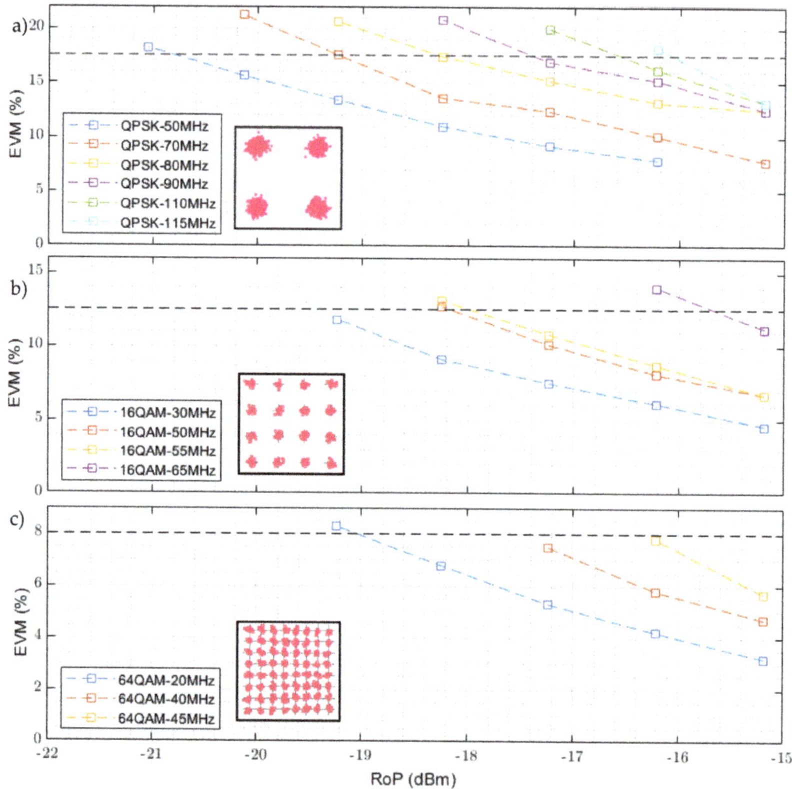

Figure 7. EVM performance vs RoP for different bandwidth signals (setup D): (**a**) QPSK modulation, (**b**) 16-QAM modulation, and (**c**) 64-QAM modulation.

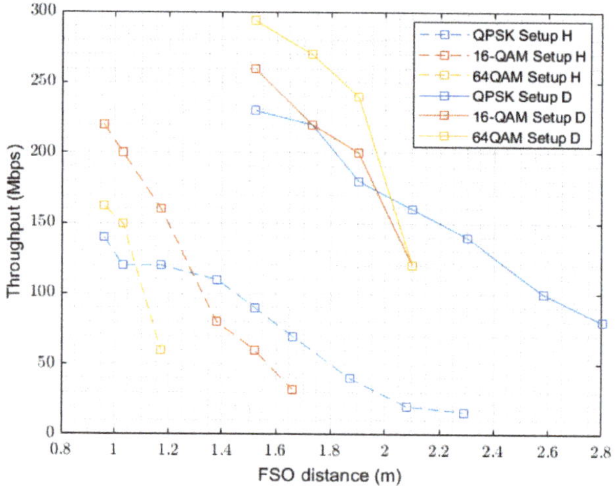

Figure 8. Throughput vs. FSO distance for different modulation formats in setups D and H.

4. Conclusions

In this paper, we have proposed the use of a centralized LED with no intermediate electro-optical conversion stages for optical fiber-wireless links. Simulations of the optics required in the interfaces have been carried out to provide the main guidelines for an optimized design, and the obtained results have been experimentally validated. As a result, we have experimentally demonstrated a 294 Mb/s VLC system using a hybrid POF/VLC link based on a single LED for indoor communications. The lack of optical sources in the fiber-to-LED interface, as well as the wired-wireless hybrid links, provides an attractive solution for massive low consumption and flexible implementation of communications links in Industry 4.0.

Author Contributions: Conceptualization, B.O. and V.A.; software, J.A.A.; validation, B.O.; data acquisition, J.A.A.; writing—original draft preparation, B.O. and J.A.A.; supervision, B.O. and V.A. All authors have read and agreed to the published version of the manuscript.

Funding: This work was supported by the Spanish Ministerio de Ciencia, Innovación y Universidades RTI2018-101658-B-I00 FOCAL project.

Institutional Review Board Statement: Not applicable.

Informed consent statement: Not applicable.

Data Availability Statement: Not applicable.

Acknowledgments: COST action NEWFOCUS (CA19111).

Conflicts of Interest: The authors declare no conflict of interest.

References

1. Schwab, K. *The Fourth Industrial Revolution*; World Economic Forum: Geneva, Switzerland, 2016.
2. Zhang, X.; Cao, Z.; Li, J.; Ge, D.; Chen, Z.; Vellekoop, I.M.; Koonen, A.M.J. Wide-Coverage Beam-Steered 40-Gbit/s Non-Line-of-Sight Optical Wireless Connectivity for Industry 4.0. *J. Light. Technol.* **2020**, *38*, 6801–6806. [CrossRef]
3. Sabella, R.; Iovanna, P.; Bottari, G.; Cavaliere, F. Optical transport for Industry 4.0. *J. Opt. Commun. Netw.* **2020**, *12*, 264–276. [CrossRef]
4. Kouhini, S.M.; Jarchlo, E.A.; Ferreira, R.; Khademi, S.; Maierbacher, G.; Siessegger, B.; Schulz, D.; Hilt, J.; Hellwig, P.; Jungnickel, V. Use of Plastic Optical Fibers for Distributed MIMO in Li-Fi Systems. In Proceedings of the 2019 Global LIFI Congress (GLC), Paris, France, 12–13 June 2019; pp. 1–5.
5. Berenguer, P.W.; Schulz, D.; Hilt, J.; Hellwig, P.; Kleinpeter, G.; Fischer, J.K.; Jungnickel, V. Optical Wireless MIMO Experiments in an Industrial Environment. *IEEE J. Sel. Areas Commun.* **2018**, *36*, 185–193. [CrossRef]
6. Haas, H. LiFi is a paradigm-shifting 5G technology. *Rev. Phys.* **2018**, *3*, 26–31. [CrossRef]
7. Bian, R.; Tavakkolnia, I.; Haas, H. 15.73 Gb/s Visible Light Communication with off-the-shelf LEDs. *J. Light. Technol.* **2019**, *37*, 2418–2424. [CrossRef]
8. Chen, Q.; Han, D.; Zhang, M.; Ghassemlooy, Z.; Boucouvalas, A.C.; Zhang, Z.; Li, T.; Jiang, X. Design and Demonstration of a TDD-Based Central-Coordinated Resource-Reserved Multiple Access (CRMA) Scheme for Bidirectional VLC Networking. *IEEE Access* **2021**, *9*, 7856–7868. [CrossRef]
9. Younus, O.I.; Le Minh, H.; Dat, P.T.; Yamamoto, N.; Pham, A.T.; Ghassemlooy, Z. Dynamic Physical-Layer Secured Link in a Mobile MIMO VLC System. *IEEE Photon. J.* **2020**, *12*, 1–14. [CrossRef]
10. Tsonev, D.; Videv, S.; Haas, H. Towards a 100 Gb/s visible light wireless access network. *Opt. Exp.* **2015**, *23*, 1627–1637. [CrossRef] [PubMed]
11. Osahon, I.N.; Pikasis, E.; Rajbhandari, S.; Popoola, W.O. Hybrid POF/VLC link with M-PAM and MLP equaliser. In Proceedings of the 2017 IEEE International Conference on Communications (ICC), Paris, France, 21–25 May 2017; pp. 1–6.
12. Lin, C.-Y.; Li, C.-Y.; Lu, H.-H.; Chang, C.-H.; Peng, P.-C.; Lin, C.-R.; Chen, J.-H. A Hybrid CATV/16-QAM-OFDM In-House Network Over SMF and GI-POF/VLC Transport. *IEEE Photon. Technol. Lett.* **2015**, *27*, 526–529. [CrossRef]
13. Joncic, M.; Kruglov, R.; Haupt, M.; Caspary, R.; Vinogradov, J.; Fischer, U.H.P. Four-Channel WDM Transmission Over 50-m SI-POF at 14.77 Gb/s Using DMT Modulation. *IEEE Photon. Technol. Lett.* **2014**, *26*, 1328–1331. [CrossRef]
14. Li, X.; Bamiedakis, N.; Wei, J.; McKendry, J.J.D.; Xie, E.; Ferreira, R.; Gu, E.; Dawson, M.D.; Penty, R.V.; White, I.H. μLED-Based Single-Wavelength Bi-directional POF Link with 10 Gb/s Aggregate Data Rate. *J. Light. Technol.* **2015**, *33*, 3571–3576. [CrossRef]
15. Correa, C.; Huijskens, F.; Tangdiongga, E.; Koonen, A. Luminaire-Free Gigabits per second LiFi Transmission employing WDM-over-POF. In Proceedings of the 2020 European Conference on Optical Communications (ECOC), Brussels, Belgium, 6–10 December 2020; pp. 1–4.

16. Nuwanpriya, A.; Ho, S.-W.; Chen, C.S. Indoor MIMO Visible Light Communications: Novel Angle Diversity Receivers for Mobile Users. *IEEE J. Sel. Areas Commun.* **2015**, *33*, 1780–1792. [CrossRef]
17. Prizmatix Home Page. Available online: https://www.prizmatix.com/Optics/Collimator.htm (accessed on 28 April 2021).
18. Mahajan, V.N. *Fundamentals of Geometrical Optics*; SPIE Press: Washington, WA, USA, 2014.
19. TS 138.101-1 V15.5.0. User equipment (UE) radio transmission and reception: Part 2: Range 2 standalone, 2019. ETSI Home Page. Available online: https://www.etsi.org/deliver/etsi_ts/138100_138199/13810102/15.05.00_60/ts_13810102v150500p.pdf (accessed on 10 April 2021).

Article

Predistortion Approaches Using Coefficient Approximation and Bidirectional LSTM for Nonlinearity Compensation in Visible Light Communication

Yun-Joong Park [1,†], Joon-Young Kim [2,†] and Jae-Il Jung [1,*]

1 Department of Electronics and Computer Engineering, Hanyang University, 17 Haengdang-dong, Seongdong-gu, Seoul 04763, Korea; pyjoong@hanyang.ac.kr
2 School of AI Convergence, Sungshin Women's University, 2 Bomun-ro 34da-gil, Seongbuk-gu, Seoul 02844, Korea; jkim@sungshin.ac.kr
* Correspondence: jijung@hanyang.ac.kr
† These authors contributed equally to this work.

Abstract: A Light-Emitting Diode (LED) has a nonlinear characteristic, and it contains fundamental limitations for the performance of Visible Light Communication (VLC) systems in indoor environments when using intensity modulation with Orthogonal Frequency Division Multiplexing (OFDM). In this paper, we investigate this nonlinear characteristic with analysis and proposal. At first, we identified the LED nonlinear characteristics in terms of bit-error performances. After analysis, we propose initial predistortion schemes to mitigate the nonlinearity matters. In the predistortion schemes, the nonlinear distortion compensation model contains predistortion features with the LED inverse characteristics. Considering a Direct-Current-biased Optical OFDM (DCO-OFDM) system, we compared the Bit-Error Rate (BER) performances with and without compensation via simulations. The performance on the LED with the compensation showed LED nonlinearity could significantly improve the bit-error performance. In addition, with consideration that the predistortion model is insufficient to represent LED distortion, we investigated possible opportunities of distortion correction using Bidirectional Long Short-Term Memory (BLSTM), one of the leading deep learning approaches. Its result showed promising improvement of the distortion compensation as well.

Keywords: nonlinearity; VLC; LED; predistortion; coefficient approximation; BLSTM

Citation: Park, Y.-J.; Kim, J.-Y.; Jung, J.-I. Novel Predistortion Approaches Using Coefficient Approximation and Bidirectional LSTM for Nonlinearity Compensation in Visible Light Communication. *Photonics* **2022**, *9*, 198. https://doi.org/10.3390/photonics9030198

Received: 28 February 2022
Accepted: 18 March 2022
Published: 20 March 2022

1. Introduction

Growing demands for high data rates and low latency communication systems, particularly in indoor and in-building environments, contribute to the significant consideration of the usage of 60 GHz and above unlicensed frequency bands in the United States [1]. Recently, optical spectrum research focused on indoor wireless communication [2]. Significant interests of Terahertz (THz) or above level communications on 6G development show the possible opportunities of the optical spectrum in the perspectives of wireless networks, interactive communications, mobility service, internet of things, and even bio-tissues [3]. Especially in the case of mobility service for in-building, its communication systems have to consider VLC for high data rates instead of current wireless systems such as NB-IoT or LoRaWAN [4].

Since the optical spectrum is in the frequency range of at least 300 GHz, multicarrier OFDM modulation can be the possible candidate for the robust modulation scheme in that spectrum for indoor visible light communications [5]. OFDM has the advantages of a high data rate and bandwidth efficiency. It also provides a scheme to mitigate intersymbol interference caused by multipath propagation. However, in VLC, the LEDs' nonlinear behavior can severely affect OFDM's performance due to its high Peak-to-Average Power Ratio.

For LEDs, the power amplifier operates up to the saturation for maximum power, and it may cause unwanted nonlinear distortion in amplitude and phase in this power operation. Signal clipping issue in the amplifier is another critical matter for OFDM [6]. The back-off to the average input power ensures that the amplifier avoids saturation, but the problem still exists that the back-off may deteriorate power efficiency. Linearization through predistortion is another method to compensate for the PA nonlinear distortion.

Since the LED is the primary source of nonlinearity, the baseband OFDM signal in VLC systems is modulated with the instantaneous power on the optical carrier called intensity modulation. Asymmetrically Clipped Optical OFDM (ACO-OFDM) and DCO-OFDM are two primary forms of OFDM with intensity modulation. In order to produce a positive signal, the bipolar OFDM signal in DCO-OFDM is superimposed on a bias point. On the other hand, the OFDM signal in ACO-OFDM is unipolar modulated by only the odd sub-carriers, and the unipolar modulation suppresses the signal at zero levels [7]. In this paper, we select DCO-OFDM for spectral efficiency.

For the investigation of the nonlinear distortion, various bias points are considered. A power back-off scheme is a possible option for the OFDM signal to set the distortion levels with LED operation near the bias point in a quasilinear segment of the LED characteristic. After sampling the LED transfer function, the predistortion method consists of the table format of the inverse of the characteristic function and compensates for the LED nonlinearity [8]. However, this method has data measurement issues in that the system must directly identify the LED data characterization. An adaptive normalized least mean square (NLMS) algorithm can be another technique to estimate correct LED bias data [9]. To compensate for the nonlinearity, it directly predicts the distortion levels given the environmental changes instead of using the fixed values of the existing Memory Look-Up-Table. High complexity in the algorithm becomes one of the remaining issues in practicality. In other words, a simple predistortion approach is necessary to resolve the distortion matters with preserving practical usability.

Especially in the case of mobility services, indoor mobility transportation becomes a possible scenario due to the LED light in the building. As shown in Figure 1, using LED lights in the hallway, the building or infrastructure can broadcast the information and data specifically designed for mobility services, indoor navigation maps, announcements, and over-the-air updates. In that communication system scene, the mobility devices such as scooters, bicycles, or autonomous robots are in a nonstationary position, and their channels have a line of sight with a few reflections, which means the channel condition is the Rician fading. In addition, those mobile devices are battery-critical, and energy efficiency is one of their top priorities. In other words, the VLC communication systems in this mobility service have to consider dynamic environmental conditions and effectiveness to combat LED distortion.

In this paper, we propose two approaches: (1) the predistortion method, effectively using the coefficient approximation without sampling the LED transfer function, and (2) the Bidirectional LSTM Approach to training the LED distortion correction without knowing the LED modeling. When we compare the BER performances of these two approaches to the case without compensation, the result confirms the possible improvement of LED distortion in the VLC OFDM system.

This paper is organized as follows. Section 2 illustrates the development procedure of the LED model and distortion. Sections 3 and 4 introduce the OFDM model, initial predistortion modeling, and the coefficient approximation scheme for LED distortion correction. Section 5 discusses the possible deep learning application for this VLC system to implement predistortion. Section 6 compares system performances of 16 Quadrature Amplitude Modulation (QAM) and 64 QAM in predistortion modeling and deep learning schemes. After presenting performance results, we conclude our paper in Section 7.

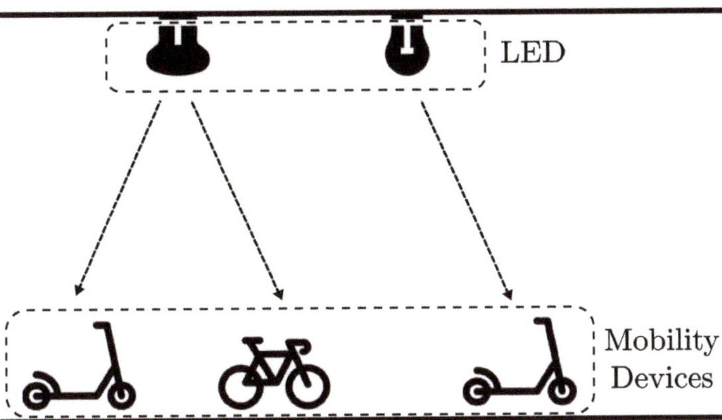

Figure 1. A general example scenario of Visible Light Communication applications for mobility services in indoor environments. The bicycle and scooters are driving or paused in the building or tunnel in this scene. Note that the line of sight between LED and mobility devices is available with multiple reflections. (Scooter icon in Vehicle and Travel Pack designed by Wishforge Games, www.wishforge.games), accessed on 3 January 2022.

2. Nonlinear Characteristics of Different LEDs in the VLC System

In the ideal condition, we consider an ideal LED as a distortion-free diode. We also define the input port signal as the driving current and the output port signal power as the emitted optical power. LEDs also exhibit nonlinearities, introducing distortions on the emitted signal [10]. Since the physical models, which include the dynamic rate equation model [10], failed to approximate practical LEDs, we model the static transfer function of the LED output power, $P_{out}(t)$, with polynomials shown in Equation (1).

$$P_{out}(t) = \sum_{n=0}^{\infty} b_n[I_{in}(t) - I_{DC}]^n \tag{1}$$

where $I_{in}(t)$ is the driving current, I_{DC} is the bias current, and b_n is the n-th order power coefficient of the transfer function. Although polynomial orders are required to be $n = 5$ to realistically model transfer functions, a second-order polynomial is proven to be a fair description [11]. The polynomial function in question is

$$P_{out}(t) = b_0 + b_1(I_{in}(t) - I_{DC}) + b_2(I_{in}(t) - I_{DC})^2 \tag{2}$$

In this paper, we set the normalized current $I_{DC} = 0.5$, and b_0, b_1, and b_2 are the Direct Current (DC) constant, the linear coefficient, and the second-order nonlinearity coefficient. Moreover, it is also known that an LED has constant behavior. As a result, the derivation of $P_{out}(t)$ with respect to $I_{in}(t)$ must be $0 \leq P_{out}(t) < 1$.

To describe the degree of nonlinearity, we define the nonlinearity parameter as ζ in the source transfer function. It is the normalized output power corresponding to the input current. For example, we assume that the LED has $\zeta = 0.5$, derived as the linear line in the transfer function shown in Figure 2. If the transfer function is concave, $\zeta > 0.5$. If it is convex, $\zeta < 0.5$. LED is the prime example that has the concave characteristic [12]. The coefficients of Equation (2) can be expressed in ζ as follows [11].

$$b_0 = \zeta, \quad b_1 = 1, \quad b_2 = -4\zeta + 2 \tag{3}$$

Figure 2 shows the concave and convex curve examples. In this figure, the concave curve is based on the red LED coefficient, but white and infrared LEDs also have concave curves. The experimental values of ζ according to LED type 85 are 0.732 in red LEDs, 0.541 in infrared LEDs, and 0.582 in white LEDs [11]. However, the examples of fixed coefficient values given LED types in the assumption are not consistent in the natural environment. Depending on types and actual production, the coefficient value can be off. For that matter, it would be reasonable to solve LED distortion issues under two assumptions: (1) The LED nonlinear parameter value is a fixed one and consistent. (2) The LED nonlinear parameter value can vary for each LED and is inconsistent. Note that our term inconsistency does mean the slight change, not significant disruption, that the LED light-emitting type may be changed. Given these two assumptions, we need to investigate the solutions for the VLC system.

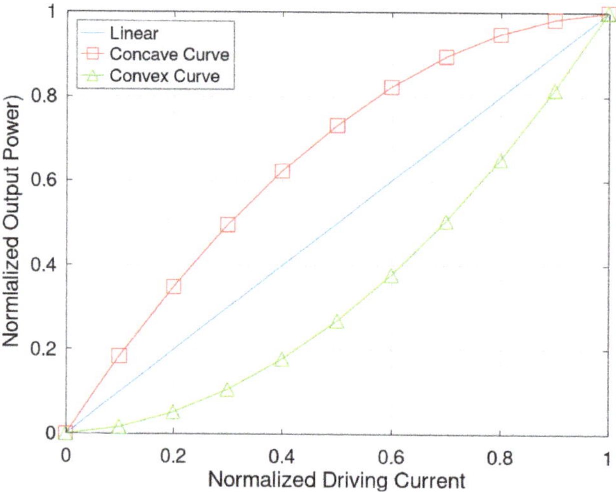

Figure 2. A comparison among concave curve with $\zeta = 0.732$, convex curve with $\zeta = 0.268$, and linear static transfer functions.

3. Initial Predistortion Modeling for LED Distortion Correction

As mentioned, Figure 3 shows the OFDM-based visible light communication transceiver system, including a predistortion module. In the existing OFDM system without predistortion, the transmitter processes signals with Inverse Fast Fourier Transform (IFFT) after QAM modulation, and it expresses signals as the orthogonal frequency components. When the system processes the IFFT output signal via LED modulation and transmits LED output, its performance deteriorates due to the nonlinearity of LEDs and the generated distortion. The distortion causes signal to become noise-sensitive and damages the orthogonality between the frequency components. The result of the distortion is the substantially high error performance at the receiver. To resolve this problem, the predistortion module, as shown in Figure 3, can compensate for LED nonlinear matters and is placed in front of the Digital-Analog Converter (DAC) for data transmission.

Since the goal is to design the predistortion module with the LED coefficient value, we describe initial predistortion modeling schemes and then coefficient approximation under the fixed value assumption for the predistortion module to compensate nonlinearity of LED luminance. Note that both models use the distortion characteristic of bias signals, and we discuss the nonfixed coefficient assumption case in the later section.

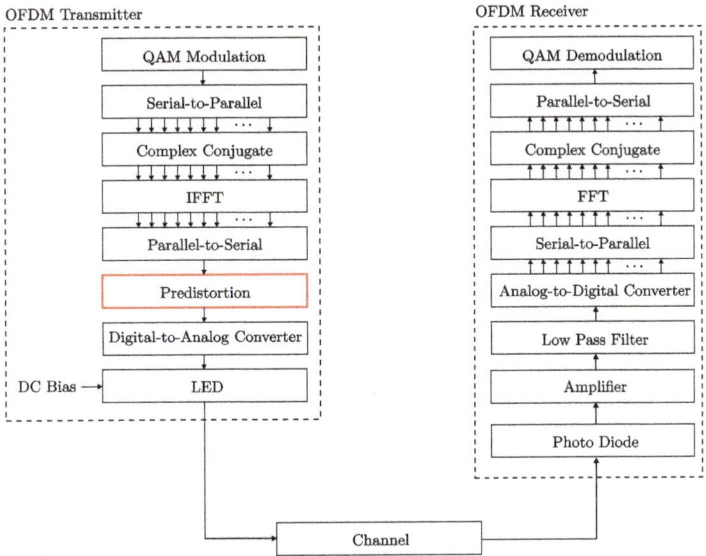

Figure 3. A block diagram of OFDM transceiver using predistorter.

In the ideal condition, the primary method is to implement a predistortion by calculating the inverse function for Equation (2). We can reformulate (2) as

$$\begin{aligned} P_{out}(t) = & b_2 I_{in}(t)^2 + (b_1 - 2b_2 I_{DC}) I_{in}(t) + \\ & b_2 I_{DC}^2 - b_1 I_{DC} + b_0 \end{aligned} \quad (4)$$

When we define $g_{inv}(t)$ as the inverse function of Equation (4), we can derive it as

$$g_{inv}(t) = \sqrt{\frac{1}{b_2} I_{in}(t) + \frac{b_1^2}{4b_2^2} - \frac{b_0}{b_2}} - \frac{b_1}{2b_2} + I_{DC} \quad (5)$$

where $I_{in}(t)$ is the input current to the predistortion module. Based on this inverse function, we can design the predistortion module and correct LED distortion.

4. Proposed Predistortion Model with Coefficient Approximation

Since the inverse method from (5) includes a root within $g_{inv}(t)$, it does require the approximation approach within hardware modules for real-time implementation. For the VLC system, the approximation approach shown in the inverse function is impractical. We introduce a simple predistortion scheme using a simple coefficient approximation to mitigate this issue. We propose implementing the predistortion using the coefficient changed according to LED color without sampling the LED transfer function.

This method uses the static transfer function of Equation (2). We modify coefficients b_0, b_1, and b_2 in (3) to the predistortion function. As mentioned, if $\zeta < 0.5$, the convex transfer function can be obtained. Therefore, the $1 - \zeta$ value is substituted instead of ζ of (3) and coefficients. Now, those coefficients can be expressed as follows.

$$b_0 = 1 - \zeta, \quad b_1 = 1, \quad b_2 = 4\zeta - 2 \quad (6)$$

When the coefficients in (6) are substituted in (2), the polynomial predistortion function $g(t)$ is

$$g_(t) = (4\zeta - 2)(I_{in}(t) - I_{DC})^2 + I_{in}(t) - I_{DC} - \zeta + 1 \quad (7)$$

Figure 4 shows the nonlinearity of the LED, predistortion function, and linearized LED transfer function. Concave curves have a characteristic of the nonlinearity of the LED. Furthermore, convex curves are based on predistortion functions. If the predistortion function $g(t)$ enters the input $I_{in}(t)$ of the LED transfer function in (2), the final transfer function becomes linear. ζ is determined according to the LED color. The proposed method can process if we know the LED color. Therefore, the sampling for making the predistortion like the conventional method is unnecessary. The proposed method has the advantage of simplicity when implementing the predistortion module.

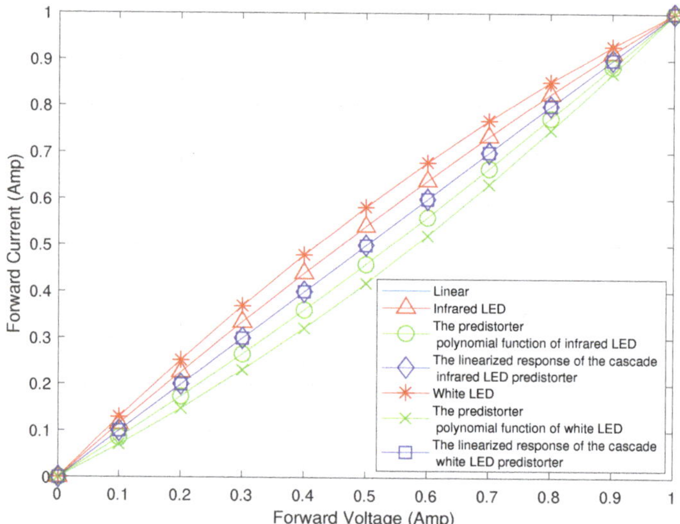

Figure 4. Nonlinear, predistortion, and linearized LED transfer function.

5. Deep Learning Approach for Possible VLC System Enhancement

In previous sections, we showed our proposed predistortion scheme with the coefficient approximation for nonlinear distortion compensation based on the performance results. We also explained that coefficient approximation could be considered the practical approach for actual implementation. However, we believe that, in Rician fading, the proposed methods did not fully address LED nonlinearity issues in the two aspects. One is the modeling of LED distortion. The LED modeling formula, including our LED modeling, can often be ill-posed and poorly represented in LED distortion patterns.

In addition to the distortion modeling, each LED produced might not follow the same distortion characteristics as theoretical LED models, and our predistortion scheme based on the theoretical model may be ill-conditioned in the actual situation. Thus, instead of the LED modeling for correction, our focus must shift to the direct correction of LED distortion. For instance, our proposed approximation modeling schemes assume that the transmitter and receiver are stationary with the same coefficients. In the practical situation, the LED distortion coefficients can be different in each produced LED case, and a slight value-off can cause significant performance degradation. Hence, this VLC also has to correct LED distortion directly with the adaptation of these coefficient changes. To resolve the correction problem, we must consider using deep learning approaches, as the possible candidates.

To find the best estimate of the model of the data and systems, deep learning algorithms, such as CNN and LSTM, gained recent popularity in various communication systems applications, including VLC research. This deep learning approach can also be appliable to solve this distortion problem. In this paper, we introduce one of the deep learning applications for possible VLC system enhancement, the so-called BLSTM.

5.1. Bidirectional LSTM Approach for the Distortion Correction

Unlike the classification and detection based on image processing, sequential data such as voice samples or text sentences correlate between present and past times. A Recurrent Neural Network (RNN) is the prime example of processing input data with network weights and structure. It also preserves the sequences of the data with the network hidden states. However, RNN has severe issues with vanishing and exploding gradient programs for the training and optimization process, resulting in the incapability of learning long-term knowledge. Long Short-Term Memory (LSTM) is one of the main algorithms to resolve these issues with RNN. LSTM contains multiple activation function modules called gates to overcome the gradient matters. This LSTM system contains memory that takes previous and current states as input, as shown in Figure 5.

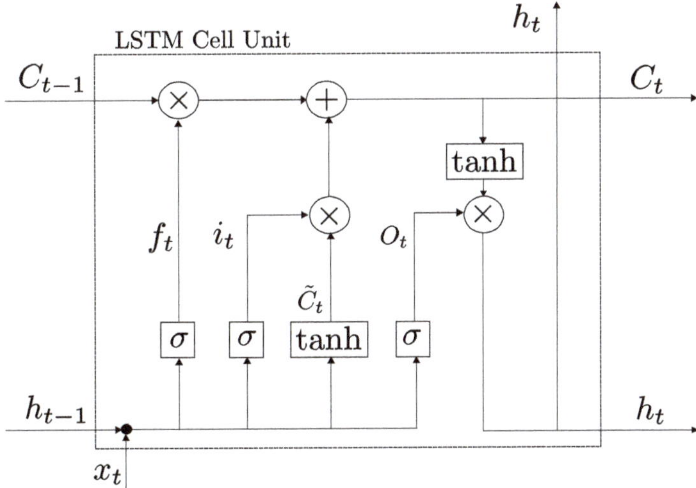

Figure 5. A basic diagram of the LSTM unit. Note that it is known to be the general structure of modern LSTM unit, and the similar and identical LSTM units are referred to in various literature [13–16].

Each element in the LSTM unit, so-called a gate, leads the LSTM unit to store and discard the data information. Apart from input and output gates, the forget gate is another key LSTM feature that controls knowledge preservation in LSTM units. Details of each parameter in Figure 5 are below [13,14]. Note that σ and tanh are defined as the sigmoid function and the hyperbolic tangent activation function.

$$i_t = \sigma(W_i x_t + U_i h_{t-1} + b_i) \tag{8}$$

$$f_t = \sigma(W_f x_t + U_f h_{t-1} + b_f) \tag{9}$$

$$C_t = f_t C_{t-1} + i_t \tilde{C}_t \tag{10}$$

$$= f_t C_{t-1} + i_t \tanh(W_c x_t + U_c h_{t-1} + b_c) \tag{11}$$

$$O_t = \sigma(W_o x_t + U_o h_{t-1} + b_o) \tag{12}$$

$$h_t = O_t \tanh(C_t) \tag{13}$$

Note that i_t, f_t, C_t, O_t, and h_t are the input gate, forget gate, memory cell, output gate, and hidden state at time step t, respectively. In addition, W_i, W_f, W_o, and W_c are the input weight matrices of parameters i, f, O, and $\tilde{C}$, and U_i, U_f, U_o, and U_c are the hidden layer weight matrices of parameters i, f, O, and $\tilde{C}$. In each formula, b_i, b_f, b_o, and b_c are the biased parameters for i, f, O, and $\tilde{C}$. Once the LSTM unit took previous memory cells and hidden states from the previous LSTM units as C_{t-1} and h_{t-1}, it processed the following sequential input at times t and x_t to produce the present memory cell and hidden state, C_t and h_t.

As illustrated in this general structure of the LSTM unit, the data sequence is a forward sequence. However, the data sequence information in forward and backward sequences can be different, and understanding the backward data sequence can produce the data prediction well. These are the main reasons why Bidirectional LSTM should be considered.

A brief block diagram of the BLSTM structure is shown in Figure 6. In this figure, the BLSTM first processes the forward LSTM before processing the backward LSTM, and the activation function produces the result [15]. Hence, this system contains the past and future input data.

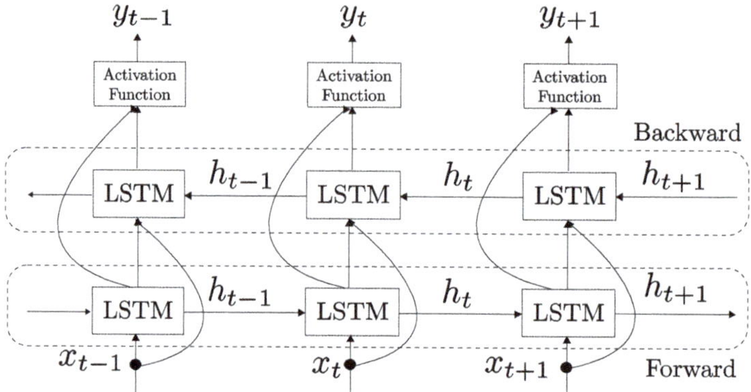

Figure 6. Bidirectional LSTM structure. Each LSTM unit in Figure 5 lines up in series in forward and backward layers in this structure. Note that the length of BLSTM is the same as the number of LSTM in the forward and backward layers.

BLSTM is a significant advancement of LSTM schemes that train the neural networks with the forward and backward data sequences [17]. Eventually, it aims to train with the data sequences to correct or predict future data. Existing example applications include speech recognition and stock index prediction [18]. Since our VLC systems process continuous time-domain signal data, LSTM becomes a suitable option for distortion correction. However, traditional LSTM trained in the forward direction of the data or signal sequence can become biased for signal correction. On the other hand, Bidirectional LSTM processes the backward and forward of the signal sequences to train both sequence directions of the signal. We consider the BLSTM for distortion correction in this paper.

5.2. *Proposed Initial BLSTM-Based VLC System*

From our VLC perspective, since the distortion occurs in the transmitter, it would be reasonable to conduct the predistortion on the transmitter instead of the receiver. Figure 7 shows that the predistortion module must be placed right before the diode. In the predistortion module, our BLSTM models are double in parallel, and each BLSTM structure is designed with two BLSTM layers and a dense layer. For the correction purpose, the output length of the BLSTM should be equivalent to the input length. Note that, due to the system complexity and hardware capability, we designed the OFDM system with an FFT size of 32.

Several properties must be carefully considered for BLSTM design and training, such as batch size, epoch, and data size. Since these parameters can impact the predistortion correction performance, our investigation extended these parameters into our scope. Once the BLSTM models are completed to train, the models are placed in the transmitter for continuous use. When the VLC system conditions and environments are changed, we need to retrain the BLSTM model, which is highly unlikely in the in-building environmental case.

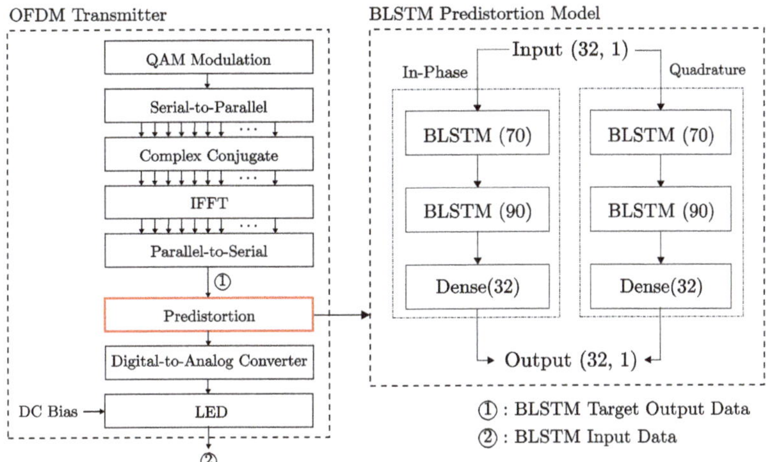

Figure 7. An Initial design of Bidirectional LSTM in predistortion module from OFDM transmitter. Note that, for training model, we consider input data as ② and target output as ①.

6. Experiment Setup and Results

For the experiment result, we operated two experiments: (1) coefficient approximation-based predistortion modeling and (2) BLSTM application. Note that we do not compare both approaches directly in this paper since each has a different assumption.

To produce the results of VLC performances with coefficient-approximation-based predistortion, given measured static transfer functions of white and infrared LEDs, a quadratic polynomial of the LED in Equation (2) was applied as the general model of the transfer function. This polynomial approximation has been widely popular to model nonlinearity characteristics of LED or laser diodes [19]. Each graph shows two QAMs in the simulation: 16 and 64.

For BLSTM experiment settings, we considered the input data as the distorted IFFT signal and target output data as the IFFT signal after the LED shown in Figure 7. We utilized our distortion model (2) to generate the distorted IFFT signal, but any distortion model or existing data based on an LED can be applied for the training. After our initial investigation, we confirmed that the BLSTM length must be more than input data size, which was 32 in our case. We also had to consider the limitation of computing resources and model complexity as well. With considering our investigation and limitation, we set the length of the Bidirectional LSTM as 70 and 90 for the first and second layers. Since we considered the indoor mobility service as an operation environmental condition, we considered the Rician fading with K factors from 4 to 20. For details of the BLSTM experiment setting, the complete list of properties and experiment settings is in Table 1. In Table 1, note that we used at least more than 25 million samples for experiment.

6.1. Experiment Results of Predistortion Modeling with Coefficient Approximation

We simulated the QAM performances of two predistortion schemes using (1) the inverse function and (2) coefficient approximation along with the conventional scheme without predistortion. Note that we set the simulation threshod as the BER reached to 10^{-5}.

Figure 8 shows OFDM's BER performance using 256 subcarriers in 16 QAM and 64 QAM for the white LED. The results showed a 3.5 dB performance gain for all predistortion schemes at BER 10^{-4} in 16 QAM. For BER 10^{-4} at 64 QAM modulation, each predistortion required us to set SNR values at 16.5 dB in the inverse function scheme and 17 dB in the coefficient modification scheme. In addition, the BER performance were deteriorated when QAM levels were increased since symbols were closer in the constellation diagram.

Table 1. Experiment properties and settings for VLC BLSTM experiments.

Properties and Settings	Types and Values
OS	Ubuntu 20.04.3 LTS
Python	3.8.10
Tensorflow (incl. Keras)	2.7.0
CUDA	11.2
CPU	Intel Xeon Silver 4208 Processor
GPU	NVIDIA GeForce RTX 3080
LED Type	White LED
FFT size	32
Modulation	16 QAM
Channel condition	Rician fading channel
Rician K factor (dB)	4, 8, 12, 16, 20
Batch size	32, 64, 128, 256, 512
Data size (per 10,000 samples)	2560, 5120, 7680, 10,240, 20,480, 25,600, 51,200
Epoch	1, 2, 3, 5

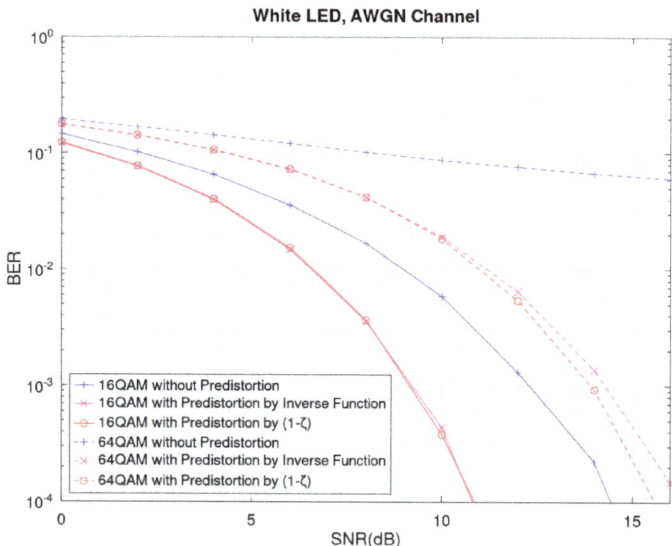

Figure 8. BER performances of the predistortion in 16/64 QAM modulation for white LED.

Figure 9 shows OFDM's BER performances using 256 subcarriers in 16 QAM and 64 QAM. In these simulations, we used the infrared LED. In this case, we also observed that both predistortion schemes gained 1.5 dB performance at BER = 10^{-4} with 16 QAM compared to the scheme without predistortion. For 64 QAM modulation to meet target BER 10^{-4}, both predistortion schemes needed to set SNR values up to 16.5 dB.

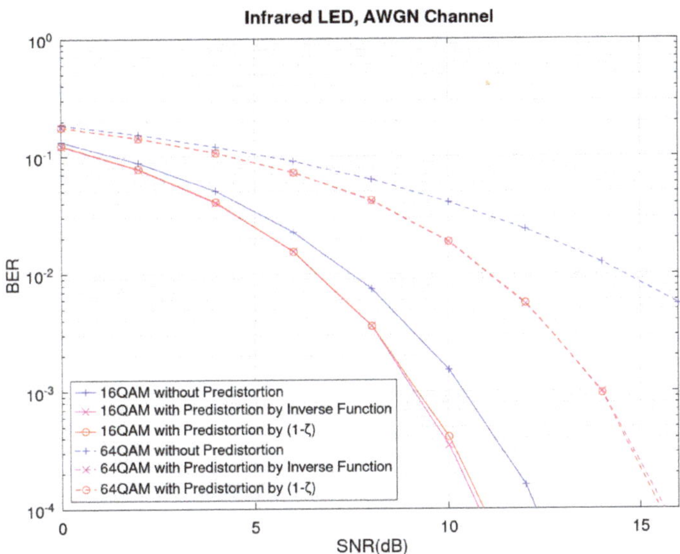

Figure 9. BER performances of the predistortion in 16/64 QAM modulation for infrared LED.

To summarize the results from Figures 8 and 9, the white LED performance was substantially improved compared to the infrared LED performance. However, the overall BER performances of the white LED with predistortion schemes were still inferior to infrared LED performance since the infrared LED operates on a lower frequency spectrum than the white LED does. In addition, infrared LED performance without predistortion had 2 dB more SNR gain than the white LED performance without predistortion.

Given the Rician fading conditions, our performance results showed that the Rician fading might alter the performance. As shown on the Rician fading results in Figure 10, the performance of the coefficient approximation predistortion method was still better than that without any predistortion in low SNRs. Our results may conclude that the predistortion modeling may mitigate the distortion matter but still can be affected by other conditions in terms of Rician fading impacts.

6.2. Experiment Results of BLSTM Application

For the BLSTM experiment, we looked for three aspects: (1) performance impacts on BLSTM parameters including batch size, data size, and epoch, (2) performance on K factor values, and (3) performance over SNR. Since the distortion correction was trained at the transmitter, we confirmed that SNR does not influence the training process but on the test. We also considered the Rician fading for our experiment due to the possible mobility service scene. The first preliminary experiment result is in Figure 11. In this graph, we can confirm that once the K factor is increased, the VLC performance is improved as well. For BLSTM-specific testing, we considered K factors 4 and 20 as the worst and best cases.

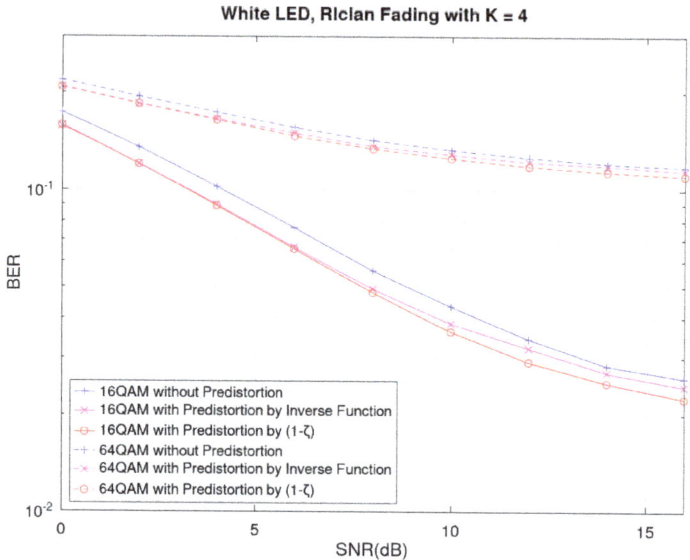

Figure 10. BER performances with Rician fading, K = 4 in 16/64 QAM modulation for white LED.

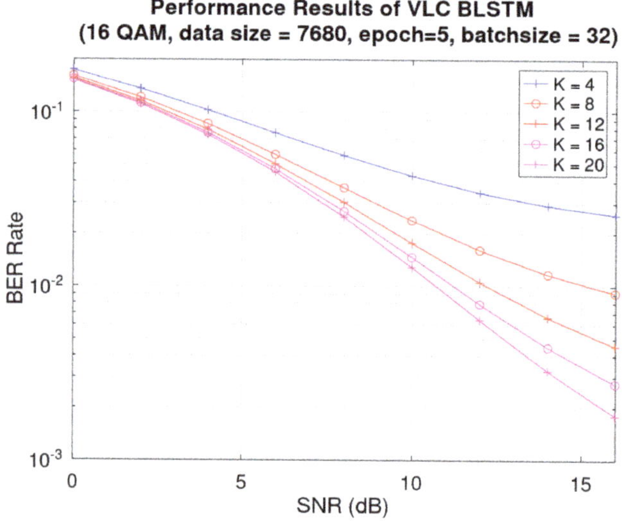

Figure 11. BER performances with various Rician fading factors, 16 QAM for white LED.

The performance results over batch size are in Figure 12. In Figure 12a,b, we observe that when batch sizes were 32 and 256, the VLC outperformed as compared to the no correction case. Note that batch size with 64 was outperformed on a K factor with 4 but underperformed on a K factor with 20. Based on the results, we may conclude that batch size is one of the critical parameters to tune the performance in a sophisticated manner.

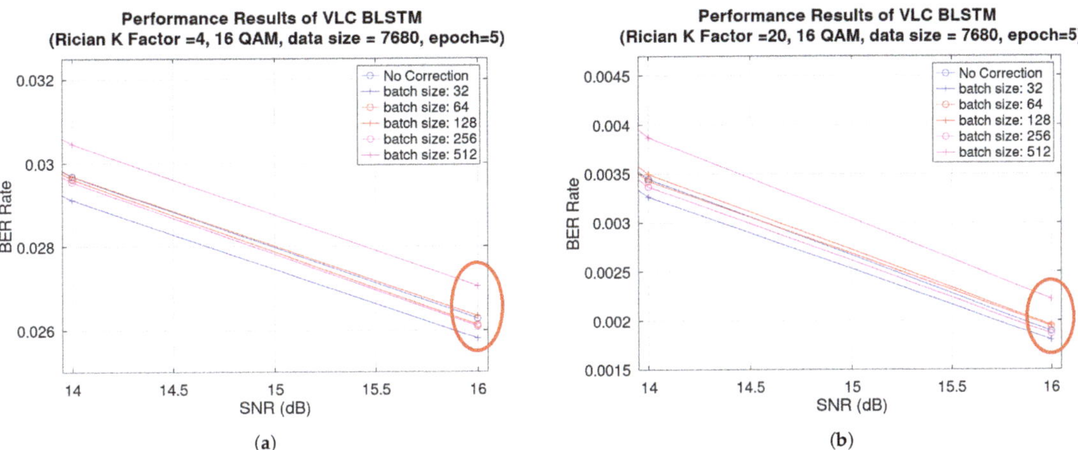

(**a**) (**b**)

Figure 12. BER performances over batch size with Rician fading channel 16 QAM for white LED: (**a**) K Factor = 4; (**b**) K Factor = 20.

The performance results over the epoch are in Figure 13. In Figure 13a,b, we observe that when the epoch was 1 and 5, the VLC outperformed as compared to the no correction case. Note that, as time goes by, the more the number of epochs increases, the more performance improvement is shown. Underperformed cases with epochs 2 and 3 proved that very few training iterations do not significantly improve performance.

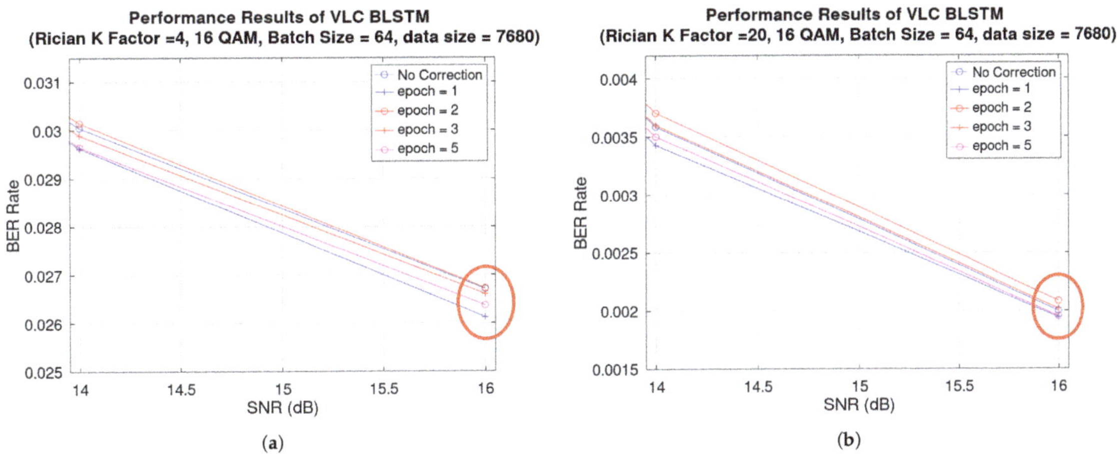

(**a**) (**b**)

Figure 13. BER performances over epoch with Rician fading channel 16 QAM for white LED: (**a**) K factor = 4; (**b**) K factor = 20.

The performance results over data size are in Figure 14. Figure 14a,b show that when data sizes were approximately over 204.8 million, the VLC outperformed much more than in the no correction case. The performance results proved that, during the training process, sufficient data must be prepared and, in our case, over about 205 million samples are required to produce better performance. Insufficient data size may cause overfitting problems, and the VLC system performs worse than in the no correction case.

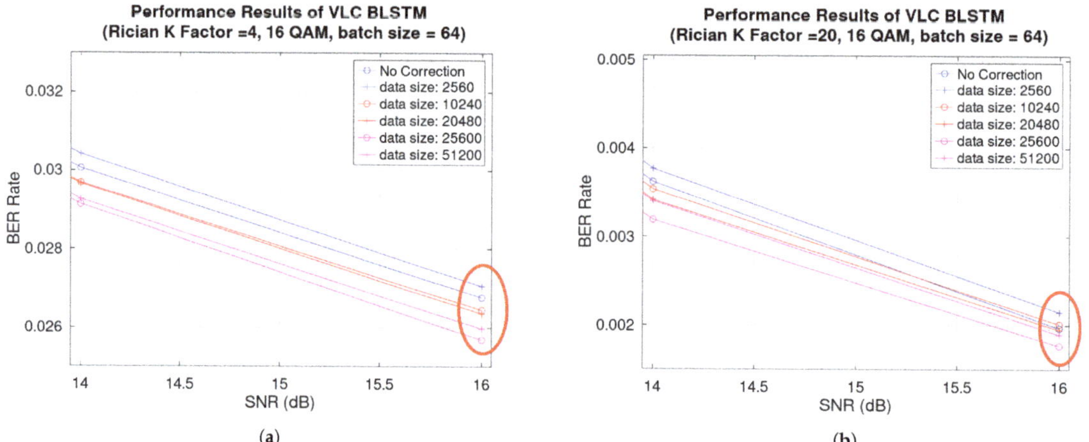

Figure 14. BER performances over data size with Rician fading channel 16 QAM for white LED: (**a**) K factor = 4; (**b**) K factor = 20.

The BER performance of each batch size over SNR is in Figure 15. The figure shows that the performance deteriorates when the batch size is increased. However, we also have to understand that the batch size is equivalent to the step size in the optimization. Given that each iteration time is fast when the batch size is large, the trade-off between batch size and iteration time must be carefully considered under limited resources and time.

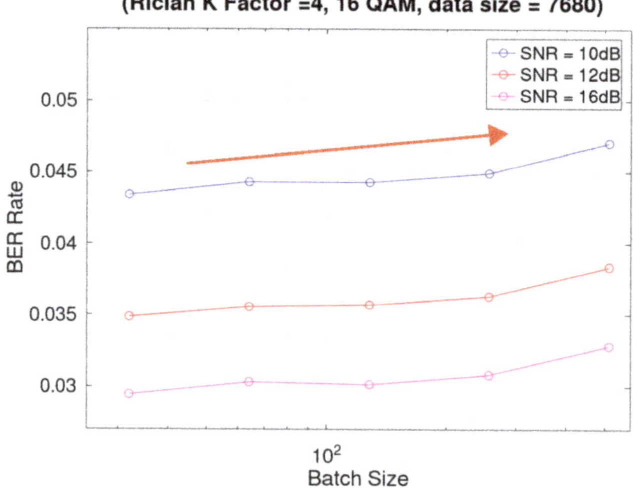

Figure 15. BER performances of batch sizes over SNR with Rician fading K factor = 4, 16 QAM.

To summarize, our BLSTM models showed the possible performance improvement of LED distortion. In addition, we only need to care for the transmitter AND no requirement for the receiver to operate this approach. As described in the in-building mobility service in Section 1, the mobility devices, including IoT applications, can use the same receiver structure while experiencing enhanced BER performance of the VLC system. In other words, they can spend less energy on the same VLC system experiences.

7. Conclusions

In our paper, we observed the nonlinearity of LEDs, and it could significantly affect the performance of optical OFDM transceiver systems. The compensated module using predistortion can substantially enhance the OFDM system performance for performance improvement. However, the existing compensation methods directly have to measure LED nonlinearity or high complex characteristics. In the OFDM-based VLC system, we proposed the practical predistortion scheme with coefficient approximation. The coefficient approximation showed the effectiveness of simple operations without a necessary sampling of the LED transfer function by using the inverse function. By simply tuning the coefficient, the OFDM system showed outperformed results compared to the system without the predistortion.

In addition, we proposed a Bidirectional LSTM model to handle the variation and distortion of LEDs without distortion modeling. We used the distortion data from our distortion models for the training purpose, and the results showed the possible promise of performance improvement. Possible discussion associated with BLSTM approaches is as follows.

- We used the simple BLSTM structure and improved distortion. However, our BLSTM structure did not address the phase issues. Since the BLSTM model cannot handle complex number data, we separated real and imaginary data and produced two models. No consensus existed between the two trained models. A significant performance improvement could be possible if we designed the BLSTM model with an interconnected structure.
- We considered the white LED case only in this experiment. To process red and infrared LEDs, BLSTM models have to be retrained. Even if we retrained the model, it is not guaranteed whether or not those models can produce the equivalent performance on those diodes. Diode-specific model structure must be explored.
- Assuming we have no prior knowledge of LEDs, the distortion correction with one model would simplify the system structure and reduce its complexity. To handle all LED distortion into one model, we may need to design a deep learning structure more complex than our tested BLSTM model.
- For our BLSTM model training, computing power is the primary factor in our proposed VLC scheme, and our initial studies encountered the limitation of the computing resources. Our future works must extend to various BLSTM layer lengths and structures, assuming we have considerable computer powers. When the VLC scheme becomes part of the 5G NR light radio family, possible opportunities for our BLSTM model are available to be trained into the cloud or multi-access edge computing (MEC) on a large scale.

In addition to our work, it would be possible to extend our BLSTM approach to the practical and effective implementation of the predistortion module and explore other deep learning models and structures dedicated to the hardware module in VLC systems for possible future work.

Author Contributions: Conceptualization, Y.-J.P. and J.-Y.K.; methodology, Y.-J.P. and J.-Y.K.; software, Y.-J.P. and J.-Y.K.; validation, Y.-J.P. and J.-Y.K.; investigation, Y.-J.P., J.-Y.K. and J.-I.J.; resources, Y.-J.P. and J.-I.J.; data curation, Y.-J.P. and J.-Y.K.; writing, Y.-J.P., J.-Y.K. and J.-I.J.; visualization, Y.-J.P. and J.-Y.K.; supervision, Y.-J.P., J.-Y.K. and J.-I.J.; project administration, Y.-J.P., J.-Y.K. and J.-I.J. All authors have read and agreed to the published version of the manuscript.

Funding: This work was supported by the Sungshin Women's University Research Grant of 2021 and the Technology Innovation Program (20013726, Development of Industrial Intelligent Technology for Manufacturing, Process, and Logistics) funded by the Ministry of Trade, Industry and Energy (MOTIE, Korea).

Conflicts of Interest: The authors declare no conflict of interest.

Abbreviations

The following abbreviations are used in this manuscript:

LED	Light-Emitting Diode
OFDM	Orthogonal Frequency Division Multiplexing
ACO-OFDM	Asymmetrically Clipped Optical OFDM
DCO-OFDM	Direct-Current-biased Optical OFDM
DC	Direct Current
AWGN	Additive White Gaussian Noise
BER	Bit-Error Rate
VLC	Visible Light Communication
QAM	Quadrature Amplitude Modulation
IFFT	Inverse Fast Fourier Transform
FFT	Fast Fourier Transform
SNR	Signal-to-Noise Ratio
RNN	Recurrent Neural Network
LSTM	Long Short-Term Memory
BLSTM	Bidirectional Long Short-Term Memory

References

1. Cabric, D.; Chen, M.; Sobel, D.; Yang, J.; Brodersen, R. Future Wireless Systems: UWB, 60 GHz, and Cognitive Radios. In Proceedings of the IEEE 2005 Custom Integrated Circuits Conference, San Jose, CA, USA, 21 September 2005; pp. 793–796.
2. Kahn, J.; Barry, J. Wireless Infrared Communications. *Proc. IEEE* **1997**, *85*, 265–298. [CrossRef]
3. Katz, M.; Ahmed, I. Opportunities and challenges for visible light communications in 6G. In Proceedings of the 2020 2nd 6G Wireless Summit (6G SUMMIT), Levi, Finland, 17–20 March 2020; pp. 1–5.
4. Santa, J.; Bernal-Escobedo, L.; Sanchez-Iborra, R. On-board unit to connect personal mobility vehicles to the IoT. *Procedia Comput. Sci.* **2020**, *175*, 173–180. [CrossRef]
5. Elgala, H.; Mesleh, R.; Haas, H.; Pricope, B. OFDM Visible Light Wireless Communication Based on White LEDs. In Proceedings of the 2007 IEEE 65th Vehicular Technology Conference—VTC2007-Spring, Dublin, Ireland, 22–25 April 2007; pp. 2185–2189.
6. Bahai, A.; Singh, M.; Goldsmith, A.; Saltzberg, B. A New Approach for Evaluating Clipping Distortion in Multicarrier Systems. *IEEE J. Select. Areas Commun.* **2002**, *20*, 1037–1046. [CrossRef]
7. Armstrong, J.; Lowery, A. Power efficient optical OFDM. *Electron. Lett.* **2006**, *42*, 370–372. [CrossRef]
8. Elgala, H.; Mesleh, R.; Haas, H. Predistortion in Optical Wireless Transmission using OFDM. In Proceedings of the 2009 Ninth International Conference on Hybrid Intelligent Systems, Shenyang, China, 12–14 August 2009; Volume 2, pp. 184–189.
9. Kim, J.K.; Hyun, K.; Park, S.K. Adaptive predistorter using NLMS algorithm for nonlinear compensation in visible-light communication system. *Electron. Lett.* **2014**, *50*, 1457–1459. [CrossRef]
10. Lee, T.P. The nonlinearity of double-heterostructure LED's for opticalcommunications. *Proc. IEEE* **1977**, *65*, 1408–1410. [CrossRef]
11. Inan, B.; Lee, S.C.J.; Randel, S.; Neokosmidis, I.; Koonen, A.M.J.; Walewski, J.W. Impact of LED Nonlinearity on Discrete Multitone Modulation. *IEEE/OSA Opt. Commun. Netw.* **2009**, *1*, 439–451. [CrossRef]
12. Ramirez-Iniguez, R.; Idrus, S.; Sun, Z. *Optical Wireless Communications-IR for Wireless Connectivity*, 1st ed.; Auerbach: Boca Raton, FL, USA, 2008; Volume 6.
13. Zhu, Y.; Gong, C.; Luo, J.; Jin, M.; Jin, X.; Xu, Z. Indoor Non-Line of Sight Visible Light Communication with a Bi-LSTM Neural Network. In Proceedings of the 2020 IEEE International Conference on Communications Workshops (ICC Workshops), Dublin, Ireland, 7–11 June 2020; pp. 1–6.
14. Smagulova, K.; James, A.P. A survey on LSTM memristive neural network architectures and applications. *Eur. Phys. J. Spec. Top.* **2019**, *228*, 2313–2324. [CrossRef]
15. Sunny, M.A.I.; Maswood, M.M.S.; Alharbi, A.G. Deep learning-based stock price prediction using LSTM and bi-directional LSTM model. In Proceedings of the 2020 2nd Novel Intelligent and Leading Emerging Sciences Conference (NILES), Giza, Egypt, 24–26 October 2020; pp. 87–92.
16. Imrana, Y.; Xiang, Y.; Ali, L.; Abdul-Rauf, Z. A bidirectional LSTM deep learning approach for intrusion detection. *Expert Syst. Appl.* **2021**, *185*, 115524. [CrossRef]
17. Amran, N.A.; Soltani, M.D.; Yaghoobi, M.; Safari, M. Deep learning based signal detection for OFDM VLC systems. In Proceedings of the 2020 IEEE International Conference on Communications Workshops (ICC Workshops), Dublin, Ireland, 7–11 June 2020; pp. 1–6.
18. Althelaya, K.A.; El-Alfy, E.S.M.; Mohammed, S. Evaluation of bidirectional LSTM for short-and long-term stock market prediction. In Proceedings of the 2018 9th International Conference on Information and Communication Systems (ICICS), Irbid, Jordan, 3–5 April 2018; pp. 151–156.
19. Daly, J.C. Fiber optic intermodulation distortion. *IEEE Trans. Commun.* **1982**, *30*, 1954–1958. [CrossRef]

 photonics

Article

The Performance Improvement of VLC-OFDM System Based on Reservoir Computing

Bingyao Cao, Kechen Yuan *, Hu Li, Shuaihang Duan, Yuwen Li and Yuanjiang Ouyang

Key Laboratory of Specialty Fiber Optics and Optical Access Networks, Shanghai Institute for Advanced Communication and Data Science, Shanghai University, Shanghai 200444, China; caobingyao@shu.edu.cn (B.C.); lihu@shu.edu.cn (H.L.); yizhi2019@shu.edu.cn (S.D.); li_yuwen@shu.edu.cn (Y.L.); oy_yuanjiang@shu.edu.cn (Y.O.)

* Correspondence: neymar@shu.edu.cn

Abstract: Nonlinear effects have been restricting the development of high-speed visible light communication (VLC) systems. Neural network (NN) has become an effective means to alleviate the nonlinearity of a VLC system due to its powerful ability to fit complicated functions. However, the complex training process of traditional NN limits its application in high-speed VLC. Without performance penalty, reservoir computing (RC) simplifies the training process of NN by training only part of the network connection weights, and has become an alternative scheme to NN. For the indoor visible light orthogonal frequency division multiplexing (VLC-OFDM) system, this paper studies the signal recovery effect of the pilot-assisted reservoir computing (PA-RC) frequency domain equalization algorithm. The pilot information is added to the feature engineering of RC to improve the accuracy of channel estimation by traditional least squares (LS) algorithm. The performance of 64 quadrature amplitude modulation (QAM) signal under different transmission rates and peak to peak voltage (Vpp) conditions is demonstrated in the experiments. Compared with the traditional frequency domain equalization algorithms, PA-RC can further expand the Vpp range that meets the 7% hard-decision forward error correction (FEC) limit of 3.8×10^{-3}. At the rate of 240 Mbps, the BER of the system is reduced by about 90%, and the utilization rate of the available frequency band of the system reaches 100%. The results show that PA-RC can effectively improve the transmission performance of VLC system well, and has strong generalization ability.

Keywords: visible light communication; orthogonal frequency division multiplexing; nonlinear equalization; reservoir computing

Citation: Cao, B.; Yuan, K.; Li, H.; Duan, S.; Li, Y.; Ouyang, Y. The Performance Improvement of VLC-OFDM System Based on Reservoir Computing. *Photonics* **2022**, *9*, 185. https://doi.org/10.3390/photonics9030185

Received: 28 February 2022
Accepted: 11 March 2022
Published: 14 March 2022

1. Introduction

It is pointed out in the potential key technologies of 6G that VLC, as a new spectrum resource technology, will become an important research direction in the future communication field [1,2]. VLC uses light-emitting diodes (LEDs) as light sources, and transmits data information with visible light while simultaneously illuminating. This means VLC has advantages such as rich spectrum resources, low cost, high transmission rate, and strong security [3]. In order to further improve the transmission rate of VLC system, relevant researchers have made a breakthrough in the research of high-order modulation format signals and advanced modulation technologies [4–7]. Among them, Orthogonal Frequency Division Multiplexing (OFDM) technology has become an important modulation technology to realize indoor high-speed VLC systems due to its advantages of high spectrum utilization and strong anti-multipath effect [8]. However, the limited modulation bandwidth and linear operating range of LEDs restrict the further development of VLC systems. The narrow modulation bandwidth causes the transmission signal on the high frequency subcarrier to suffer severe channel impairment [9]. With the increase of signal modulation order and communication rate, the nonlinear characteristics of LED will have a more significant impact on the BER performance of VLC systems [10,11]. In addition, the inherent high peak to average power ratio (PAPR) of OFDM system makes the system more sensitive to the nonlinear effect of LED [12].

In order to alleviate the distortion of the received signal caused by the nonlinearity of VLC system, researchers mainly correct the distorted signal from the perspectives of pre- and post-distortion [13–17]. Compared with the pre-distortion algorithms, the post-distortion compensation algorithms have better performance by mitigating nonlinear interference existing in the entire communication system and transmission channel [18,19]. Post-distortion methods based on polynomial models usually require high model accuracy. The polynomial coefficients are difficult to determine in strong nonlinear and complex communication scenarios [20]. Traditional machine learning (ML) algorithms such as K-means clustering [21], support vector machine (SVM) [22], and K nearest neighbors (KNN) [23] generate nonlinear decision boundaries by learning the received data, which suppresses the influence of constellation distortion. Gaussian Mixture Model (GMM) divides data into several probability models based on Gaussian distribution, overcoming the limitation of Kmeans, taking distance as the only reference [24]. However, the above methods usually perform decision optimization on the equalized signal constellation. When the signal constellation diagram is seriously distorted, the traditional ML algorithms will not be able to achieve the expected performance.

With rapid development of integrated circuits and constant improvement of computer's computing power, deep neural network (DNN) with stronger fitting ability of nonlinear function has become an important method to improve the transmission performance of VLC systems. Among them, a series of DNN models, such as convolutional neural network (CNN) [25], dual-branch multilayer perceptron (MLP) [26] and long short-term memory network (LSTM) [27] with memory ability, are applied in VLC systems to overcome nonlinear effects. A deep learning model based on end-to-end optimization is proposed in [28], and the damage caused by each link in the VLC-OFDM system is reduced by introducing neural network models in different processing steps of the system. Ref. [29] takes channel impairment in the VLC-OFDM system as a learning task, and recovers distorted signals by introducing two DNN networks in the channel equalization and signal decision steps. In order to reduce the complexity of DNN, a dual network structure based on constellation decision is proposed in [30] to equalize the inner and outer ring signals of the constellation diagram, respectively. However, complex network structures are usually accompanied by higher training costs. A large number of parameters make the network converge slowly and easily fall into local minimum, resulting in poor generalization ability of the model. Reservoir Computing (RC) has recently become an alternative to traditional NN due to its excellent nonlinear processing capability and low training complexity [31–33]. RC, also known as Echo State Network (ESN), replaces the hidden layers of NN with a large-scale sparsely randomly connected network (reservoir). By training partial weights of the network, the training process of the algorithm is greatly simplified, and the shortcomings of the traditional recurrent neural network (RNN) structure are difficult to determine and the training process is too complicated. As a new type of random weight recursive network, it has been proved that RC can reduce the training complexity of NN and improve the performance of optical communication system [34].

In this paper, a pilot-assisted RC (PA-RC) nonlinear equalization algorithm is proposed to improve the transmission performance of the system. The shortcomings of the traditional least squares (LS) algorithm are improved by introducing PA-RC into the channel estimation stage. We analyze in detail several key parameters that affect the performance of the algorithm. The trend of bit error rate (BER) curve of 64QAM signal at different rates and different signal Vpp in the VLC-OFDM system and the performance improvement brought by the algorithm are experimentally studied. The results indicate that compared with zero forcing equalization (ZFE) algorithm, PA-RC can significantly improve BER performance of the system. At the rate of 240 Mbps, the Vpp range of the signal meeting the FEC threshold is expanded by more than 0.5 V. When the signal Vpp is 1.2 V, the system BER is reduced by 90%, and the available frequency band utilization rate reaches 100%. In addition, we have proved through experiments that PA-RC has strong generalization ability, and the model trained under the condition of fixed signal-to-noise ratio (SNR) can be applied to the system under different transmission conditions. To the best of our knowledge, this is the first time that RC is used for equalization in VLC system.

2. The Proposed Scheme

2.1. Nonlinear Equalizer Based on Reservoir Computing

The PA-RC structure adopted in this paper is shown in Figure 1, which mainly includes the input layer, reservoir, and output layer. The input vector consists of valid data and received pilots within a frame. In a VLC-OFDM system, the received pilot data directly reflects the real situation of the channel. Due to the non-ideal characteristics of transmission system and channel, these pilots are inevitably subject to nonlinear interference. Therefore, the traditional LS channel estimation algorithm has high sensitivity to noise. In the proposed algorithm, the received pilot data is used as one of the features to learn the mapping relationship between input and output, thereby further improving the accuracy of channel estimation. The middle layer of the network uses a random sparsely connected network (reservoir) to replace the hidden layer of the traditional NN. The random connection between neurons makes the reservoir have short-term memory capacity. In this paper, the leaky integral neurons are used to replace ordinary neurons to optimize the performance of the RC algorithm. The output layer uses a linear activation function to linearly combine the neurons in the reservoir to output the recovered signal. The state and output equations of the proposed algorithm are shown in Equations (1) and (2):

$$\mathbf{S}(n+1) = (1-a)\mathbf{S}(n) + f(\mathbf{W}_{in}\mathbf{X}(n+1) + \mathbf{W}\mathbf{S}(n) + \mathbf{W}_{back}\mathbf{Y}(n)) \quad (1)$$

$$\mathbf{Y}_{n+1} = f_{out}(\mathbf{W}_{out}(\mathbf{x}_{n+1}, \mathbf{S}_{n+1}, \mathbf{Y}_n)) \quad (2)$$

where $\mathbf{W}_{in}$, $\mathbf{W}$ and $\mathbf{W}_{back}$ represent the input, the state variables, and output connection weight matrices to the state variables, respectively; $\mathbf{X}(n) = [x_1(n), x_2(n), \ldots, x_K(n)]^T$ is the input vector at time n; $\mathbf{Y}(n) = [y_1(n), y_2(n), \ldots, y_L(n)]^T$ is the output vector at time n; $\mathbf{S}(n) = [s_1(n), s_2(n), \ldots, s_N(n)]^T$ is the state of N internal neurons at time n; $\mathbf{W}_{out}$ represents the connection weight matrix of the reservoir, input, and output to output; a is the leakage rate; f and f_{out} represent two different activation functions.

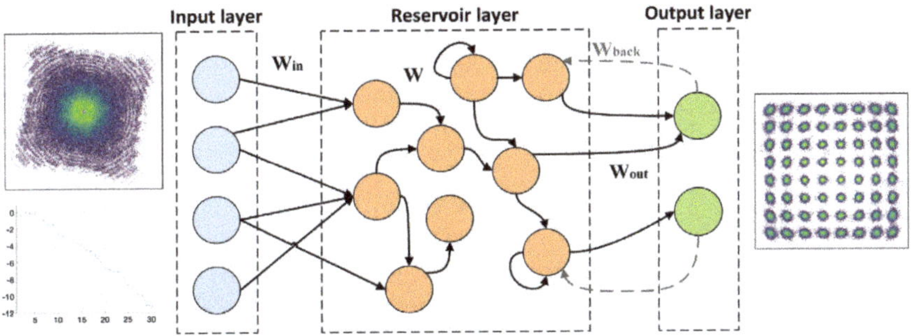

Figure 1. Diagram of PA-RC nonlinear equalizer.

In order not to lose phase information, the length of the input vector in the network is 4 (K = 4), which mainly includes the real and imaginary parts of the received data and pilot information. The output is the recovered signal constellation points, so the size of L is 2. The size of the reservoir is discussed in the experimental section. Equation (1) indicates that the state of the reservoir at a certain time is determined by the input vector at the current time, the state of the reservoir, and the output vector at the previous time. a controls the retention of the reservoir's state at the last moment, which determines the long-term memory capability of the network. When a is 1, the model degenerates back to the ordinary ESN. f usually chooses *tanh* as the activation function to improve the nonlinear representation ability of the network. The state matrix $\boldsymbol{W}$ represents the richness of the internal network of the reservoir. In order to reduce the computational complexity as much as possible, the sparsity of $\boldsymbol{W}$ is generally maintained at 1% to 5%. Equation (2) shows that we need to train $\boldsymbol{W}_{out}$ according to the input and output data of the system to obtain an output that is closer to the expected value. For the signal equalization problem, f_{out} can take the linear identity function.

2.2. Training Process

Generally, the training process of RC includes two stages: states selection and weights calculation. In the state selection stage, the initial state of the reservoir needs to be determined first, and the initial state is usually assumed to be 0. The update process of the state is shown in formula (1). $\boldsymbol{W}_{in}$, $\boldsymbol{W}$, and $\boldsymbol{W}_{back}$ are randomly generated and kept constant throughout the training phase. For simplicity, $\boldsymbol{W}_{back}$ is assumed to be 0. The input vector and the state vector are spliced in a column to form a one-dimensional column vector $\boldsymbol{U}$, and the $\boldsymbol{U}$ obtained during the sampling period is formed into a state matrix $\boldsymbol{B}$ by columns:

$$\boldsymbol{B} = [\boldsymbol{U}(m), \boldsymbol{U}(m+1), ..., \boldsymbol{U}(M)] \tag{3}$$

where $\boldsymbol{U}(n) = [x_1(n), x_2(n), ..., x_K(n), s_1(n), s_2(n), ..., s_N(n)]^T$; m is the sampling time; M is the size of the training set.

In the weights calculation stage, the matrix $\boldsymbol{W}_{out}$ needs to be calculated according to the collected system state matrix $\boldsymbol{B}$ and training data, so that the actual output of the network is close to the expected output result. As shown in formula (4):

$$\boldsymbol{Y}_{expect} \approx \boldsymbol{Y}(n) = \boldsymbol{W}_{out}^T \boldsymbol{B} \tag{4}$$

where $\boldsymbol{Y}_{expect}$ is the expected output.

To make the error between $\boldsymbol{Y}(n)$ and $\boldsymbol{Y}_{expect}$ as small as possible, we define the mean squared error (MSE) as the target loss function in the training phase and solve this linear regression problem using the least squares method. Computationally, this problem can be further processed as a pseudo-inverse problem of $\boldsymbol{B}$:

$$\boldsymbol{W}_{out} = \boldsymbol{Y}_{expect} \boldsymbol{B}^T (\boldsymbol{B}\boldsymbol{B}^T + \zeta \boldsymbol{I})^{-1} \tag{5}$$

Since $\boldsymbol{B}$ may be ill-conditioned in practical applications, a pseudo-inverse algorithm or a regularization technique needs to be used to ensure that the above problems are solved. In Equation (5), we use ridge regression to process $\boldsymbol{B}$. Improve the stability and reliability of traditional LS by adding a regularization term (ζ). The completion of $\boldsymbol{W}_{out}$ calculation represents the end of RC training process.

3. Experimental Setup

The VLC-OFDM transmission system based on 64QAM signals is shown in Figure 2. In order to ensure high frequency band utilization, DC-biased Optical Orthogonal Frequency Division Multiplexing (DCO-OFDM) modulation technology is adopted to generate positive real signals. Figure 2a shows the experimental setup of the indoor VLC system. At the transmitter, the offline random sequence is generated by MATLAB. The input data generates parallel data streams with the same number of sub-carriers by operations such as serial-to-parallel conversion, QAM mapping, and Hermitian symmetry. In order to avoid the influence of DC signal, the first subcarrier is set to 0. Before inverse fast Fourier transform (IFFT), a certain length of training sequence is added to the data frame. The training sequence is also called pilot data, and with the pilot information, the received signals can be synchronized and the channel response can be estimated in real time. In order to mitigate the inter-symbol interference (ISI) caused by the multipath effect, a cyclic prefix of a certain length needs to be added before the OFDM signal. By adding the pilot signal and the cyclic prefix, a complete OFDM signal is generated. The time domain signals after IFFT and clipping are stored in the random access memory (RAM) of the field programmable gate array (FPGA). After that, the transmitted signal is converted into an analog signal by a digital-to-analog converter (DAC-AD9708), and positive real signals are generated by an adjustable attenuator and an amplifier (AMP-OPA657) followed by a DC offset, thereby Drive blue LED (LSLED405-5) with wavelength of 405 nm for electro-optical conversion to send optical signals into 1.2 m indoor free space.

A lens is placed at the receiver of the system to collect the beam to improve the SNR of the receiver. The collected received light signals are converted into electrical signals by a photodetector (Hamamatsu-S10784). Signals are then captured by a digital phosphor oscilloscope (Tektronix 7354C) at a rate of 10× upsampling. In the signal off-line processing stage, the received signals are synchronized, and parallel-serial conversion and fast Fourier transform (FFT) are used to obtain the frequency-domain OFDM data. The pilot data and valid data information are sent in a frame into the PA-RC network to estimate the channel and recover the distorted signal. Finally, the BER performance of the received data demodulated by QAM is analyzed. The spectrum of the transmitter and receiver of the system are shown in Figure 2b,c.

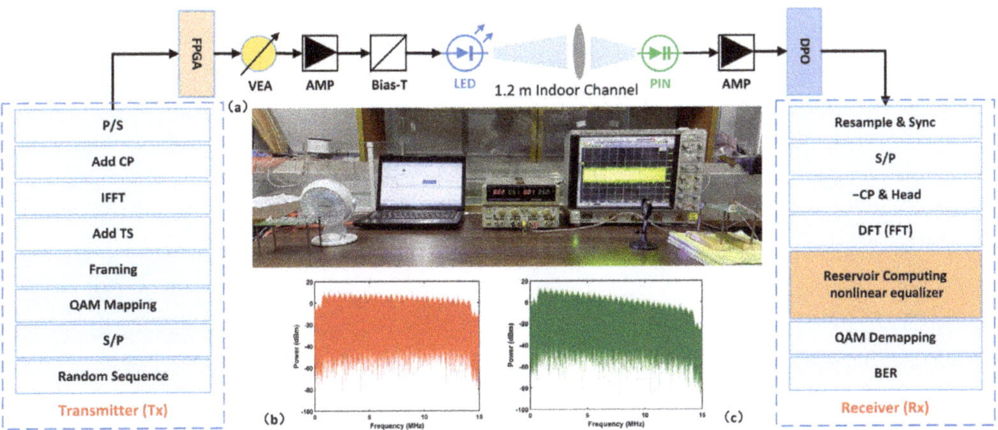

Figure 2. System structure diagram: (**a**) The experimental bench; (**b**) the transmitted spectra of the transmitter; (**c**) the received spectra of the receiver.

4. Experimental Results and Analysis

The OFDM related parameters used in the experiment are shown in Table 1. We use DCO-OFDM modulation technology to generate positive real signals that can drive LEDs to work properly. Therefore, the valid data sub-carriers are half of the total number of subcarriers. The performance of the first subcarrier is often poor because it is close to DC. Therefore, the first sub-carrier does not transmit valid data. A frame transmission sequence includes 100 OFDM symbols, the first of which is a training symbol. The transmitted sequences are generated completely randomly, and the training and test sets are 8000 and 10,000 in size, respectively. To demonstrate the adaptability to environmental changes of the model, training and test sets are collected at two different time periods.

We first discuss several important parameters related to the performance of PA-RC, namely the number of neurons in the reservoir, sampling time, spectral radius, leakage rate, scaling factor and sparsity, and take the system BER as the index to measure the performance of the algorithm. During the experiment, the default values of the above six parameters are 100, 1000, 0.1, 0.9, 0.1, and 0.3, respectively. We verify the effect of parameter changes on the algorithm performance under the condition that the signal Vpp is 1.2 V. In order to follow the principle of the control variable method, when a parameter is changed, other fixed parameters use default values. A set of parameters with stronger generalization ability is selected by comparing the BER under different system transmission rates.

Table 1. Parameters of OFDM.

Parameter	Value
Number of data-carrying subcarriers	31
FFT size	64
Length of cyclic prefix (CP)	16
Number of training symbols	1
Number of subcarriers near DC	1
QAM signal modulation order	64
Length of one frame	100

The number of neurons in PA-RC determines the size of the reservoir. Figure 3a shows that PA-RC performance increases with the increase of the reservoir and gradually tends to be stable. It is not difficult to understand that when the number of neurons in the reservoir is small, it will lead to insufficient representation ability of the model for the nonlinear system. As the number of neurons increases, the network describes the system more accurately. When the number of neurons increases to a certain extent, the training complexity of the model increases, and the network may overfit. The function of the sampling time is to avoid the influence of the reservoir's initial state on the network performance. As shown in Figure 3b, on the nonlinear characterization problem of VLC system, a small sampling time can make the model achieve better performance. The performance of the model starts to deteriorate after the 4000th sample point. A later sampling time means a smaller number of reservoir states and expected outputs to be collected. The linear regression process may increase the training error of RC due to insufficient number of samples. Equation (5) indicates that the size of the state matrix $\boldsymbol{B}$ directly determines the computational complexity of $\boldsymbol{W}_{out}$, and the size of $\boldsymbol{B}$ is determined by the number and sampling time of reservoir neurons. In order to balance model performance and training complexity, the number of neurons and sampling time are taken as 100 and 4000, respectively.

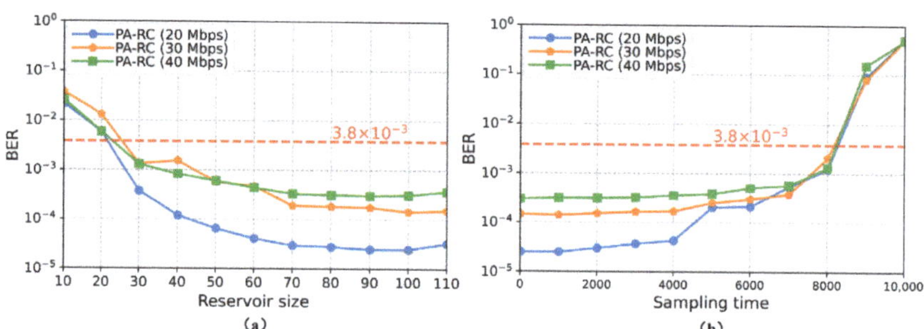

Figure 3. BER vs. RC parameters. (**a**) Neurons in the reservoir; (**b**) sampling time.

The spectral radius is defined as the absolute value of the largest eigenvalue of the connection weight matrix $\boldsymbol{W}$, which is an important condition to ensure the stability of the algorithm. When spectral radius is less than 1, the network has the property of echo. As shown in Figure 4a, the system BER increases gradually with the increase of spectral radius. In the range of 0 to 0.2, the performance of the network is optimal. The proposed algorithm uses leaky integral neurons to replace ordinary neurons in the reservoir. Therefore, we discuss the impact of leakage rate on the performance of PA-RC. As shown in Figure 4b, a small value of leakage rate results in a slow response of the reservoir to the input signal. By increasing the leakage rate, the performance of the algorithm is significantly improved.

In the range of 0.9 to 1, the performance of PA-RC tends to be stable. When the leakage rate is 1, the leaky integral neurons degenerate into ordinary neurons. Theoretically, leaky integral neurons can improve the performance of the reservoir. Without loss of generality, the values of spectral radius and leakage rate are 0.1 and 0.98, respectively.

The scaling factor is mainly used to adjust the feature scale of the input signal. When the value of scaling factor is small, the network element works around the linear center of activation function. As the scaling factor increases, the internal units of the network gradually approach the saturation point of f, and the nonlinearity of the model is stronger. As shown in Figure 5a, a value between 1 and 2 is more appropriate for the scaling factor. Sparsity represents the proportion of non-zero elements to the total elements in the neuron connection matrix of the reservoir. Generally, the smaller the sparsity is, the simpler the internal structure of the reservoir is, but its nonlinear representation ability is weakened. From Figure 5b, we find that for the problem of the nonlinear equalization of the VLC system, the sparsity within a fixed range (1∼5%) has little effect on the performance of the algorithm. In order to reduce the network complexity of the reservoir layer as much as possible, we set the sparsity as 0.01. Since then, we have identified all parameters that affect the computational performance of the reservoir. The parameters related to PA-RC are shown in Table 2.

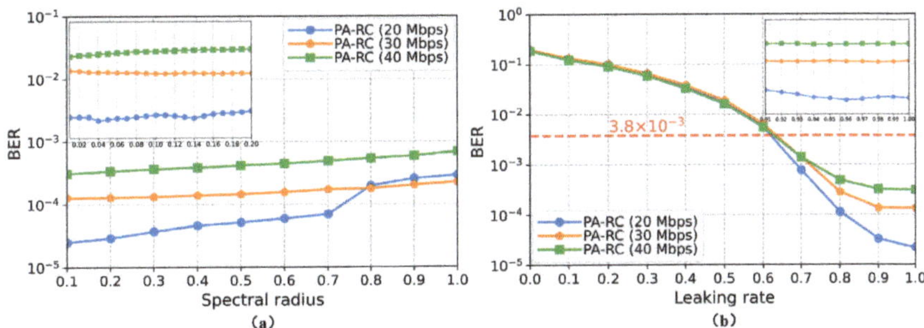

Figure 4. BER vs. RC parameters. (**a**) Spectral radius; (**b**) leaking rate.

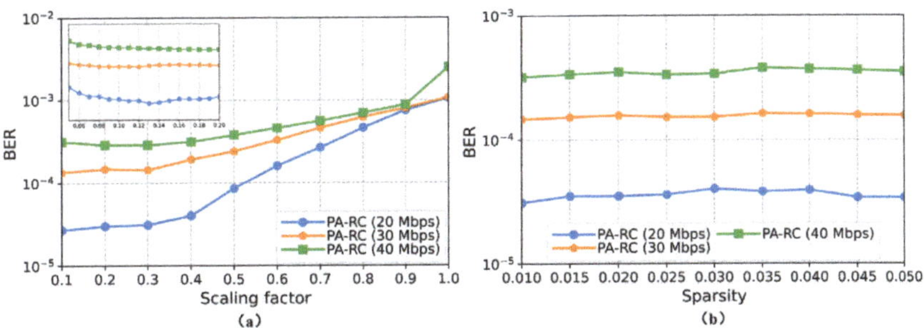

Figure 5. BER vs. parameters. (**a**) Scaling factor; (**b**) sparsity.

Table 2. Parameters of PA-RC.

Parameter	Value
Number of neurons in the reservoir	100
Sampling time	4000
Spectral radius	0.1
Leakage rate	0.98
Scaling factor	0.12
Sparsity	1%
Training set size	8000
Test set size	10,000

We increase the nonlinearity of the system by changing the signal Vpp and compare the effect of system performance at different rates. The signal Vpp directly reflects the size of the current system SNR. The LS curve represents the system BER recovered by LS channel estimation combined with ZFE. The LS + RC curve represents the BER of the signal demapped by RC after ZFE. PA-RC represents the BER curve of the signal recovered by the algorithm used in this paper. As shown in Figure 6, it is obvious that under different transmission rates, the system BER shows a trend of first decreasing and then increasing. When Vpp is small, due to the low SNR of the system, Gaussian White Noise (GWN) interferes greatly with the transmitted signals. With the increase of Vpp, BER decreases. When Vpp increases to a certain extent, the system performance begins to deteriorate. This is easy to understand, because the linear operating range of non-linear devices such as LED is limited. When the signal power is high, the LED will enter the non-linear operating range and the signal will be clipped and distorted.

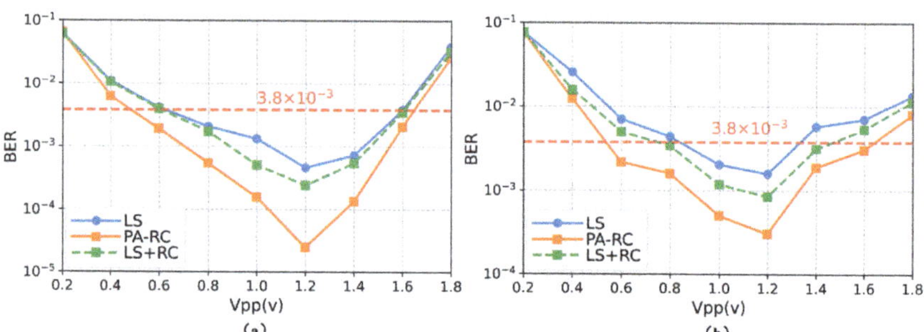

Figure 6. BER vs. signal Vpp. (**a**) The data is 120 Mbps; (**b**) the data is 240 Mbps.

As shown in Figure 6a, under the rate of 120 Mbps, with the help of the ZFE to equalize received signals, the Vpp range of the signal whose system BER meets the FEC threshold (3.8×10^{-3}) is about 1.0 V. By using the RC algorithm to further restore the ZFE equalized signals, the system BER is reduced, but the range of Vpp is not expanded. because the recovery effect of ZFE completely depends on the accuracy of the channel estimation by LS. Due to the serious attenuation of the channel response of high-frequency subcarriers, the channel information estimated by the pilot is strongly interfered by noise. Therefore, RC cannot further recover the signals. PA-RC can better recover received signals by correcting the channel information estimated by the pilot. The BER is greatly reduced, and the Vpp range is expanded to about 1.2 V. When the Vpp is about 1.2 V, the system BER performance reaches the best, and the proposed algorithm reduces the system BER by more than 95%. When the system transmission rate increases to 240 Mbps, the BER performance deteriorates, as shown in Figure 6b. This is because as the data rate increases, the number of light emissions per unit time increases, resulting in enhanced nonlinearity. Using ZFE to recover the received signals, the Vpp range that satisfies the FEC threshold is about 0.5 V. After using RC to demap the signal based on the LS channel estimation, the Vpp range of the signal satisfying the FEC threshold is expanded to 0.6 V. The influence of channel impairment is compensated by PA-RC, and the Vpp range of the signal satisfying FEC is expanded to about 1.1 V. From the results, the system transmission performance has been significantly improved. When the signal Vpp is 0.6 V and 1.6 V, the BER is still below the FEC threshold. It shows that RC can still bring some improvement to the system under the condition of low SNR and strong nonlinear interference. When Vpp is 1.2 V, the system BER is reduced by about 90%.

In order to reflect the improvement of the utilization rate of the available frequency band of the system brought by PA-RC, we draw the BER curves of high frequency subcarriers under different SNR. As shown in Figure 7, the BER of the subcarriers gradually deteriorates with the increase of frequency. The main reason is that the modulation bandwidth of nonlinear devices such as LED is limited, resulting in the gradual attenuation of VLC channel response with the increase of frequency. When Vpp is 0.6 V, the transmitted signals are seriously interfered by GWN, and the transmission performance of high-frequency subcarrier deteriorates seriously. As shown in Figure 7a, starting from the 23rd subcarrier, the BER is already above the FEC threshold. After using PA-RC to restore the received signals, the number of available sub-carriers reaches 28. When Vpp is 1.2 V, the number of available subcarriers in the system increases significantly under optimal transmission conditions. The PA-RC makes the utilization rate of the available frequency band of the system reach 100%. When Vpp is 1.5 V, the transmitted signals suffer severe clipping distortion. By using PA-RC to suppress nonlinear effects in the system, the available frequency band utilization of the system is improved by about 16%. This indicates that the proposed algorithm can further improve the transmission performance of high-frequency subcarriers without losing the system transmission capacity.

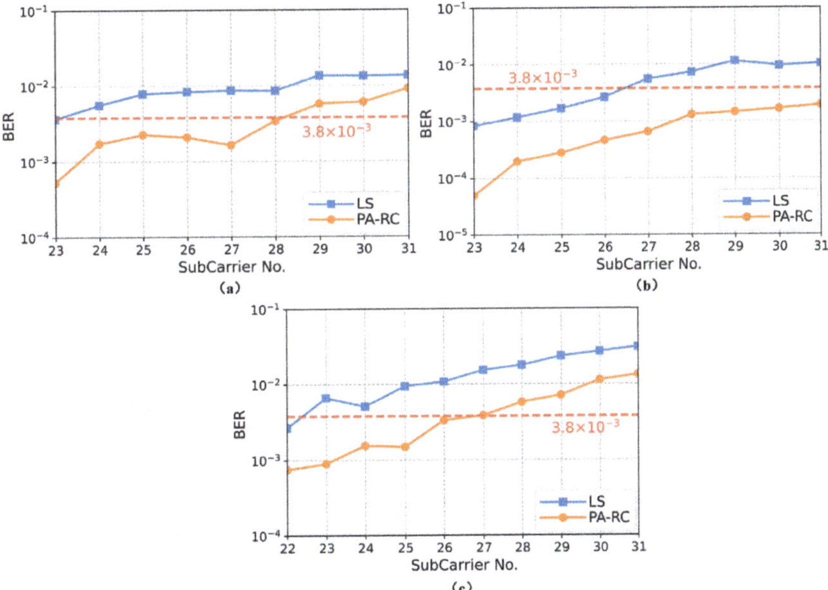

Figure 7. BER vs. Subcarrier number under the rate of 240 Mbps. (**a**) The Vpp is 0.6 V; (**b**) the Vpp is 1.2 V; (**c**) the Vpp is 1.5 V.

We verified the adaptability and sensitivity of PA-RC to changes in the system transmission environment at the transmission rate of 240 Mbps. The received signals under other SNR conditions are recovered by the model trained under the fixed SNR, and the system BER is used as the judgment index of the algorithm performance. As shown in

Figure 8, PA-RC (0.6 V), PA-RC (1.2 V), and PA-RC (1.5 V) represent the models trained with signal Vpp of 0.6 V, 1.2 V, and 1.5 V, respectively. PA-RC (0.6 V) has poor performance when Vpp is greater than 1 V. When Vpp is less than 1.2 V, the equalization performance of the PA-RC (1.5 V) is inferior to that of PA-RC (0.6 V) and PA-RC (1.2 V). It can be seen from the previous analysis that different Vpp conditions represent different interferences on the signals. Therefore, when the signal Vpp is quite different from the conditions during model training, the algorithm will not be able to achieve the best equalization performance. Note that PA-RC (1.2 V) has better performance under different transmission conditions, which indicates that the model trained around 1.2 V has the best generalization ability. Figure 8b shows the comparison of the BER performance of PA-RC (0.6 V) and LS under the optimal transmission conditions of the high-frequency sub-carriers of the system. Due to the great changes in transmission conditions, the performance of the PA-RC algorithm is degraded, but the recovery effect of the signal is still better than that of LS. The results in Figure 8 show that even the models trained under low SNR (0.6 V) and high SNR (1.5 V) conditions still have excellent equalization performance in changing environments.

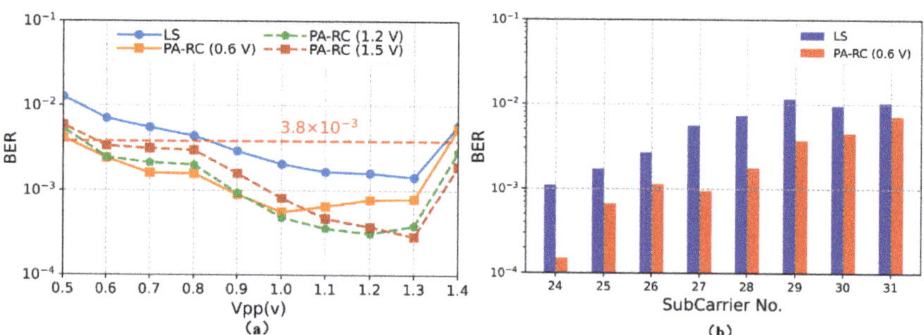

Figure 8. The data rate is 240 Mbps: (**a**) BER vs. Vpp; (**b**) BER vs. Subcarrier number when Vpp is 1.2 V.

In order to verify that RC has the same performance as the traditional NN while simplifying the model training, we choose two representative algorithms, the fully connected neural network (FCNN) and the long short-term memory network (LSTM) trained under different iterations, as the comparison objects. For the convenience of comparison, both FCNN and LSTM adopt a single hidden layer structure. By changing the number of neurons in the middle layer, the algorithm performance under different network scales can be compared. When Vpp is 1.2 V, as shown in Figure 9a, FCNN (10 epochs) and LSTM (10 epochs) have poor signal recovery performance. When N is 100, the two networks have only the same performance as LS (1.92×10^{-3}). When the epochs is 30, the performance of FCNN and LSTM is close to RC. It shows that traditional NN needs more iterations to converge. However, PA-RC only needs to solve a simple linear regression problem during the training process, so the convergence rate is fast. We note that the PA-RC algorithm outperformed LS with 30 neurons, while FCNN (10 epochs) and LSTM (10 epochs) required 40 neurons. When Vpp is 1.5 V, the system suffers from strong nonlinear disturbance. As shown in Figure 9b, under the condition of fewer neurons, FCNN and LSTM improve the performance through more iterations. When N is 30, LSTM (30 epochs) and FCNN (30 epochs) outperform LS (5.82×10^{-3}). With the increase of neurons in the middle layer, the PA-RC converges rapidly and achieves the same performance as the two NNs. The results in Figure 9 indicate that PA-RC has the same excellent nonlinear mitigation ability as traditional NN, and training is simple and the convergence rate is fast. Furthermore, we find that LSTM has similar performance to FCNN. Mainly because the objects of equalization in this paper are the frequency domain signals. Therefore, LSTM does not exert its advantages in time series problems. Compared with [29] using two NNs, We use PA-RC to directly equalize the received signals, which greatly reduces the complexity of the receiver.

In addition, the number of neurons in the reservoir of PA-RC is small, which is the most essential difference from the NN equalizer used in current VLC systems.

Finally, we show the received signal constellations on the 10th and 31st sub-carriers of OFDM symbols under transmission rate of 240 Mbps. Figure 10a,c show the restored constellations of the received signals on the 10th and 30th subcarriers by ZFE, respectively. We find that the signal constellations on the high frequency subcarriers produce obvious rotation. This indicates that there is phase noise in the system. At high frequencies, due to more serious channel impairments, the estimation of the channel by the traditional LS algorithm will become inaccurate, so the effect of phase noise is more obvious. Figure 10b,d show the recovery effect of PA-RC on the signals. Compared with ZFE, the signal distribution after PA-RC restoration is more dense. It indicates that the proposed algorithm has a better equalization performance on received signals. The channel information estimated by pilots is corrected by PA-RC, and the common phase error (CPE) of the received signals is effectively eliminated. We know that the constellation points of the 64QAM signal are distributed on seven circles of constant modulus length. It is not difficult to find in Figure 10 that the signals on the outer ring are affected by more serious nonlinearity. Mainly because the signals on the outer ring have higher power, in addition to amplifying the multiplicative noise in the system, they will make it easier for nonlinear devices such as LEDs to enter the nonlinear working area. The signals are corrected by PA-RC, and the CPE and amplitude noise of the high frequency part are effectively suppressed. However, there are still some phase noise that cannot be completely eliminated. In future work, we will combine some commonly used phase noise suppression algorithms to better recover the signals.

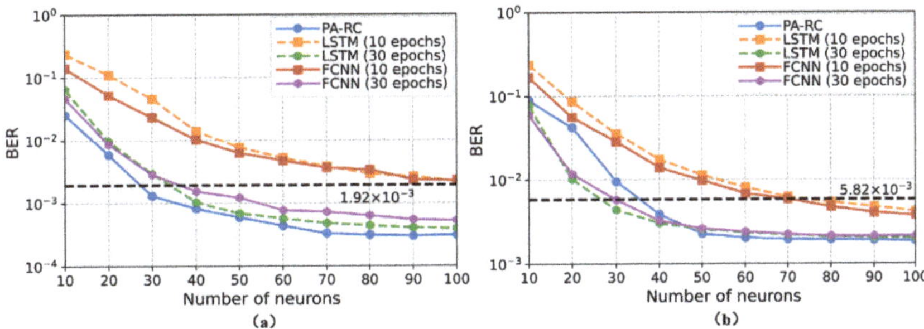

Figure 9. BER vs. Number of neurons. (**a**) The Vpp is 1.2 V; (**b**) the Vpp is 1.5 V.

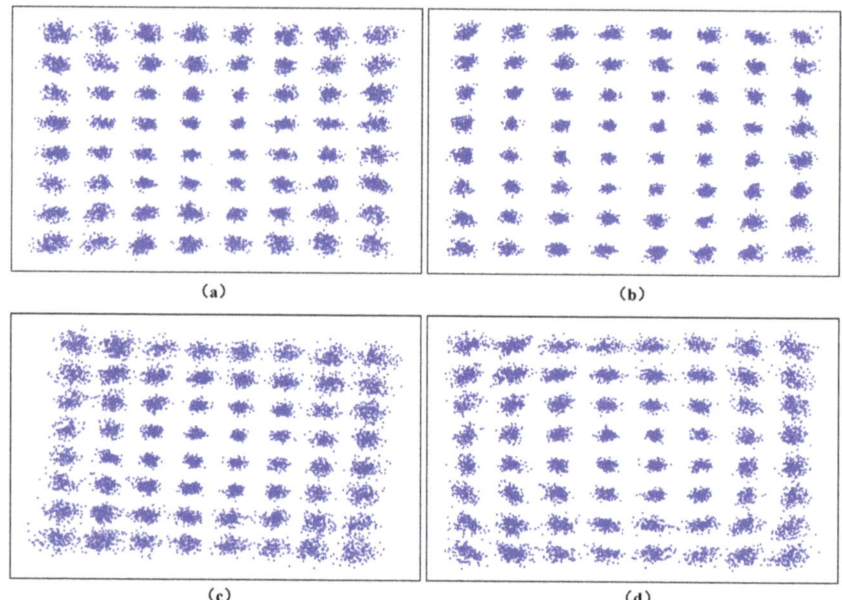

Figure 10. Constellation diagram of received signal on the 10th subcarrier recovered by (**a**) LS and (**b**) RC; constellation diagram of received signal on the 31th subcarrier recovered by (**c**) LS and (**d**) RC.

5. Conclusions

In this paper, we adopt PA-RC to compensate for the signals subjected to nonlinear interference in a VLC-OFDM system. Taking the received pilot information as one of the features of the input vector enables the model to obtain the true state of the current channel, thereby better recovering signals. The experiment compares the signal recovery effect of PA-RC and traditional frequency domain equalization algorithm under strong nonlinear effects conditions. We study the relationship between system BER performance and signal Vpp. When Vpp is too small or too large, the signals are severely distorted. The results show that PA-RC can significantly reduce the system BER. At the transmission rate of 120 Mbps, compared with ZFE, PA-RC expands the Vpp range by 0.2 V. When Vpp is 1.2 V, the system BER is reduced by more than 95%. As the transmission rate increases, the system BER performance drops significantly. When the transmission rate is 240 Mbps, even under the optimal transmission conditions, the high-frequency subcarriers still cannot meet the transmission conditions. By utilizing PA-RC, the range of signal Vpp meeting FEC threshold is expanded from 0.5 V to 1.1 V. When Vpp is 1.2 V, the utilization rate of the available frequency band of the system reaches 100%, and the system BER is reduced by about 90%. Furthermore, we investigate the effect of system environment changes on PA-RC performance. The trained PA-RC model has stronger generalization ability when Vpp is 1.2 V. In order to simplify the training process of the RC, the reservoir used in this paper is randomly generated. The optimal reservoir structure should match the specific problem. In the future, it is believed that more diverse RC network structures will be applied in the VLC system to further alleviate nonlinear effects.

Author Contributions: Conceptualization, B.C. and K.Y.; methodology, K.Y.; software, K.Y. and S.D.; validation, K.Y. and S.D.; formal analysis, B.C.; investigation, K.Y.; resources, B.C.; data curation, K.Y. and H.L.; writing—original draft preparation, B.C.; writing—review and editing, K.Y., Y.L. and Y.O.; visualization, K.Y. All authors have read and agreed to the published version of the manuscript.

Funding: This work was supported in part by National Key Research and Development Program of China (2021YFB2900800 2021YFB2900802), the Science and Technology Commission of Shanghai Municipality (Project No. 20511102400, 20ZR1420900) and 111 project (D20031).

Institutional Review Board Statement: Not applicable.

Informed Consent Statement: Informed consent was obtained from all subjects involved in the study.

Conflicts of Interest: The authors declare no conflict of interest.

References

1. Yang, P.; Xiao, Y.; Xiao, M. 6G Wireless Communications: Vision and Potential Techniques. *IEEE Netw.* **2019**, *33*, 70–75. [CrossRef]
2. Ariyanti, S.; Suryanegara M. Visible Light Communication (VLC) for 6G Technology: The Potency and Research Challenges. In Proceedings of the 2020 Fourth World Conference on Smart Trends in Systems Security and Sustainablity, London, UK, 27–28 July 2020.
3. Wang, Y.; Huang, X.; Tao, L. 45-Gb/s RGB-LED based WDM visible light communication system employing CAP modulation and RLS based adaptive equalization. *Opt. Express* **2015**, *23*, 13633–13686. [CrossRef] [PubMed]
4. Noshad, M.; Brandt-Pearce, M. Hadamard Coded Modulation for Visible Light Communications. *Hadamard Coded Modul. Visible Light Commun.* **2016**, *64*, 1167–1175. [CrossRef]
5. Huang, X.; Chen, S.; Wang Z. 2.0-Gb/s Visible Light Link Based on Adaptive Bit Allocation OFDM of a Single Phosphorescent White LED. *IEEE Photonics J.* **2017**, *7*, 1–8. [CrossRef]
6. Chi, N.; Zhao, H.; Wang Z. Bandwidth-efficient visible light communication system based on faster-than-Nyquist pre-coded CAP modulation. *Chin. Opt. Lett.* **2017**, *15*, 6–11.
7. Dissanayake, S.; Armstrong, J. Comparison of ACO-OFDM, DCO-OFDM and ADO-OFDM in IM/DD System. *J. Lightwave Technol.* **2013**, *31*, 1063–1072. [CrossRef]
8. Mohammed, M.; He, C.; Armstrong, J. Mitigation of Side-Effect Modulation in Optical OFDM VLC Systems. *IEEE Access* **2018**, *6*, 58161–58170. [CrossRef]
9. Dimitrov, S.; Haas, H. Information rate of OFDM-based optical wireless communication systems with nonlinear distortion. *J. Lightwave Technol.* **2013**, *28*, 213–227. [CrossRef]
10. Zou, P.; Hu F.; Li G. Optimized QAM order with probabilistic shaping for the nonlinear underwater VLC channel. In Proceedings of the Optical Fiber Communication Conference, San Diego, CA, USA, 8–12 March 2020.
11. Ahmed, M.; Elsherbini, M.; Abdelkader, H. Exploring the effect of LED nonlinearity on theperformance of layered ACO-OFDM. *Appl. Opt.* **2020**, *59*, 7343–7351.
12. Sharifi, A.A.; Emami, H. PAPR reduction of asymmetrically clipped optical OFDM signals: Optimizing PTS technique using improved flower pollination algorithm. *Opt. Commun.* **2020**, *474*, 126057. [CrossRef]
13. Lu, X.; Zhao M.; Qiao L. Non-linear Compensation of Multi-CAP VLC System Employing Pre-Distortion Base on Clustering of Machine Learning. In Proceedings of the Optical Fiber Communication Conference, San Diego, CA, USA, 11–15 March 2018.
14. Mitra, R.; Bhatia, V. Chebyshev Polynomial-Based Adaptive Predistorter for Nonlinear LED Compensation in VLC. *IEEE Photonics Technol. Lett.* **2016**, *28*, 1053–1056. [CrossRef]
15. Abd Elkarim, M.; Aly M.; AbdelKader, H. LED nonlinearity mitigation in LACO-OFDMoptical communications based on adaptive predistortion and postdistortion techniques. *Appl. Opt.* **2021**, *60*, 7279–7289. [CrossRef]
16. Lain, J.; Chen, Y. An ANN-Based Adaptive Predistorter for LED Nonlinearity in Indoor Visible Light Communications. *Electronics* **2021**, *10*, 948. [CrossRef]
17. Chen, C.; Nie, Y.; Liu, M. Digital pre-equalization for OFDM-based VLC systems: Centralized or distributed?. *IEEE Photonics Technol. Lett.* **2021**, *33*, 1081–1084. [CrossRef]
18. Lu, H.; Jin, J.; Wang, J. Alleviation of LED nonlinearity impact in visible light communication using companding and predistortion. *IET Commun.* **2019**, *13*, 818–821. [CrossRef]
19. Miao, P.; Chen, G.; Wang, X.; Yao, Y.; Chambers, J.A. Adaptive Nonlinear Equalization Combining Sparse Bayesian Learning and Kalman Filtering for Visible Light Communications. *J. Lightwave Technol.* **2020**, *38*, 6732–6745. [CrossRef]
20. Niu, Y.; Deng X.; Fan W. LED Nonlinearity Post-compensator with Legendre polynomials in Visible Light Communications. In Proceedings of the IEEE Conference on Industrial Electronics and Applications, Chengdu, China, 1–4 August 2021.
21. Wu, X.; Chi, N.; Zwierko, P. The phase estimation of geometric shaping 8-QAM modulations based on K-means clustering in underwater visible light communication. *Opt. Commun.* **2019**, *444*, 147–153. [CrossRef]
22. Sun, H.; Zhang, Y.; Wang, F. SVM Aided Signal Detection in Generalized Spatial Modulation VLC System. *IEEE Access* **2021**, *9*, 80360–80372. [CrossRef]
23. Yuan K.; Cao, B. Nonlinear mitigation of VLC-OFDM based on low complexity DW-KNN. In Proceedings of the 2021 International Conference on Signal Processing and Communication Technology, Tianjin, China, 24–26 December 2021; *accepted*.
24. Wu, X.; Hu, F.; Zou, P. The performance improvement of visible light communication systems under strong nonlinearities based on Gaussian mixture model. *Microw. Opt. Technol. Lett.* **2019**, *62*, 547–554. [CrossRef]

25. Gao, Y.; Wu, Z.; Wang, J. Convolution neural network-based time-domain equalizer for DFT-Spread OFDM VLC system. *Opt. Commun.* **2019**, *435*, 35–40. [CrossRef]
26. Zhao, Y.; Zou, P.; Yu, W. Two tributaries heterogeneous neural network based channel emulator for underwater visible light communication systems. *Opt. Express* **2019**, *27*, 22532–22541. [CrossRef] [PubMed]
27. Lu, X.; Lu, C.; Yu, W. Memory-controlled deep LSTM neural network post-equalizer used in high-speed PAM VLC system. *Opt. Express* **2019**, *27*, 7822–7833. [CrossRef] [PubMed]
28. Miao, P.; Zhu, B.; Qi, C. A model-driven deep learning method for LED nonlinearity mitigation in OFDM-based optical communications. *IEEE Access* **2019**, *7*, 71436–71446. [CrossRef]
29. Miao, P.; Yin, W.; Peng, H.; Yao, Y. Study of the Performance of Deep Learning-Based Channel Equalization for Indoor Visible Light Communication Systems. *Photonics* **2021**, *8*, 453. [CrossRef]
30. Chen, H.; Niu, W.; Zhao, Y.; Zhang, J.; Chi, N.; Li, Z. Adaptive deep-learning equalizer based on constellation partitioning scheme with reduced computational complexity in UVLC system. *Opt. Express* **2021**, *29*, 21773–21782. [CrossRef] [PubMed]
31. Verstraeten, D.; Schrauwen, B.; Haene, M.; Stroobandt, D. An experimental unification of reservoir computing methods. *Neural Netw.* **2007**, *20*, 391–403. [CrossRef]
32. Duport, F.; Schneider, B.; Smerieri, A.; Haelterman, M.; Massar, S. All-optical reservoir computing. *Opt. Express* **2012**, *20*, 22783–22795. [CrossRef] [PubMed]
33. Feng, X.; Zhang, L.; Pang, X.; Gu, X.; Yu, X. Numerical Study of Parallel Optoelectronic Reservoir Computing to Enhance Nonlinear Channel Equalization. *Photonics* **2021**, *8*, 406. [CrossRef]
34. Wang, S.; Fang, S.; Wang, L. Signal recovery based on optoelectronic reservoir computing for high speed optical fiber communication system. *Opt. Commun.* **2021**, *495*, 127082. [CrossRef]

MDPI

Article

Neural Network-Based Transceiver Design for VLC System over ISI Channel

Lin Li, Zhaorui Zhu * and Jian Zhang

National Digital Switching System Engineering and Technological Research Center, Zhengzhou 450000, China; wclilin@163.com (L.L.); zhang_xinda@126.com (J.Z.)
* Correspondence: zhaoruizhu_xyz@126.com

Abstract: In this letter, we construct the neural network (NN)-based transceiver to compensate for the varying inter-symbol-interference (ISI) effect in visible light communication (VLC) systems. For processing variable-length sequences, the convolution neural network (CNN) is utilized, and then the residual network structure is further leveraged at the receiver part to enhance the performance. To cope with varying ISI, the pilot sequence, instead of channel side information (CSI) obtained by an additional module, is integrated into the framework to recover the data sequence directly. Simulation results show that the symbol error rate (SER) performance of the proposed NN-based transceiver can outperform separately designed transceiver schemes and approach the ideal perfect CSI (PCSI) case with a few pilot symbols or even no pilot.

Keywords: visible light communication (VLC); neural network (NN); deep learning; autoencoder (AE); transceiver design

Citation: Li, L.; Zhu, Z.; Zhang, J. Neural Network-Based Transceiver Design for VLC System over ISI Channel. *Photonics* **2022**, *9*, 190. https://doi.org/10.3390/photonics9030190

Received: 13 February 2022
Accepted: 14 March 2022
Published: 16 March 2022

1. Introduction

Visible light communication (VLC) [1] has recently been widely researched by academia and industry, due to its advantage of simultaneous lighting [2] and communication. For low complexity and cost, intensity modulation using light emitting diodes (LED) and direct detection with a photodetector (PD) is commonly employed in the VLC system. However, there are some challenges that create an obstacle to the development of VLC. First, the nonlinearity characteristic of LEDs is significant. Thus, the transmitted signal usually satisfies the peak intensity constraint. Second, due to the limited bandwidth of LED, efficient constellation modulation [3] is applied for a higher data rate. Third, the reflected signal leads to the inter-symbol-interference (ISI), which should be compensated by extra equalization [4] or error correction code.

Machine learning, especially deep learning [5], is now penetrating every facet of wireless communication [6]. The general method of deep learning is comprised of two types: in one respect, the individual parts of communication systems, such as pilot design [7], channel estimation and detection [8], are replaced by a learned efficient neural network (NN); in the other respect, the end-to-end learning of the whole communication system creates a new paradigm for joint optimization of transceivers [9]. Due to the similarity of autoencoder (AE) and communication systems, several works have been completed on transceiver design using AE to further promote the performance of transceivers. The early works of NN-based transceivers focus on the additive white Gaussian noise (AWGN) channel model, which deserves more practical consideration for application. In [10], the authors focus on the continuous data transmission and synchronization issue in the receiver. However, the ISI is neglected, and a stacked fully connected neural network is inefficient. In [11], the pilot and data are trained together in the AE; however, the fading channel is the only single path and the sequence length is limited due to the fully connected structure. In [12], Zhu et al. provide the convolution neural network (CNN) structure, which is adequate for sequence training. Still, the simulated ISI channel is static and not practical for varying models.

During the next several years, similar approaches are migrated to enhance the performance of transceivers in the VLC domain, where typical characteristics, such as unipolarity of signal [13], illumination requirements [14,15] and nonlinearity of channel [16], should be sufficiently considered. In [17], for higher bandwidth efficiency, the VLC orthogonal frequency division multiplexing (OFDM) system with the stochastic ISI model is optimized with the AE approach. However, the ISI is mainly eliminated by the cyclic prefix, which can be further modified through the deep learning method. Meanwhile, in [18], to explicitly integrate the channel side information (CSI) into AE VLC system, the classic model-based method is required to estimate CSI in the receiver, which adds extra complexity to the whole system. Similarly, for more real-life application constraints, Ref. [19] proposed VLCnet, which takes into account illumination level, flicker influence and channel impulse response. However, an additional minimum mean square error (MMSE) equalizer with real CSI is still required in practical implementation.

In this letter, we propose an NN-based transceiver for the VLC system over the ISI channel with the modified AE model, which extends the work [20] focusing on the single path channel and CSI obtained by the traditional method. The contribution of this paper can be summarized as follows:

(1) To handle the sequence input issue, we propose the AE framework with a 1-D convolution (Conv1D) layer structure. Meanwhile, the residual network structure is utilized at the receiver to improve training performance. The whole architecture is flexible for processing continuous transmission signals, which is prevalent in current communication systems.

(2) To the best of the author's knowledge, it is the initial work to integrate the pilot sequence and data sequence into the transceiver design with NN in the VLC domain. This joint structure enables the optimization of transmitter and receiver with implicit channel estimation in the whole system. Thus, the additional channel estimation part using the traditional method can be eliminated. The pilot-assisted transceiver enables the receiver to recover the data sequence directly without explicit CSI.

The simulation results demonstrate that the symbol error rate (SER) performance of the proposed transceiver can outperform the individually designed transceiver scheme and approach the ideal perfect CSI (PCSI) case with a few pilot symbols for above 2 level modulation or even no pilot for a 2 level case.

2. VLC System Model With ISI

The common ISI VLC model can be given as in Figure 1, where the light-of-sight (LOS) path is the dominant one and the reflected path by the wall causes the multipath distortion. The validation of the model has been experimentally illustrated in [21] under the off-the-shelf devices and standard indoor environments. Due to the slowly changing property (compared to the baud rate), the ISI VLC model is usually regarded as a linear time-invariant system during multiple symbols duration. For the validity of channel estimation, data and pilot sequence experience the same channel condition. Referring to [22], the received signal containing data and pilot can be written as:

$$y(t) = h^{(1)}(t) \otimes x(t) + h^{(2)}(t) \otimes x(t - \tau_d) + w(t), \tag{1}$$

where $h^{(1)}(t)$ and $h^{(2)}(t)$ are the impulse response for the LOS path and the reflected path, respectively, $x(t)$ is the transmitted signal, τ_d is the transmission delay for the second path, $w(t)$ is the AWGN.

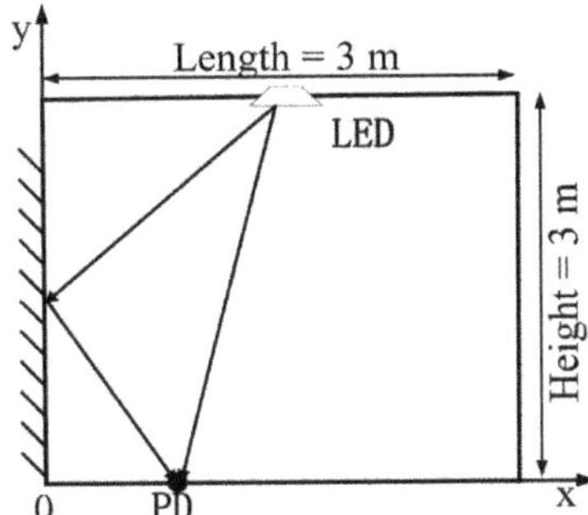

Figure 1. Two paths ISI model in VLC system.

Assuming a matched filter with the impulse response of the LOS path, then we can obtain the discrete time model as [18]:

$$\mathbf{y} = \mathbf{H}\mathbf{x} + \mathbf{w}, \tag{2}$$

where $\mathbf{y} = (y_1, y_2, \cdots, y_N)^T$ is the received signal, $\mathbf{x} = (x_1, x_2, \cdots, x_N)^T$ is the transmitted signal, $\mathbf{w} = (w_1, w_2, \cdots, w_N)^T$ is the AWGN with mean 0 and variance σ^2. $\mathbf{H}$ is a Toeplitz matrix containing the shifted two path channel coefficients and (i,j)-th element $[\mathbf{H}]_{ij}$ is expressed as:

$$[\mathbf{H}]_{ij} = \begin{cases} 1 + \gamma(1 - \Delta), & \text{for } j = i, \\ \gamma\Delta, & \text{for } j = i - 1, \\ 0, & \text{else}, \end{cases} \tag{3}$$

where $\Delta = \tau_d / T = (d_{\mathrm{ISI}} - d)/(cT)$ is the normalized delay, d is the LOS path transmission distance, d_{ISI} is the reflected path transmission distance, c is the speed of light and T is the symbol time interval. In the indoor VLC model, the channel DC gain ratio γ can be calculated as:

$$\gamma = h^{(2)} / h^{(1)} = \rho d^4 / d_{\mathrm{ISI}}^4, \tag{4}$$

where ρ is the walls reflectivity factor. The channel DC gain h of an optical link can be obtained as:

$$h = \begin{cases} \frac{(m+1)Sn^2}{2\pi d^2 \sin^2(\Psi_c)} \cos^m(\phi) T_s(\Psi) \cos(\Psi), & 0 \le \Psi \le \Psi_c, \\ 0, & \Psi > \Psi_c. \end{cases} \tag{5}$$

The order of Lambertian emission is $m = -\ln 2 / \ln(\cos \Phi_{1/2})$, ϕ is the angle of irradiance, Ψ is the angle of incidence and other parameters are introduced in Table 1.

Table 1. VLC system parameter.

Parameter	Value
Room Dimension (Length × Height)	3 m × 3 m
LED position	(1.5 m, 3 m)
LED beam width $\Phi_{1/2}$	60°
PD detector area S	0.1 m × 0.1 m
PD field of view (FOV) Ψ_c	90°
Refractive index of a lens at a PD n	1.5
Optical filter gain $T_s(\Psi)$	1
Walls reflectivity factor ρ	0.53
Symbol time interval T	10^{-8} s

3. Autoencoder Model

Here, we propose the NN framework in Figure 2 to solve the transceiver design issues in the above section. For the flexibility of the processing sequence, we leverage Conv1D layers in the AE model. To compensate for the varying ISI, the pilot sequence is creatively incorporated into the NN-based transceiver.

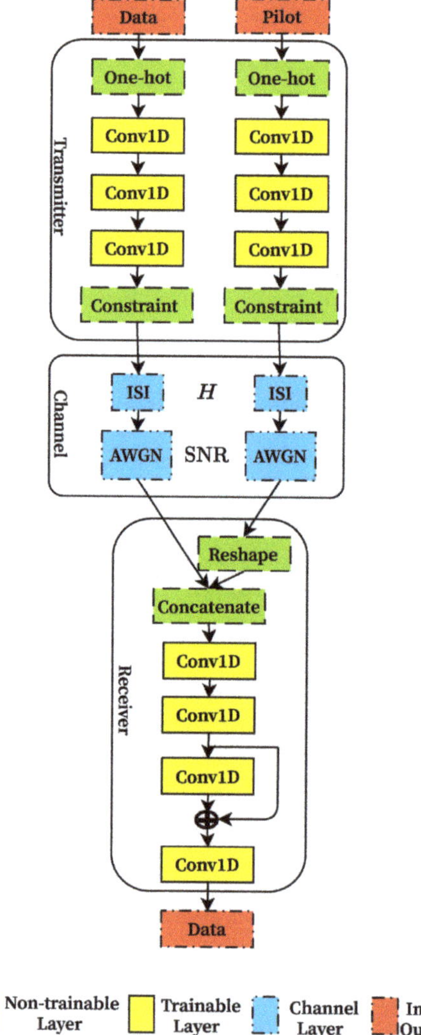

Figure 2. NN-based transceiver framework for VLC system.

In the transmitter part, data and pilot, which represent random and fixed value sequences, respectively, are input into the corresponding transmitter NNs. All the data sequences $\mathbf{s}_d \in \{1, \cdots, M\}^{N_d}$ and pilot sequences $\mathbf{s}_p = \mathbf{1}_{N_p \times 1}$ are mapped into one-hot vectors, whose index value is 1 and other values are all 0. The Conv1D layer enables the input vectors to be convolved and added by trainable convolution kernels and biases, respectively, which offer efficient mutual operation between the kernel-size-dependent nearby independent symbols instead of redundant distant symbols in a fully connected structure. Detailed parameters of all Conv1D layers, including filters number, kernel size and activation function, are provided in Table 2. The last Conv1D layers in the pilot transmitter, data transmitter and receiver employ a 'valid' padding scheme, while other Conv1D layers utilize the 'same' scheme. The strides are configured as 1. The rectified linear unit (ReLU) activation function used in majority layers produces nonlinearity and superior convergence performance, while the linear function employed in the layers before constraint layers guarantees that the signal space will not be adjusted. For satisfying the non-negative and peak power constraint of the sent light signal, that is, $x_i \in [0, A], i = 1, \cdots, N$, a weighted sigmoid activation function is utilized in the constraint layer, that is, $A \times \text{sigmoid}(\cdot)$, where A is the peak power constraint.

Table 2. Structrue of Autoencoder.

Part	Layer	Filters	Kernel Size	Activation
Pilot transmitter	Conv1D	N_p	l_p	ReLU
	Conv1D	M	l_p	ReLU
	Conv1D	1	1	Linear
	Constraint	-	-	Sigmoid
Data transmitter	Conv1D	N_d	l_p	ReLU
	Conv1D	M	l_p	ReLU
	Conv1D	1	1	Linear
	Constraint	-	-	Sigmoid
Receiver	Conv1D	N_d	l_p	ReLU
	Conv1D	M	l_p	ReLU
	Conv1D	M	l_p	ReLU
	Conv1D	M	1	Softmax

In the channel layer, both data and pilot sequences are firstly multiplied by an ISI Toeplitz matrix, whose row vectors are the shifted multipath channel vectors. Then, the noise generated from a standard normal distribution with a fixed variance $\sigma^2 = 1/\text{SNR}$, where SNR is the signal-to-noise ratio, is added into the distorted sequences.

At the receiver part, if the noisy pilot sequence is only concatenated with the data sequence, it is notoriously hard for the receiver NN to treat pilot and data distinctively based on the provided modest NN scale. Therefore, for the sake of addressing the influence of the pilot, the noisy pilot sequence $\mathbf{y}_p$ is reshaped into pilot matrix $\mathbf{Y}_p = \mathbf{1}_{N_d \times 1} {\mathbf{y}_p}^T \in \mathbb{R}^{N_d \times N_p}$, and then data sequence $\mathbf{y}_d$ is concatenated together into the matrix $[\mathbf{y}_d, \mathbf{Y}_p] \in \mathbb{R}^{N_d \times (1+N_p)}$. An intuitive explanation of these operations is illustrated in Figure 3. In the leftmost part of Figure 3, the three boxes represent the pilot sequence while the next six bold boxes represent the data sequence. With the proposed concatenation method, in the following convolution operations, all the pilot symbols can influence the detection of the data symbol by implicit joint channel estimation and equalization. The concrete performances depend on the eventually learnable NN parameters. To enhance the capability of the NN receiver and accelerate its convergence, we leverage the residual network structure, which means the inputs and outputs of specific layers are added together. We use the softmax activation function in the last layer to transform the input values into a probability vector over all possible messages. The loss function of the training process is the categorical cross-entropy of a data sequence, which is given as:

$$L_{loss} = -\sum_{m=1}^{M} \mathbf{u}_m \log \hat{\mathbf{u}}_m, \tag{6}$$

where $\mathbf{u}_m$ represents the mth index value of one-hot vector $\mathbf{u}$ and $\hat{\mathbf{u}}_m$ is the corresponding estimated value.

The NN structure and parameters are conceived empirically. Extensive hyper-parameters searching, which might enhance the eventual performance, is not under consideration for brevity. The NN is trained using the back-propagation algorithm. The data and corresponding label are the same randomly generated signal sequence in the AE unsupervised learning strategy. Once the training process of NN is completed, the transmitter part can send the data and pilot sequence using the NN or simplified lookup table. Based on the perfectly synchronized pilot and data, the receiver can recover the data sequence straightforwardly without estimating the CSI explicitly.

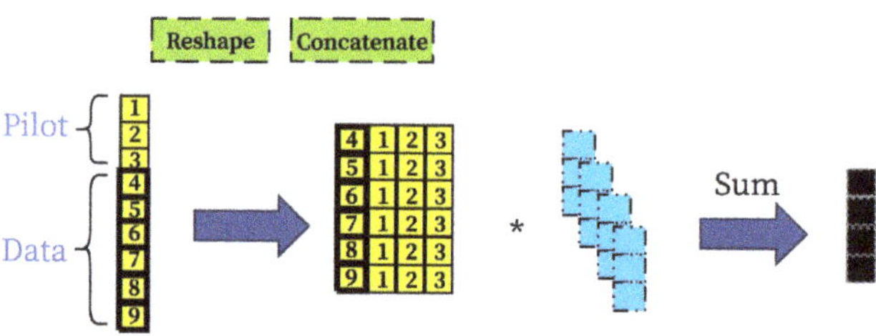

Figure 3. Visualization explanation of 'Reshape' and 'Concatenate' layers.

4. Simulation Results

Based on the channel model in Figure 1, we simulate two typical PD positions (around the corner/in the middle of the room) and the average performance to verify the effectiveness of our proposed method. The basic parameter configuration for the VLC system is given in Table 1. The PD's positions and corresponding parameters are presented in Table 3.

Table 3. PD positions and corresponding parameters.

PD Position (m)	d (m)	d_{ISI} (m)	γ	Δ
(0.1, 0)	3.31	3.4	0.48	0.03
(1.5, 0)	3	4.24	0.13	0.41

We set the sequence length of data and pilot as $N_d = 100$ and $N_p \in \{3, 10\}$, respectively. The tested constellation set cardinality $M \in \{2, 4\}$. The SNR is defined as $1/\sigma^2$ here and peak power as $A = 1$. The multipath channel coefficients are generated randomly considering the PD uniformly appears in the x axis, that is, the PD's coordinate $(x_{\text{PD}}, y_{\text{PD}})$ satisfies $x_{\text{PD}} \sim \mathcal{U}[0, 3m]$, $y_{\text{PD}} = 0$.

In our baseline method, we use M-PAM for data and pilot sequences with length $N_d = 100$ and pilot value $\mathbf{x}_p \in \{[0, 1, 0]^T, [1, 1, 1]^T\}$ at the transmitter. As for the receiver, minimum mean square error (MMSE) channel estimation result or PCSI is provided, and then maximum likelihood sequence estimation (MLSE), using the Viterbi algorithm [23], is utilized.

In our simulation, we use TensorFlow 2.0 and Python 3.6. During the training process, 100,000 samples are employed for 50 epochs. For every 10 epochs, the progressively increasing batch size from set $\{64, 128, 256, 512, 1024\}$ is employed. The Adam optimizer is used and the learning rate decreases with the 'loss' monitor factor 0.1, the patience 2, the initial learning rate 0.001 and the minimum learning rate 0.00001. For training effectiveness, the early-stopping strategy is applied with the 'loss' monitor and the patience 5.

In Figure 4, we consider the case $M = 2$. 'AE NOCSI' means that no pilot or CSI is input into the receiver NN. The training SNR is given in the caption of Figure 4 (In the 'NOCSI' case, if the training SNR is too low or the same as using the pilot case, the final SER performance will converge to a constant value in the high SNR domain; therefore, we configure the training SNR slightly higher than using the pilot case). Once the NN converges steadily, the learned data constellation $\mathcal{S} = \{0, 1\}$, the same as the 2-PAM scheme, for all the symbols in the sequence and the pilot sequence is $\mathbf{1}_{N_p}$. It can be seen from Figure 4a that when the PD locates around the corner of the room, the SER performance of AE schemes using pilot or not is inferior to MLSE schemes with excellent CSI conditions. However, Figure 4b demonstrates that AE schemes with a few pilot symbols can approach the optimal MLSE with PCSI. The average SER performance in Figure 4c further clarifies that in the majority positions in a room, AE can learn efficient transceiver and baseline methods, and even without CSI, the receiver can still compensate for the detrimental effect of ISI and can finally recover the sequence despite the limited decline in the high SNR domain.

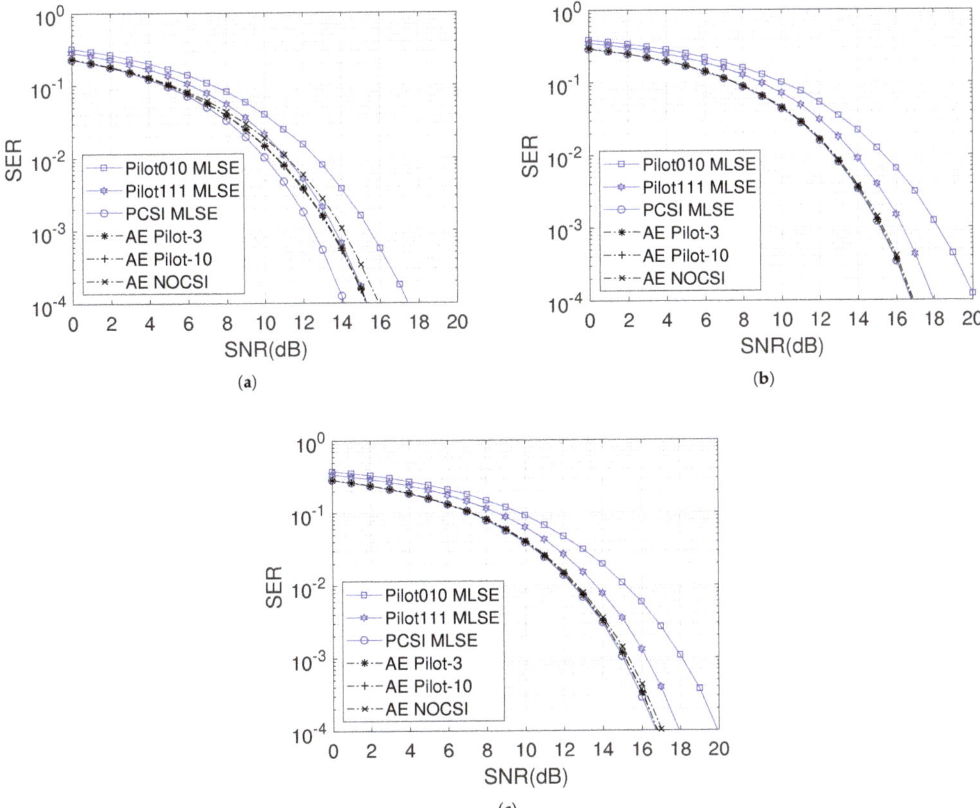

Figure 4. The SER performance of input symbol constellation cardinality $M = 2$. The training SNR = 16 dB for 'NOCSI' case and SNR = 13 dB for using pilot case. (**a**) PD position $(0.1, 0)$ result. (**b**) PD position $(1.5, 0)$ result. (**c**) Average result of PD position following uniform distribution.

The AE results are slightly distinctive when $M = 4$. Similar to the $M = 2$ case, the learned pilot sequence is $\mathbf{1}_{N_p}$. Nonetheless, the learned data constellation sets illustrated in Table 4 indicate that the constellation sets of schemes (Almost all the learned symbols in the sequence are mapped into the same constellation set. Only the last two or three symbols are mapped into different sets, but the values are still similar. Here, we only focus on the majority cases) using pilot converge to the equal-interval 4-PAM but the interval is irregular without CSI, which agrees with the PCSI and noisy CSI case in [20]. It can be seen from Figure 5 that the SER performance of 'AE NOCSI' significantly degrades compared with AE using pilot. Thus, the importance of CSI is obvious for above 2 level modulation. With the aid of the pilot sequence, the SER curve can approach the results of MLSE with PCSI in Figure 5b,c, which shows that the joint design of the transceiver using AE can reduce the power consumption with fewer pilot symbols to meet the desired SER performance. However, the performance is unfavorable in Figure 5a, similar to in Figure 4a, especially in the high SNR case, where the ISI occupies a more important position than noise. To handle these issues, one feasible strategy is to enlarge the pilot symbol number and leverage a deeper NN structure, which deserves a delicate experimental validation in our future work.

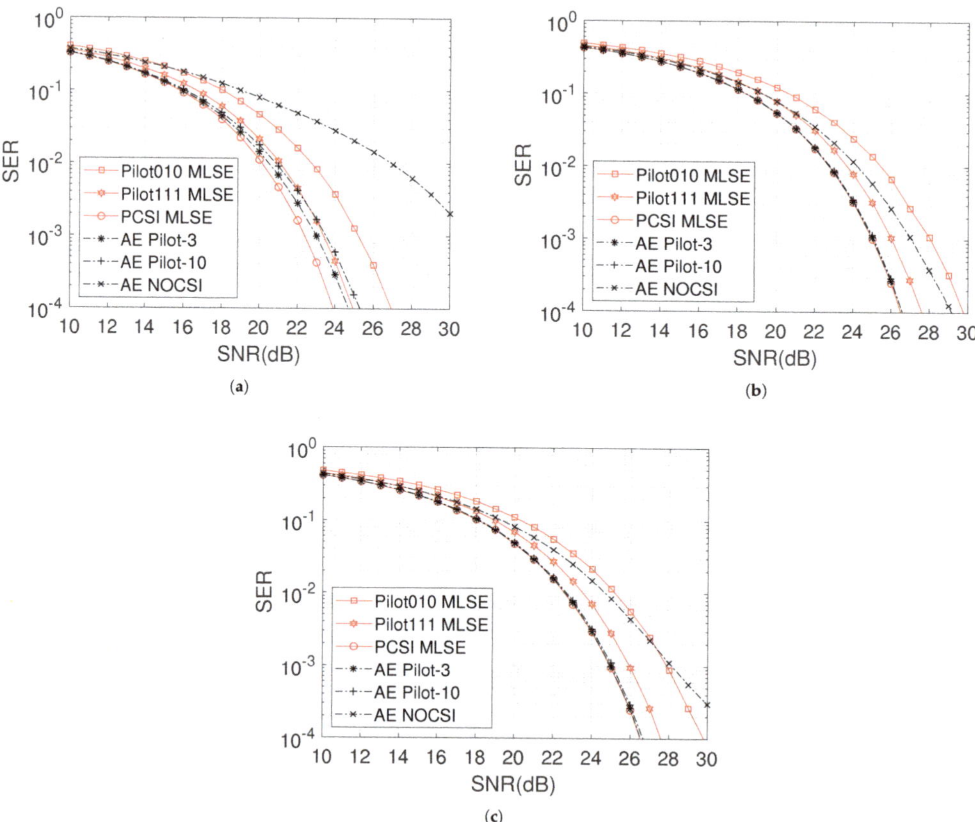

Figure 5. The SER performance of input symbol constellation cardinality $M = 4$. The training SNR = 26 dB for 'NOCSI' case and SNR = 23 dB for using pilot case. (**a**) PD position $(0.1, 0)$ result. (**b**) PD position $(1.5, 0)$ result. (**c**) Average result of PD position following uniform distribution.

Table 4. Learned constellation and baseline constellation for $M = 4$.

Case	Constellation SET
4-PAM	[0, 0.33, 0.67, 1]
AE NOCSI	[0, 0.25, 0.56, 1]
AE Pilot-3	[0, 0.32, 0.65, 1]
AE Pilot-10	[0, 0.32, 0.66, 1]

5. Conclusions and Future Work

In this letter, we propose an NN-based transceiver design scheme to compensate for the varying ISI effect in the VLC system. By leveraging the AE and CNN structure, the data and pilot sequence constellation mapping strategy at the transmitter, and the direct sequence estimator at the receiver can simultaneously be obtained. Simulation results demonstrate that the proposed transceiver can outperform the model-based method especially when imperfect CSI is available with limited pilot symbols. The NN-based transceiver design paradigm paves a new way for a bandwidth-efficient VLC system as the high-speed information transmission requirement increased.

Three potential future works, which extend the application of the NN-based transceiver for VLC systems, are discussed below. First, the channel factors adopted in the validation process are relatively simple; thus, more complicated issues, such as hardware imperfection and interference by environmental light, are suitable for sufficiently utilizing the NN's capability. Second, prevalent OFDM schemes are extensively applied for high-speed VLC systems. An NN-based OFDM system desiring less pilot consumption is attractive. Eventually, the powerful Transformer [24] structure witnesses substantial breakthroughs in almost all deep learning domains. A substitute with Transformer blocks for CNN layers might lead to unexpected performance.

Author Contributions: Conceptualization, L.L.; methodology, Z.Z.; software, Z.Z.; validation, L.L.; formal analysis, L.L.; investigation, L.L.; resources, L.L.; data curation, Z.Z.; writing—original draft preparation, L.L. and Z.Z.; writing—review and editing, J.Z.; visualization, Z.Z.; supervision, J.Z.; project administration, J.Z.; funding acquisition, J.Z. All authors have read and agreed to the published version of the manuscript.

Funding: This research was funded by the National Natural Science Foundation of China (NSFC) under Grant (62071489, 61901524).

Institutional Review Board Statement: Not applicable.

Informed Consent Statement: Not applicable.

Data Availability Statement: Not applicable.

Conflicts of Interest: The authors declare no conflict of interest.

References

1. Karunatilaka, D.; Zafar, F.; Kalavally, V.; Parthiban, R. LED based indoor visible light communications: State of the art. *IEEE Commun. Surv. Tutor.* **2015**, 17, 1649–1678. [CrossRef]
2. Guo, J.N.; Zhang, J.; Xin, G.; Li, L. Constant transmission efficiency dimming control scheme for vlc systems. *Photonics* **2021**, 8, 7. [CrossRef]
3. Chen, C.; Zhong, W. De; Yang, H.; Du, P.; Yang, Y. Flexible-Rate SIC-Free NOMA for Downlink VLC Based on Constellation Partitioning Coding. *IEEE Wirel. Commun. Lett.* **2019**, *8*, 568–571. [CrossRef]
4. Chen, C.; Zhong, W.-D.; Wu, D. Indoor OFDM visible light communications employing adaptive digital pre-frequency domain equalization. In Proceedings of the 2016 Conference on Lasers and Electro-Optics (CLEO), San Jose, CA, USA, 5–10 June 2016._at.2016.jth2a.118. [CrossRef]
5. Lecun, Y.; Bengio, Y.; Hinton, G. Deep learning. *Nature* **2015**, *521*, 436–444. [CrossRef] [PubMed]
6. Gunduz, D.; De Kerret, P.; Sidiropoulos, N.D.; Gesbert, D.; Murthy, C.R.; Van Der Schaar, M. Machine Learning in the Air. *IEEE J. Sel. Areas Commun.* **2019**, *37*, 2184–2199. [CrossRef]
7. Xu, J.; Zhu, P.; Li, J.; You, X. Deep Learning-Based Pilot Design for Multi-User Distributed Massive MIMO Systems. *IEEE Wirel. Commun. Lett.* **2019**, *8*, 1016–1019. [CrossRef]
8. Ye, H.; Li, G.Y.; Juang, B.H. Power of Deep Learning for Channel Estimation and Signal Detection in OFDM Systems. *IEEE Wirel. Commun. Lett.* **2018**, *7*, 114–117. [CrossRef]
9. O'Shea, T.; Hoydis, J. An Introduction to Deep Learning for the Physical Layer. *IEEE Trans. Cogn. Commun. Netw.* **2017**, *3*, 563–575. [CrossRef]
10. Dorner, S.; Cammerer, S.; Hoydis, J.; Brink, S. Ten Deep Learning Based Communication over the Air. *IEEE J. Sel. Top. Signal Process.* **2018**, *12*, 132–143. [CrossRef]
11. Aoudia, F.A.; Hoydis, J. Model-Free Training of End-to-End Communication Systems. *IEEE J. Sel. Areas Commun.* **2019**, *37*, 2503–2516. [CrossRef]
12. Zhu, B.; Wang, J.; He, L.; Song, J. Joint Transceiver Optimization for Wireless Communication PHY Using Neural Network. *IEEE J. Sel. Areas Commun.* **2019**, *37*, 1364–1373. [CrossRef]
13. Soltani, M.; Fatnassi, W.; Aboutaleb, A.; Rezki, Z.; Bhuyan, A.; Titus, P. Autoencoder-Based Optical Wireless Communications Systems. In Proceedings of the 2018 IEEE Globecom Work, Abu Dhabi, United Arab Emirates, 9–13 December 2018. [CrossRef]

14. Lee, H.; Lee, S.H.; Quek, T.Q.S.; Lee, I. Deep Learning Framework for Wireless Systems: Applications to Optical Wireless Communications. *IEEE Commun. Mag.* **2019**, *57*, 35–41. [CrossRef]
15. Zhang, D.F.; Yu, H.Y.; Zhu, Y.J.; Zhu, Z.R. A Transceiver Design Based on an Autoencoder Network for Multi-Color VLC Systems. *IEEE Photonics J.* **2020**, *12*, 7902716. [CrossRef]
16. Si-Ma, L.H.; Zhu, Z.R.; Yu, H.Y. Model-Aware End-to-End Learning for SISO Optical Wireless Communication over Poisson Channel. *IEEE Photonics J.* **2020**, *12*, 7907115. [CrossRef]
17. Miao, P.; Zhu, B.; Qi, C.; Jin, Y.; Lin, C. A Model-Driven Deep Learning Method for LED Nonlinearity Mitigation in OFDM-Based Optical Communications. *IEEE Access* **2019**, *7*, 71436–71446. [CrossRef]
18. Lee, H.; Quek, T.Q.S.; Lee, S.H. A Deep Learning Approach to Universal Binary Visible Light Communication Transceiver. *IEEE Trans. Wirel. Commun.* **2020**, *19*, 956–969. [CrossRef]
19. Ulkar, M.G.; Baykas, T.; Pusane, A.E. VLCnet: Deep learning based end-to-end visible light communication system. *J. Light. Technol.* **2020**, *38*, 5937–5948. [CrossRef]
20. Zhu, Z.R.; Zhang, J.; Chen, R.H.; Yu, H.Y. Autoencoder-based transceiver design for OWC systems in log-normal fading channel. *IEEE Photonics J.* **2019**, *11*, 7905912. [CrossRef]
21. Komine, T.; Nakagawa, M. Fundamental analysis for visible-light communication system using LED lights. *IEEE Trans. Consum. Electron.* **2004**, *50*, 100–107. [CrossRef]
22. Wang, H.; Kim, S. Decoding of polar codes for intersymbol interference in visible-light communication. *IEEE Photonics Technol. Lett.* **2018**, *30*, 1111–1114. [CrossRef]
23. Omura, J.K. On the Viterbi Decoding Algorithm. *IEEE Trans. Inf. Theory* **1969**, *15*, 177–179. [CrossRef]
24. Vaswani, A.; Shazeer, N.; Parmar, N.; Uszkoreit, J.; Jones, L.; Gomez, A.N.; Kaiser, Ł.; Polosukhin, I. Attention is all you need. *Adv. Neural Inf. Process. Syst.* **2017**, *2017*, 5999–6009.

Article

Demonstration of Performance Improvement in Multi-User NOMA VLC System Using Joint Transceiver Optimization

Tong Wu [1], Zixiong Wang [2,*], Shiying Han [3], Jinlong Yu [2] and Yang Jiang [4]

1 Tianjin International Engineering Institute, Tianjin University, Tianjin 300072, China; wutong1996@tju.edu.cn
2 School of Electrical and Information Engineering, Tianjin University, Tianjin 300072, China; yujinlong@tju.edu.cn
3 College of Electronic Information and Optical Engineering, Nankai University, Tianjin 300350, China; syhan@nankai.edu.cn
4 College of Physics, Guizhou University, Guiyang 550025, China; jiangy@gzu.edu.cn
* Correspondence: wangzixiong@tju.edu.cn

Abstract: The bit error ratio (BER) performance of a non-orthogonal multiple access (NOMA) visible light communication (VLC) system is poor due to the unequal distances between adjacent points in the superposition constellation (SC). In this paper, we propose a novel scheme to improve the BER performance by adjusting parameters to change the shape of SC at the transmitter and by adjusting the parameters of successive interference cancellation (SIC) decoding at the receiver simultaneously, which is called a SC and SIC adjustment (SC-SIC-A) scheme. For multi-user NOMA VLC system, we derive the closed-form BER expression for each user, where the modulation format is four-quadrature amplitude modulation. According to the derived BER expressions, we formulate an optimization problem that minimizes the average BER for all users by adjusting the obtained parameters of SC and SIC decoding via differential evolution algorithm. The improvement of capacity performance is investigated consequently. In order to verify the feasibility and effectiveness of the proposed SC-SIC-A scheme, we carried out theoretical analysis, Monte Carlo simulation and experiments of two-user and three-user NOMA VLC systems. Results show that the SC-SIC-A scheme outperforms the existing schemes in NOMA VLC system, where the signal-to-noise ratio (SNR) reductions to achieve BER of 10^{-3} are 1.3 dB and 0.8 dB for both users in the two-user NOMA VLC system, respectively, and the SNR reductions to achieve BER of 10^{-3} are 5.7 dB, 4.3 dB and 4.6 dB for all users in the three-user NOMA VLC system, respectively.

Keywords: non-orthogonal multiple access (NOMA); visible light communication (VLC); superposition constellation adjustment; successive interference cancellation; bit error ratio; NOMA triangle

Citation: Wu, T.; Wang, Z.; Han, S.; Yu, J.; Jiang, Y. Demonstration of Performance Improvement in Multi-User NOMA VLC System Using Joint Transceiver Optimization. *Photonics* **2022**, *9*, 168. https://doi.org/10.3390/photonics9030168

Received: 6 January 2022
Accepted: 6 March 2022
Published: 9 March 2022

Publisher's Note: MDPI stays neutral with regard to jurisdictional claims in published maps and institutional affiliations.

1. Introduction

As fifth generation (5G) mobile communication and the Internet of things emerge, a large number of data-intensive applications have been born, resulting in the exponential growth of data traffic [1,2]. The conventional radio frequency (RF) spectrum cannot meet the requirement of high data transmission due to its limited spectrum resource [3]. As a good solution for satisfying the spectrum requirement, visible light communication (VLC) has attracted considerable attention for the wireless network due to its abundant unlicensed spectrum, immunity to electromagnetic interference, low energy consumption and effective frequency and spatial reuse [4–6]. VLC is carried out by using the illumination infrastructure with light-emitting diode (LED), where illumination and communication are implemented simultaneously. With the popularity of LED as a light source in indoor and outdoor environments, VLC has become a green and energy-saving communication method [7]. However, the modulation bandwidth of LED is only tens of megahertz (MHz), which limits the capacity of a VLC system [8]. The capacity of VLC system can be increased by adopting advanced modulation formats, which is carried out under orthogonal frequency division multiplexing (OFDM) [9,10]. It has been demonstrated that the spectral efficiency, fairness and capacity performance of multi-user OFDM VLC systems could be further improved by employing non-orthogonal multiple access (NOMA). Meanwhile, NOMA could significantly improve the quality and reliability of reception in OFDM VLC systems due to different power allocation among all users [11,12].

NOMA is an attractive and efficient access method for wireless network, which allows multiple terminals to use the same resources, such as time and frequency [13]. NOMA has the advantages of high spectral efficiency, high flexibility, low transmission latency and improved fairness [14–16]. Superposition coding and successive interference cancellation (SIC) decoding are two key techniques in the NOMA scheme. The performance analysis and optimization of NOMA VLC system have been widely studied. The theoretical bit error analysis of a downlink NOMA VLC system based on a high-order modulation scheme is investigated in [17]. In [18], the bit error ratio (BER) performance of a two-user NOMA VLC system using different modulation formats was investigated. In order to improve the system's performance, the method of adjusting superposition constellation (SC) has been proposed in the literature. The poor SC due to different arrival times is compensated by phase predistortion method [19]. Moreover, SC can be adjusted based on convex optimization [20]. The above works are the research of the two-user NOMA VLC system. To improve system capacity and spectral efficiency, NOMA technology should be used to serve more users in one resource block. However, with the increase in the number of users, the interference among users becomes severe, resulting in the deterioration of system performance. Thus, the performance of NOMA VLC system supporting three or more users has been investigated in [21,22]. The BER performance under noisy channel state information is researched in [21], while in [22], the capacity region of a practical uplink NOMA for multiple users is investigated. Nevertheless, for multiuser NOMA VLC system using 4-quadrature amplitude modulation (QAM), the closed-form BER expressions for all users as well as the improvement of BER and capacity performance have not been studied.

In this paper, we propose a joint transceiver optimization scheme to improve the BER and capacity performance of a multi-user NOMA VLC system using 4-QAM-OFDM. Since the distances between adjacent points in the SC are quite different, the BER performance of multi-user NOMA VLC system is poor. Therefore, adjustment parameters are introduced to change the shape of SC at the transmitter [20]. This scheme is called the superposition constellation adjustment (SC-A) scheme. We notice that, under the SC-A scheme, the power allocation coefficients are no longer optimal for SIC decoding at the receiver. Thus, the joint transceiver optimization scheme was proposed, which is called the SC and SIC adjustment (SC-SIC-A) scheme. In this scheme, we adjust the parameters of the SC at the transmitter and that of the SIC decoding at the receiver to improve the performance of the downlink multi-user NOMA VLC system without increasing the overall transmitted power of signal. More specifically, we derive the closed-form BER expression of each user in detail under the SC-SIC-A scheme. Meanwhile, we find the rule for the number of terms in each user's BER expression, which is named as NOMA triangle regarding the terms of Q-function. According to the derived BER expressions, an optimization problem is formulated to minimize the average BER for all users by adjusting the parameters of SC at the transmitter and that of SIC decoding at the receiver. The optimization problem is solved by the differential evolution (DE) algorithm to obtain the optimal adjustment parameters. In addition to BER performance, the expression of channel capacity for each user under the proposed SC-SIC-A scheme is also derived. When the optimal adjustment parameters are applied to the SC-SIC-A scheme, capacity performance is also improved. In order to demonstrate the effectiveness and feasibility of the proposed SC-SIC-A scheme, we carry out theoretical analysis, Monte-Carlo (MC) simulation, and experiment, where the results match very well. The results show that compared with the SC-A scheme, the SC-SIC-A scheme provides 1.3-dB and 0.8-dB signal-to-noise ratio (SNR) reductions to achieve BER of

10^{-3} for two-user NOMA VLC system, respectively, and provides 5.7-dB, 4.3-dB and 4.6-dB SNR reductions to achieve the BER of 10^{-3} for three-user NOMA VLC system, respectively.

This paper extends its conference version [23] from the following aspects. Firstly, we derive the closed-form BER expression for each user in the multi-user NOMA VLC system under the SC-SIC-A scheme. Secondly, the capacity expression for each user under the proposed scheme is derived. Thirdly, the BER and capacity performance of NOMA VLC system are analyzed when the number of users is two and three. Finally, the indoor NOMA VLC experiment is carried out in this paper to demonstrate the feasibility of the proposed scheme.

The remainder of the paper is organized as follows. Section 2 describes the scheme of SC-SIC-A for multi-user NOMA. The performance analysis and optimization by using joint SC and SIC decoding adjustment are presented in Section 3. Section 4 shows the results and discussions, and the conclusion is finally provided in Section 5.

2. System Model

The schematic of multi-user NOMA VLC system using joint transceiver optimization is shown in Figure 1. We assume that there are K users in the NOMA VLC system. LED transmits signals to the K users simultaneously. The data of each user contain $\log_2 4$ bits, which are mapped to the constellation point $S_k \in \mathcal{S}_k$ by using 4-QAM. K signals are superposed, where the power allocation coefficients for K users are denoted by $\alpha_1, \alpha_2, \ldots, \alpha_K$. Note that $\sum_{k=1}^{K} \alpha_k = 1$. Under the K-user NOMA scheme, users with lower channel gain will be allocated higher power to ensure fairness. In this paper, we assume that channel gains are identical, i.e., $|h_1|^2 = |h_2|^2 = \cdots = |h_K|^2$, and the power allocation coefficients follow $\alpha_1 \geq \alpha_2 \geq \cdots \geq \alpha_K$. Thus, the original superposed NOMA signal can be written as follows.

$$S = \sum_{k=1}^{K} \sqrt{\alpha_k} S_k. \tag{1}$$

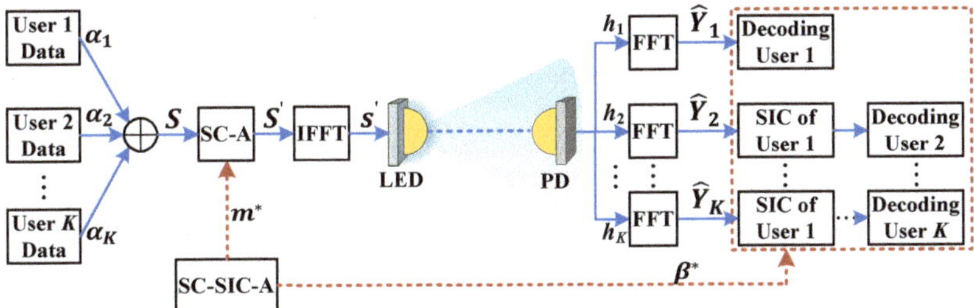

Figure 1. The schematic of the downlink multi-user NOMA VLC system using joint transceiver optimization.

Regarding the constellation of the superposed signal S, the distances between the adjacent points are quite different from each other. We take the SC-A scheme to adjust SC and, therefore, to improve BER performance. The illustration of the SC-A process is shown in Figure 2, where the first and second quadrants of SC are taken into consideration. Since the SC is symmetric, we need to use 2^{K-1} scale parameters to adjust the SC for K-user NOMA VLC system. The SC scale parameter vector $\boldsymbol{m} = [m_1, m_2, \ldots, m_{2^{K-1}}]^{\mathrm{T}}$ is identical for horizontal and vertical directions. Under the SC-A scheme, the original point (x_i, y_j) is adjusted to $\left(x_i', y_j'\right) = (m_i x_i, m_j y_j)$. Accordingly, the original superposed signal S is converted to S'. The total transmitted power of the signal is converted from P to P'.

$$P = \frac{1}{2^{K-1}} \left(\sum_{i=1}^{2^{K-1}} x_i^2 + \sum_{j=1}^{2^{K-1}} y_j^2 \right), \tag{2}$$

$$P' = \frac{1}{2^{K-1}}\left[\sum_{i=1}^{2^{K-1}} x_i'^2 + \sum_{j=1}^{2^{K-1}} y_j'^2\right] = \frac{1}{2^{K-1}}\left[\sum_{i=1}^{2^{K-1}} (m_i x_i)^2 + \sum_{j=1}^{2^{K-1}} (m_j y_j)^2\right]. \quad (3)$$

After taking inverse fast Fourier transform (IFFT), the frequency-domain signal S' is converted to the time-domain OFDM signal s', which drives the LED under direct current (DC) bias [24]. Thus, the modulated light x at the output of LED is as follows:

$$x = \eta(s' + s_{DC}), \quad (4)$$

where η is the current-to-light conversion efficiency of LED, and s_{DC} is the DC bias.

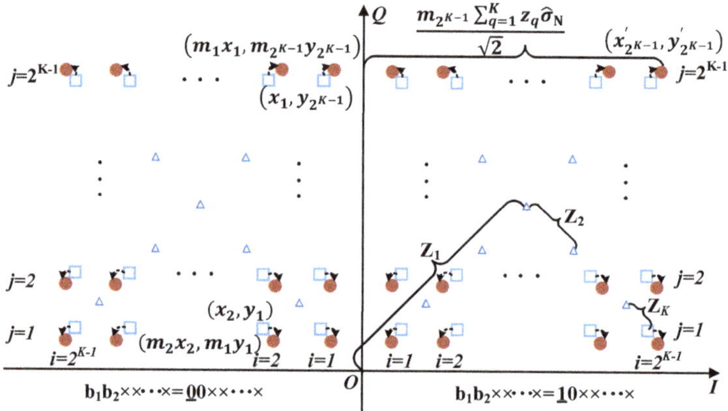

Figure 2. The demonstration of SC-A in the first and second quadrants. Triangle: the intermediate points of superposition signals in constellation before SC-A; square: the points of the SC before SC-A; circle: the points of the SC after SC-A.

The LED output light x propagates through an indoor channel. If only line-of-sight (LoS) propagation is considered, the channel gain h_k between the LED and the user k is given by the following [25].

$$h_k = \begin{cases} \frac{A(m+1)}{2\pi d_k^2}\cos^m(\phi_k)\cos(\psi_k)T_S(\psi_k)g(\psi_k), & |\psi_k| \le \psi_c \\ 0, & |\psi_k| > \psi_c \end{cases}. \quad (5)$$

In Equation (5), A is the physical area of photo-detector (PD), and $m = -\ln(2)/\ln(\cos\phi_{1/2})$ is the order of Lambertian emission, where $\phi_{1/2}$ is the semiangle at half power. d_k is the distance between the LED and user k, ϕ_k is the emission angle, ψ_k is the angle of incidence, and $T_S(\psi_k)$ is the gain of an optical filter. $g(\psi_k) = n_s^2/\sin^2(\psi_c)$ is the gain of an optical concentrator, where ψ_c is the concentrator field of view (FOV) semiangle and n_s denotes the refractive index.

The modulated light is collected by PD, which converts it into an electrical signal. After removing the DC component, the signal for user k is as follows:

$$y_{R,k} = \eta R h_k s' + n_k, \quad (6)$$

where R is the responsivity of PD, and n_k is the additive white Gaussian noise (AWGN) with a mean of zero and the variance of σ_N^2. $y_{R,k}$ can be normalized as follows:

$$\hat{y}_{R,k} = \frac{y_{R,k}}{\eta R h_k} = s' + \hat{n}_k, \quad (7)$$

where $\hat{n}_k = n_k/(\eta R h_k)$ is the normalized noise with the mean of zero and the variance of $\hat{\sigma}_{\mathrm{N}}^2 = \sigma_{\mathrm{N}}^2/(\eta R h_k)^2$.

The fast Fourier transform (FFT) module converts the normalized signal $\hat{y}_{R,k}$ to the frequency-domain signal $\hat{Y}_{R,k} = S' + \hat{N}_k$, where $\hat{N}_k$ is the normalized noise in frequency domain. $\hat{N}_k = \hat{N}'_{\mathrm{H}} + j\hat{N}'_{\mathrm{V}}$, where $\hat{N}'_{\mathrm{H}}$ and $\hat{N}'_{\mathrm{V}}$ are the real part and imaginary part of $\hat{N}_k$, respectively. After that, the SIC algorithm is used to decode each user's data from $\hat{Y}_{R,k}$. Since the SC-A scheme is used at the transmitter, the original power allocation coefficients $\alpha_1, \alpha_2, \ldots, \alpha_K$ are no longer optimal for SIC decoding. We should find the optimal parameters $\beta_1, \beta_2, \ldots, \beta_K$ to replace $\alpha_1, \alpha_2, \ldots, \alpha_K$ in the procedure of SIC decoding. The entire process that contains the parameter adjustment for SC at the transmitter and for SIC at the receiver is called an SC-SIC-A scheme.

3. Performance Analysis and Optimization under SC-SIC-A Scheme

In this section, we investigate the BER and capacity performance of the multi-user NOMA VLC system under SC-SIC-A scheme. We derive the BER expressions of all K users, where the rule for the number of terms in each user's BER expression is found. We formulate an optimization problem to minimize the average BER by jointly adjusting the parameters of SC and SIC decoding. Based on the same optimized parameters, the capacity performance of multi-user NOMA VLC systems under SC-SIC-A scheme is also studied.

3.1. Derivation of BER

We derive the BER expressions of all the K users under the SC-SIC-A scheme. The SNR of user k is defined as $\mathrm{SNR}_k = \gamma_k = \frac{\alpha_k P h_k^2}{\sigma_{\mathrm{N}}^2} = \frac{\alpha_k P}{\hat{\sigma}_{\mathrm{N}}^2}$. As depicted in Figure 2, each point of SC represents $2K$ bits, which can be denoted by $b_1, b_2, \ldots, b_{2K}$. $\{b_1, b_2\}, \{b_3, b_4\}, \ldots$ and $\{b_{2K-1}, b_{2K}\}$ are the information bits of 4-QAM signals for all the K users, respectively. The original scales of the K users' 4-QAM signals are denoted by $Z_1, Z_2, \ldots, Z_K$, respectively. We have $Z_k^2 = \alpha_k P = z_k^2 \hat{\sigma}_{\mathrm{N}}^2$, where $z_k = \sqrt{\frac{\alpha_k}{\alpha_1}\gamma_1}$, $k \in \{1, 2, \ldots, K\}$. Then, the BER expressions of all the K users can be derived as follows.

3.1.1. BER Expression of User 1

Firstly, we derive the BER expression of bits $\{b_1, b_2\}$ for user 1. Due to the symmetry of the adjusted SC, the BER of b_1 for user 1 in the first quadrant is the same as that in the rest three quadrants. Since the detection of b_2 for user 1 in the vertical direction follows the same principle as that of b_1 in the horizontal direction, we have $\mathrm{BER}_{(K,1)} = \frac{1}{2}(\mathrm{BER}_{1,b_1} + \mathrm{BER}_{1,b_2}) = \mathrm{BER}_{1,b_1}$. As shown in Figure 2, each quadrant of the SC can be divided into 2^{K-1} columns. The BER for the in-phase components of the points in each column are the same, while the BERs for the in-phase components in various columns are different from each other. Thus, the BER expression is as follows:

$$\mathrm{BER}_{(K,1)} = \frac{1}{2}(\mathrm{BER}_{1,b_1} + \mathrm{BER}_{1,b_2}) = \mathrm{BER}_{1,b_1} = \mathop{\mathbb{E}}_{i \in I}\left[\mathrm{BER}_{1,x'_i}\right], \tag{8}$$

where the set $I = \{1, 2, \ldots, 2^{K-1}\}$ represents the indices of columns in any quadrant as shown in Figure 2, $\mathbb{E}$ stands for the operation of expectation and BER_{1,x'_i} is the error probability of b_1 for user 1 at the adjusted SC point $\left(x'_i, y'_j\right)$. Therefore, the BER expression of user 1 in Equation (8) is derived as follows:

$$\mathrm{BER}_{(K,1)} = \frac{1}{2^{K-1}} \sum_{\ell_1=0}^{1} \sum_{\ell_2=0}^{1} \cdots \sum_{\ell_{K-1}=0}^{1} \left\{ Q\left[m_{(\sum_{a=1}^{K-1} 2^{a-1}\ell_a)+1} \left(z_1 + \sum_{q=2}^{K} (-1)^{\ell_{q-1}} z_q \right) \right] - Q(\infty) \right\}, \tag{9}$$

where the detailed derivation process of $\mathrm{BER}_{1,x'_{2^{K-1}}} = Q\left(m_{2^{K-1}} \sum_{q=1}^{K} z_q\right) - Q(\infty)$ is shown in Appendix A.1.

3.1.2. BER Expression of User 2

Secondly, we derive the BER expression of bits $\{b_3, b_4\}$ for user 2. The BER performance of user 2 is affected by the previous bits detection of user 1 under the SIC scheme. Erroneous detection for user 2's bits can be found by both the conditions that user 1's bits are detected correctly and incorrectly. Thus, based on the derivation principle of user 1's BER expression, we can obtain the BER expression of user 2 as follows:

$$\begin{aligned}\mathrm{BER}_{(K,2)} &= \frac{1}{2}(\mathrm{BER}_{2,b_3} + \mathrm{BER}_{2,b_4}) = \mathrm{BER}_{2,b_3} = \mathop{\mathbb{E}}_{i\in I}\left[\mathrm{BER}_{2,x_i'}\right] \\ &= \mathop{\mathbb{E}}_{i\in I}\left[\mathrm{BER}_{2,x_i'\cap \mathrm{U}_1 \text{ is correct}}\right] + \mathop{\mathbb{E}}_{i\in I}\left[\mathrm{BER}_{2,x_i'\cap \mathrm{U}_1 \text{ is wrong}}\right] \\ &= \mathop{\mathbb{E}}_{i\in I}\left[\sum_{c_1=0}^{1} \mathrm{BER}_{2,x_i'\cap \mathrm{U}_{1,c_1}}\right],\end{aligned} \tag{10}$$

where $\mathrm{BER}_{2,x_i'}$ is the error probability of b_3 for user 2 at the adjusted SC point $\left(x_i', y_j'\right)$, U_1 represents user 1 and c_1 stands for the detection result of user 1's bit. The index $c_1 = (0,1)$ represents the conditions that U_1's bits are detected correctly and wrongly, respectively. Thus, the BER expression of user 2 in Equation (10) can be written as follows:

$$\begin{aligned}\mathrm{BER}_{(K,2)} =& \frac{1}{2^{K-1}} \sum_{\ell_1=0}^{1}\sum_{\ell_2=0}^{1}\cdots\sum_{\ell_{K-1}=0}^{1}(-1)^{\ell_1}\left\{\sum_{r_{1,1}=0}^{1} Q\left[m_{\sum_{a=1}^{K-1}2^{a-1}\ell_a+1}\left(z_1+\sum_{q=2}^{K}(-1)^{\ell_{q-1}}z_q\right)\right.\right. \\ &\left.+(-1)^{r_{1,1}}z_1'\right] - Q\left[m_{\sum_{a=1}^{K-1}2^{a-1}\ell_a+1}\left(z_1+\sum_{q=2}^{K}(-1)^{\ell_{q-1}}z_q\right)\right] - Q\left[(-1)^{\ell_{k-1}}\infty\right]\Bigg\},\end{aligned} \tag{11}$$

where the detailed derivation processes of $\mathrm{BER}_{2,x_{2K-1}'\cap \mathrm{U}_{1,1}} = Q\left(m_{2^{K-1}}\sum_{q=1}^{K} z_q + z_1'\right) - Q(\infty)$ and $\mathrm{BER}_{2,x_{2K-1}'\cap \mathrm{U}_{1,0}} = Q\left(m_{2^{K-1}}\sum_{q=1}^{K} z_q - z_1'\right) - Q\left(m_{2^{K-1}}\sum_{q=1}^{K} z_q\right)$ are shown in Appendix A.2.

3.1.3. BER Expression of User k

Finally, we derive the BER expression of bits $\{b_{2k-1}, b_{2k}\}$ for user k. The BER of user k is affected by the previous bits detection of $k-1$ users. Thus, there are 2^{k-1} conditions for the detection of user k. Thus, based on the derivation principles of b_1 for user 1 and b_3 for user 2 at the adjusted SC point (x_{2K-1}', y_{2K-1}'), the BER of user $k(\geq 2)$ is as follows:

$$\begin{aligned}\mathrm{BER}_{(K,k)} =& \frac{1}{2}(\mathrm{BER}_{k,b_{2k-1}} + \mathrm{BER}_{k,b_{2k}}) = \mathrm{BER}_{k,b_{2k-1}} \\ =& \mathop{\mathbb{E}}_{i\in I}\left[\sum_{c_1=0}^{1}\sum_{c_2=0}^{1}\cdots\sum_{c_{k-1}=0}^{1}\mathrm{BER}_{k,x_i'\cap \mathrm{U}_{1,c_1}\cap \mathrm{U}_{2,c_2}\cdots\cap \mathrm{U}_{k-1,c_{k-1}}}\right] \\ =& \frac{1}{2^{K-1}}\sum_{\ell_1=0}^{1}\sum_{\ell_2=0}^{1}\cdots\sum_{\ell_{K-1}=0}^{1}(-1)^{\ell_{k-1}}\left\{\sum_{r_{1,1}=0}^{1}\sum_{r_{1,2}=0}^{1}\cdots\sum_{r_{1,k-1}=0}^{1} Q\left[m_{\sum_{a=1}^{K-1}2^{a-1}\ell_a+1}\right.\right. \\ &\left.\left(z_1+\sum_{q=2}^{K}(-1)^{\ell_{q-1}}z_q\right)+\sum_{t=1}^{k-1}(-1)^{r_{1,t}}z_t'\right] \\ &-\sum_{r_{2,1}=0}^{1}\sum_{r_{2,2}=0}^{1}\cdots\sum_{r_{2,k-2}=0}^{1} Q\left[m_{\sum_{a=1}^{K-1}2^{a-1}\ell_a+1}\left(z_1+\sum_{q=2}^{K}(-1)^{\ell_{q-1}}z_q\right)+\sum_{t=1}^{k-2}(-1)^{r_{2,t}}z_t'\right] \\ &\qquad\vdots \\ &-\sum_{r_{k-1,1}=0}^{1} Q\left[m_{(\sum_{a=1}^{K-1}2^{a-1}\ell_a)+1}\left(z_1+\sum_{q=2}^{K}(-1)^{\ell_{q-1}}z_q\right)+(-1)^{r_{k-1,1}}z_1'\right] \\ &\left.-Q\left[m_{\sum_{a=1}^{K-1}2^{a-1}\ell_a+1}\left(z_1+\sum_{q=2}^{K}(-1)^{\ell_{q-1}}z_q\right)\right] - Q\left[(-1)^{\ell_{k-1}}\infty\right]\right\},\end{aligned} \tag{12}$$

where $z'_t = \sqrt{\frac{\beta_t \gamma_1}{\alpha_1}}$, $t \in \{1, 2, \ldots, k-1\}$. $c_\ell(\ell = 1, 2, \ldots, k-1)$ stands for the previous detection results of $k-1$ users' bits, which represent the conditions that user ℓ's bit is detected correctly $(c_\ell = 0)$ or incorrectly $(c_\ell = 1)$. Let $\boldsymbol{\alpha} = [\alpha_1, \ldots, \alpha_K]^{\mathrm{T}}$ be the power allocation coefficient vector and $\boldsymbol{\beta} = [\beta_1, \ldots, \beta_K]^{\mathrm{T}}$ be the SIC decoding parameter vector. When $\boldsymbol{\alpha} = \boldsymbol{\beta}$, Equations (9) and (12) reduce to the BER expressions for K-user NOMA VLC system under the SC-A scheme.When $\boldsymbol{\alpha} = \boldsymbol{\beta}$ and SC scale parameter vector $\boldsymbol{m} = [1, 1, \ldots, 1]^{\mathrm{T}}$, Equations (9) and (12) reduce to the BER expressions for the conventional NOMA VLC system without any adjustments.

Note that as the number of users increases, i.e., $K = 4$, the BER expression of user 4 is affected by SIC decoding parameters at the receiver. The reason for how SIC decoding parameters at the receiver affect the BER expression is described in Appendix A.3, where we find that the bit error does not exist under certain SIC decoding parameters. Therefore, Equation (12) holds under the following restrictions

$$
\begin{cases}
(-1)^{r_{1,1}} \sum\limits_{t=u}^{k-1} (-1)^{r_{1,t}} z'_t \geq 0, \quad u \in \{1, 2, \ldots, k-1\}. \\
(-1)^{r_{2,1}} \sum\limits_{t=u}^{k-2} (-1)^{r_{2,t}} z'_t \geq 0, \quad u \in \{1, 2, \ldots, k-2\}. \\
\quad \vdots \\
(-1)^{r_{k-1,1}} \sum\limits_{t=u}^{1} (-1)^{r_{k-1,t}} z'_t \geq 0, \quad u \in \{1\}.
\end{cases}
\tag{13}
$$

When the restrictions in Equation (13) are not satisfied, the overlapping part of the decision regions for all users is affected by $z'_1, z'_2, \ldots$ and z'_{k-1}, which is consistent with Case II in Appendix A.3. Thus, some parts in Equation (12) become zero under certain z'_1, $z'_2, \ldots$, and z'_{k-1}, i.e., when $(-1)^{r_{1,1}} \sum_{t=1}^{k-1} (-1)^{r_{1,t}} z'_t < 0$, the $\sum_{t=1}^{k-1} (-1)^{r_{1,t}} z'_t$ in Equation (12) becomes zero.

3.2. NOMA Triangle Regarding the Terms of Q-Function

When the modulation format is 4-QAM, the BER expressions of user k under K-user NOMA VLC system are shown in Equations (9) and (12), where Equations (9) and (12) apply to user 1 and user $k(\geq 2)$, respectively. As shown in Equations (9) and (12), the BER expression for a particular user under the multi-user NOMA VLC system is the summation of the Q-function. Based on the derivation principles of BER expressions in the Section 3.1, the BER of user k is affected by the previous detection results of $k-1$ users. Thus, there are 2^{k-1} conditions for the detection of user k. As the conditions number of detection increases, the number of terms for BER expressions in terms of Q-function also grows.

In order to investigate the BER expression for each user, we list the number of terms for each user's BER expression in Table 1, where N_K stands for the number of users in NOMA VLC system, N_T represents the number of terms for a particular user's BER expression and U_K represents a particular user under K-user NOMA scheme. The value of N_T for each user's BER expression does not take $Q(-\infty)$ and $Q(\infty)$ into account, since $Q(-\infty) = 1$ and $Q(\infty) = 0$. Note that, in the vertical direction of Table 1, when the number of superimposed users increases by one under NOMA scheme, the number of terms for the BER expressions doubles under the same U_K. We can see that, under the K-user NOMA VLC system, the number of terms of BER expressions for K users includes $\left[2^{K-1}, 2^{K-1}\sum_{i=0}^{1} 2^i, 2^{K-1}\sum_{i=0}^{2} 2^i, \ldots, 2^{K-1}\sum_{i=0}^{k-1} 2^i, \ldots, 2^{K-1}\sum_{i=0}^{K-1} 2^i\right]^{\mathrm{T}}$, respectively. According to the rule shown in the Table 1, we can obtain the number of terms for each user's BER expression under K-user NOMA scheme. Since the number of terms for BER expressions for K users is distributed as a triangle in Table 1, we name this rule as NOMA triangle regarding the terms of Q-function.

Table 1. NOMA triangle regarding the terms of Q-function for K-user NOMA VLC system using 4-QAM.

N_T \ U_K / N_K	1	2	3	4	5	⋯	k	⋯	K
1	2^0								
2	2^1	$2^1\sum_{i=0}^{1}2^i$							
3	2^2	$2^2\sum_{i=0}^{1}2^i$	$2^2\sum_{i=0}^{2}2^i$						
4	2^3	$2^3\sum_{i=0}^{1}2^i$	$2^3\sum_{i=0}^{2}2^i$	$2^3\sum_{i=0}^{3}2^i$					
5	2^4	$2^4\sum_{i=0}^{1}2^i$	$2^4\sum_{i=0}^{2}2^i$	$2^4\sum_{i=0}^{3}2^i$	$2^4\sum_{i=0}^{4}2^i$				
⋮	⋮	⋮	⋮	⋮	⋮	⋱			
k	2^{k-1}	$2^{k-1}\sum_{i=0}^{1}2^i$	$2^{k-1}\sum_{i=0}^{2}2^i$	$2^{k-1}\sum_{i=0}^{3}2^i$	$2^{k-1}\sum_{i=0}^{4}2^i$	⋯	$2^{k-1}\sum_{i=0}^{k-1}2^i$		
⋮	⋮	⋮	⋮	⋮	⋮	⋮	⋮	⋱	
K	2^{K-1}	$2^{K-1}\sum_{i=0}^{1}2^i$	$2^{K-1}\sum_{i=0}^{2}2^i$	$2^{K-1}\sum_{i=0}^{3}2^i$	$2^{K-1}\sum_{i=0}^{4}2^i$	⋯	$2^{K-1}\sum_{i=0}^{k-1}2^i$	⋯	$2^{K-1}\sum_{i=0}^{K-1}2^i$

3.3. Optimization of BER Performance

We would like to find the minimum average BER under the SC-SIC-A scheme. The BER expressions are a function of SC scale parameters $m_1, m_2, \ldots, m_{2K-1}$ and the SIC decoding parameters $\beta_1, \beta_2, \ldots, \beta_K$. Thus, we establish an optimization problem to improve BER performance of a multi-user NOMA VLC system by joint optimization of the vector of SC scale parameters $\boldsymbol{m}$ and the vector of SIC decoding parameters $\boldsymbol{\beta}$. The minimization of average BER among K users is used as the objective function, which can be formulated as follows:

$$\begin{aligned} \mathrm{P}: &\underset{\boldsymbol{m},\,\boldsymbol{\beta}}{\text{minimize}} \quad \frac{1}{K}\sum_{k=1}^{K}\mathrm{BER}_{(K,k)}, \\ &s.t., \quad \mathrm{C1}: P' \le P, \\ &\qquad\quad \mathrm{C2}: \beta_K \le \beta_{K-1} \le, \ldots, \le \beta_1, \end{aligned} \tag{14}$$

where (C1) and (C2) are the constraints of problem (14), and (C1) guarantees that the total transmitted power of signal after adjustment is no larger than that before the adjustment and (C2) is used to maintain the same decoding order as that of the conventional NOMA VLC system without any adjustments. Due to the robustness of DE algorithm in optimization problems, the problem (14) is solved by DE algorithm [26]. By using the DE algorithm, the optimal SC scale parameters $m_1^*, m_2^*, \ldots, m_{2K-1}^*$ and the SIC decoding parameters $\beta_1^*, \beta_2^* \ldots, \beta_K^*$ are obtained. When only (C1) is considered, the optimization problem (14) for the proposed SC-SIC-A scheme is simplified to that for the SC-A scheme.

3.4. Derivation of Capacity

We derive the capacity expression of each user in the K-user NOMA VLC system under the SC-SIC-A scheme. The channel capacity of user 1 can be written as $C_{(K,1)} = \max I(\hat{Y}_{R,1}; S_1)$, where I represents the mutual information, S_1 stands for the transmitter data of user 1 and $\hat{Y}_{R,1}$ is the superposition signal at the receiver of user 1 under the SC-SIC-A scheme, which is the function of SC scale parameters $m_1, m_2, \ldots, m_{2K-1}$. Then, $I(\hat{Y}_{R,1}; S_1)$ can be derived as follows:

$$
\begin{aligned}
I(\hat{Y}_{R,1};S_1) &= I(S'+\hat{N}_1;S_1) = \int f(\hat{Y}_{R,1})\log_2\frac{f(\hat{Y}_{R,1}|S_1)}{f(\hat{Y}_{R,1})}d\hat{Y}_{R,1} \\
&= \log_2 4 \times \frac{1}{4^K}\sum_{S_1\in\mathcal{S}_1}\cdots\sum_{S_K\in\mathcal{S}_K}\int f(\hat{Y}_{R,k}|S_k,\ldots,S_K)d\hat{Y}_{R,1} \\
&- \frac{1}{4^K}\sum_{S_1\in\mathcal{S}_1}\sum_{S_2\in\mathcal{S}_2}\cdots\sum_{S_K\in\mathcal{S}_K}\int f(\hat{Y}_{R,1}|S_1,S_2,\ldots,S_K) \\
&\times \log_2\frac{\sum_{S_1\in\mathcal{S}_1}\sum_{S_2\in\mathcal{S}_2}\cdots\sum_{S_K\in\mathcal{S}_K} f(\hat{Y}_{R,1}|S_1,S_2,\ldots,S_K)}{\sum_{S_2\in\mathcal{S}_2}\cdots\sum_{S_K\in\mathcal{S}_K} f(\hat{Y}_{R,1}|S_1,S_2,\ldots,S_K)}d\hat{Y}_{R,1},
\end{aligned}
\tag{15}
$$

where the following is the case.

$$
\begin{aligned}
f(\hat{Y}_{R,1}) &= \frac{1}{4^K}\sum_{S_1\in\mathcal{S}_1}\sum_{S_2\in\mathcal{S}_2}\cdots\sum_{S_K\in\mathcal{S}_K} f(\hat{Y}_{R,1}|S_1,S_2,\ldots,S_K), \\
f(\hat{Y}_{R,1}|S_1) &= \frac{1}{4^{K-1}}\sum_{S_2\in\mathcal{S}_2}\cdots\sum_{S_K\in\mathcal{S}_K} f(\hat{Y}_{R,1}|S_1,S_2,\ldots,S_K), \\
f(\hat{Y}_{R,1}|S_1,S_2,\ldots,S_K) &= \frac{1}{\pi}e^{-\left|\hat{Y}_{R,1}-\sum_{t=1}^{K}\sqrt{\alpha_t}S_t\right|^2}.
\end{aligned}
\tag{16}
$$

Based on the same principle, the capacity of user k can be derived as follows:

$$
\begin{aligned}
C_{(K,k)} &= \max I(\hat{Y}_{R,k};S_k|S_1,S_2,\ldots,S_{k-1}) \\
&= \max I\left(\hat{Y}_{R,k}-\sum_{t=1}^{k-1}\sqrt{\beta_t}S_t+\hat{N}_k;S_k\right) \\
&= \max I(\tilde{Y}_{R,k};S_k),
\end{aligned}
\tag{17}
$$

where $\hat{Y}_{R,k}$ is the superposition signal at the receiver of user k, S_k represents the transmitter data of user 1 and $\tilde{Y}_{R,k} = \hat{Y}_{R,k} - \sum_{t=1}^{k-1}\sqrt{\beta_t}S_t + \hat{N}_k$. Then, $I(\tilde{Y}_{R,k};S_k)$ can be written as follows:

$$
\begin{aligned}
I(\tilde{Y}_{R,k};S_k) &= \int f(\tilde{Y}_{R,k})\log_2\frac{f(\tilde{Y}_{R,k}|S_k)}{f(\tilde{Y}_{R,k})}d\tilde{Y}_{R,k} \\
&= \log_2 4 \times \frac{1}{4^{K-k}}\sum_{S_k\in\mathcal{S}_k}\cdots\sum_{S_K\in\mathcal{S}_K}\int f(\tilde{Y}_{R,k}|S_k,\ldots,S_K)d\tilde{Y}_{R,k} \\
&- \frac{1}{4^{K-k}}\sum_{S_k\in\mathcal{S}_k}\sum_{S_{k+1}\in\mathcal{S}_{k+1}}\cdots\sum_{S_K\in\mathcal{S}_K}\int f(\tilde{Y}_{R,k}|S_k,\ldots,S_K) \\
&\times \log_2\frac{\sum_{S_k\in\mathcal{S}_k}\cdots\sum_{S_K\in\mathcal{S}_K} f(\tilde{Y}_{R,k}|S_k,\ldots,S_K)}{\sum_{S_{k+1}\in\mathcal{S}_{k+1}}\cdots\sum_{S_K\in\mathcal{S}_K} f(\tilde{Y}_{R,k}|S_k,\ldots,S_K)}d\tilde{Y}_{R,k},
\end{aligned}
\tag{18}
$$

where the following is the case.

$$
\begin{aligned}
f(\tilde{Y}_{R,k}) &= \frac{1}{4^{K-k}}\sum_{S_k\in\mathcal{S}_k}\sum_{S_{k+1}\in\mathcal{S}_{k+1}}\cdots\sum_{S_K\in\mathcal{S}_K} f(\hat{Y}_{R,k}-\sum_{t=1}^{k-1}\sqrt{\beta_t}S_t+\hat{N}_k|S_k,\ldots,S_K), \\
f(\tilde{Y}_{R,k}|S_k) &= \frac{1}{4^{K-k-1}}\sum_{S_{k+1}\in\mathcal{S}_{k+1}}\cdots\sum_{S_K\in\mathcal{S}_K} f(\hat{Y}_{R,k}-\sum_{t=1}^{k-1}\sqrt{\beta_t}S_t+\hat{N}_k|S_k,\ldots,S_K), \\
f(\tilde{Y}_{R,k}|S_k,\ldots,S_K) &= \frac{1}{\pi}e^{-\left|\hat{Y}_{R,k}-\sum_{t=1}^{k-1}\sqrt{\beta_t}S_t+\hat{N}_k-\sum_{t=k}^{K}\sqrt{\alpha_t}S_t\right|^2}.
\end{aligned}
\tag{19}
$$

From Equations (15)–(19), we can find that capacity expressions are a function of SC scale parameters $m_1, m_2, \ldots, m_{2K-1}$ and the SIC decoding parameters $\beta_1, \beta_2, \ldots, \beta_K$. The same optimized parameters obtained by solving the problem (P) are also suitable for investigation on the capacity performance of the SC-SIC-A scheme.

4. Results and Discussion

In this section, without loss of generality, we investigate the BER and capacity performance for multi-user NOMA VLC system by letting $K = 2$ and $K = 3$. We present the results of theoretical analyses, MC simulations and experiments compared with other benchmark schemes to show the effectiveness and feasibility of the proposed SC-SIC-A scheme in Figure 1. The parameters of DE algorithm, MC simulation and experiment in NOMA VLC system are listed in Tables 2–4.

Table 2. DE algorithm parameters.

Parameter	Value
Population size	500
Generation number	1000
Mutation factor	0.5
Crossover probability	0.2

Table 3. Simulation parameters [25,27,28].

Parameter	Value
FOV at a receiver, ψ_c	60°
PD detection area, A	1 cm^2
Optical filter, $T_S(\psi_i)$	1
Semiangle at half power, $\phi_{1/2}$	60°
Refractive index, n_s	1.5
Responsivity of PD, R	1 A/W
Current-to-light conversion efficiency, η	1 W/A
Power allocation coefficient vector when $K = 2$, $\boldsymbol{\alpha}$	$[0.6, 0.4]^{\mathrm{T}}$
Power allocation coefficient vector when $K = 3$, $\boldsymbol{\alpha}$	$[0.6, 0.3, 0.1]^{\mathrm{T}}$

Table 4. Experiment parameters.

Parameter	Value
Power allocation coefficient vector when $K = 2$, $\boldsymbol{\alpha}$	$[0.6, 0.4]^{\mathrm{T}}$
Power allocation coefficient vector when $K = 3$, $\boldsymbol{\alpha}$	$[0.6, 0.3, 0.1]^{\mathrm{T}}$
Data rate	10 M symbol/s
Semiangle at half power of VLC transmitter module	60°
Optical wavelength of VLC transmitter module	459.3 nm
Active area of VLC receiver module	1 cm^2

4.1. Benchmark Schemes

The BER and capacity performance of the proposed SC-SIC-A scheme in Figure 1 is compared with two benchmark schemes, which are called the no adjustment scheme and the SC-A scheme. The no adjustment scheme superimposes user signals by using NOMA without adjusting the parameters of SC or SIC decoding, while the SC-A scheme adjusts the parameters of SC at the transmitter. Note that the SC-A scheme for two-user NOMA VLC system was proposed in [20]. In this work, for comparison, the SC-A scheme has been applied to two-user and three-user NOMA VLC systems, respectively.

4.2. Theoretical Analysis and MC Simulation

We analyze the BER performance of the multi-user NOMA VLC system by letting $K = 2$. The SCs under the three schemes are depicted in Figure 3, where the adjusted parameters are obtained according to the optimization problem (P), when γ_1 is 21 dB. We can see that, under the SC-SIC-A scheme, all four points with the same color are confined in one quadrant and the distances between adjacent points in constellation are almost identical compared with that under the SC-A scheme and under the no adjustment scheme. Thus, an improved BER performance for all users can be expected under the SC-SIC-A scheme.

The BER performance is shown in Figure 4, where the results of theoretical analysis and MC simulation match very well. Note that under the no adjustment scheme, user 1 and user 2 can achieve the BER of 10^{-3} when γ_1 is 23.9 dB, for which its corresponding BER curves basically coincide. Under the SC-A scheme, γ_1 to achieve BER of 10^{-3} decreases from 23.9 dB to 14.8 dB for user 1, and the value of γ_1 to achieve BER of 10^{-3} decreases from 23.9 dB to 15.7 dB for user 2 at the same time. Under the proposed SC-SIC-A scheme in Figure 1, the obtained vectors of optimal adjustment parameters are $\boldsymbol{m}^* = [3.1644, 0.9529]^{\mathrm{T}}$ and $\boldsymbol{\beta}^* = [0.8015, 0.1947]^{\mathrm{T}}$, when γ_1 is 21 dB. With $\boldsymbol{m}^*$ and $\boldsymbol{\beta}^*$, the equality in (C1) holds, and $\sum_{k=1}^{2} \beta_k^*$ is smaller than and very close to 1. Moreover, we can see that the optimal power allocation coefficients $\beta_1^*, \beta_2^*, \ldots, \beta_K^*$ for SIC decoding are different from the original power allocation coefficients $\alpha_1, \alpha_2, \ldots, \alpha_K$. The values of γ_1 to achieve the BER of 10^{-3} are 13.5 dB and 14.9 dB for the two users, respectively. Compared with the no adjustment scheme, the corresponding reductions in γ_1 to achieve the BER of 10^{-3} are 9.1 dB and 8.2 dB for the two users, respectively. Meanwhile, the corresponding reductions in γ_1 to achieve the same level of BER are 1.3 dB and 0.8 dB, respectively, compared with the SC-A scheme.

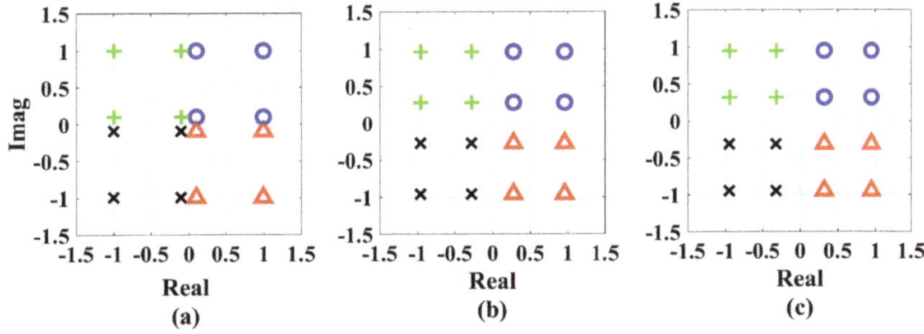

Figure 3. The SCs of (**a**) the no adjustment scheme, (**b**) the SC-A scheme and (**c**) the SC-SIC-A scheme. The parameters of (**b**,**c**) are obtained, when $K = 2$ and $\gamma_1 = 21$ dB. Colours and markers are used to distinguish the data of user 1.

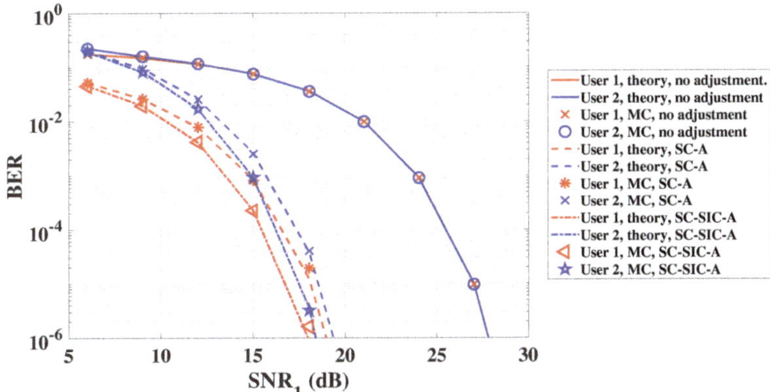

Figure 4. The BER performance for each user of multi-user NOMA VLC system, when $K = 2$.

Then, we analyze the performance of the multi-user NOMA VLC system by letting $K = 3$. The SCs under three schemes when γ_1 is 21 dB are shown in Figure 5. We can see that, under the no adjustment scheme, some of the points with one colour enter the adjacent quadrants with another colour, which deteriorates BER performance and leads to error floors of all users. Meanwhile, under the proposed SC-SIC-A scheme, all 16 points with the same colour are confined in the corresponding quadrants and the distances between adjacent points in constellation are almost identical compared to that under the no adjustment scheme and SC-A scheme. Thus, a better BER performance can be achieved by jointly adjusting the parameters of SC at the transmitter and SIC decoding at the receiver.

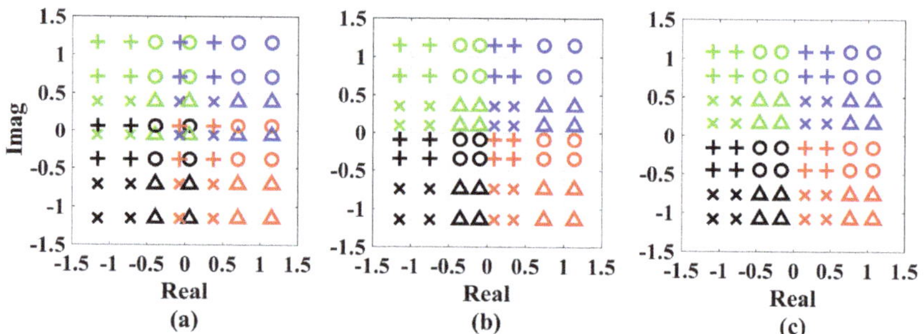

Figure 5. The SCs of (**a**) the no adjustment scheme, (**b**) the SC-A scheme and (**c**) the SC-SIC-A scheme. The parameters of (**b**,**c**) are obtained, when $K = 3$ and $\gamma_1 = 21$ dB. Colours are used to distinguish the data of user 1, and markers are used to distinguish the data of user 2 when user 1's data are detected.

The BER performance of three-user NOMA VLC system is shown in Figure 6, where the results of theoretical analysis and MC simulation match very well. We can find that all three users cannot achieve the BER of 10^{-3}, and the error floors for all users can be observed under the no adjustment scheme. Under the SC-A scheme, the error floors are eliminated, and the values of γ_1 to achieve the BER of 10^{-3} are 24.3 dB, 24.3 dB and 25.8 dB for the three users, respectively. Under the SC-SIC-A scheme in Figure 1, the obtained vector of optimal adjustment parameters are $\boldsymbol{m}^* = [-2.5583, 1.1801, 1.0851, 0.9344]^{\mathrm{T}}$ and $\boldsymbol{\beta}^* = [0.7607, 0.1940, 0.0451]^{\mathrm{T}}$, when γ_1 is 21 dB. Moreover, note that the optimal adjustment parameters $\beta_1^*, \beta_2^*, \ldots, \beta_K^*$ for SIC decoding are different from the original power allocation coefficients $\alpha_1, \alpha_2, \ldots, \alpha_K$ due to the SC adjustment at the transmitter. With joint transceiver optimization, the values of γ_1 to achieve the BER of 10^{-3} are 18.6 dB, 20.0 dB and 21.2 dB for the three users, respectively. The reductions in γ_1 to achieve BER of 10^{-3} are 5.7 dB, 4.3 dB and 4.6 dB for the three users, respectively, comparing to the BER performance under SC-A scheme. When γ_1 is 21 dB, the SCs in the three schemes are shown in the Figures 3 and 5, by letting $K = 2$ and $K = 3$. It can be found that the adjusted SC is clearer. Therefore, BER performance is improved. The γ_1 reductions to achieve BER of 10^{-3} by using the SC-SIC-A scheme compared with SC-A scheme under various power allocation coefficients are shown in Table 5, from which we can observe that the SC-SIC-A scheme performs almost better than the SC-A scheme under various power allocation coefficients. We can conclude that whether it is the two-user or the three-user NOMA VLC system, BER performance is significantly improved under the SC-SIC-A scheme.

The sum capacity performance of the multi-user NOMA VLC system with $K = 2$ and $K = 3$ is shown in Figure 7. To begin with, we analyze the sum capacity performance of the two-user NOMA VLC system. We can observe that the sum capacity increases very slowly and achieves its maximum value when γ_1 is 27.0 dB under the no adjustment scheme. Under the SC-A scheme, the value of γ_1 making the sum capacity achieve 4 bps/Hz is 15.0 dB. The reduction in γ_1 to achieve sum capacity of 4 bps/Hz is 12.0 dB. Under the SC-SIC-A scheme, the value of γ_1 is 15.0 dB, when the sum capacity achieves 4 bps/Hz, which is the same as the SC-A scheme. Then, we analyze the sum capacity performance of NOMA VLC system, when $K = 3$. We can find that, when SC is not adjusted, sum capacity increases very slowly and reaches its maximum value until γ_1 is 29.8 dB. Under the SC-A scheme, the value of γ_1 making the sum capacity achieve 6 bps/Hz is 25.0 dB, which is smaller than that under the no adjustment scheme. However, the value of γ_1 to achieve the maximum of sum capacity is 24.0 dB under the SC-SIC-A scheme. Meanwhile, when γ_1 is more than 13.7 dB, the sum capacity curve acquired by the SC-SIC-A scheme performs better than others. Therefore, the sum capacity performance is improved.

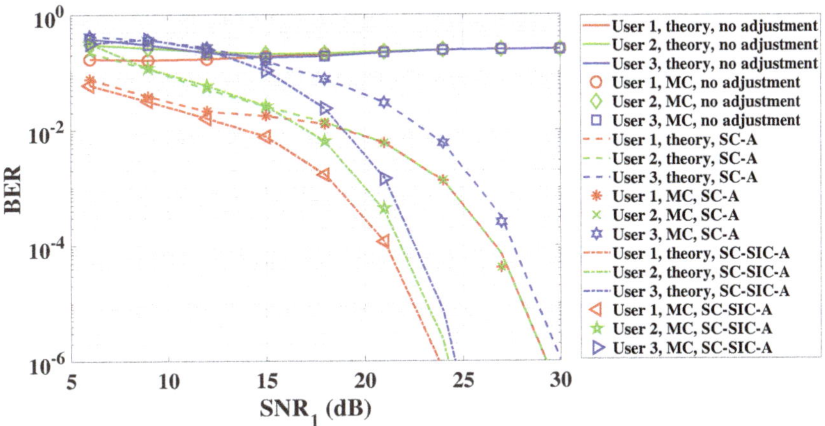

Figure 6. The BER performance for each user of multi-user NOMA VLC system, when $K = 3$.

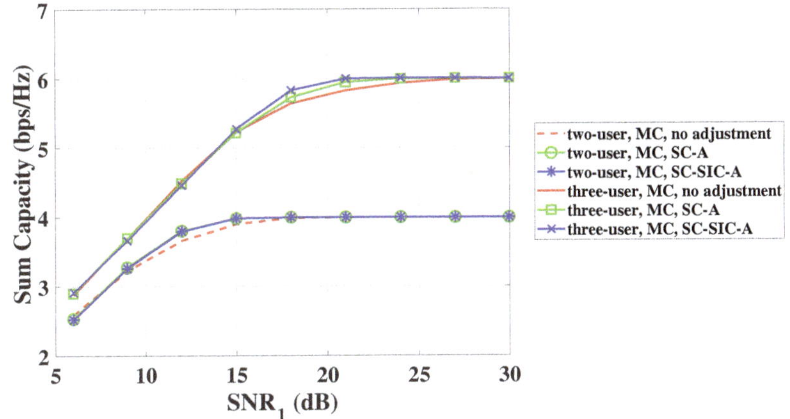

Figure 7. The sum capacity of multi-user NOMA VLC system, when $K = 2$ and $K = 3$.

Table 5. The γ_1 reductions to achieve BER of 10^{-3} by using the SC-SIC-A scheme compared with SC-A scheme under various power allocation coefficients.

Power Allocation Coefficient Vector α	γ_1 Reduction for User 1	γ_1 Reduction for User 2	γ_1 Reduction for User 3
$[0.6, 0.4]^T$	1.3 dB	0.8 dB	-
$[0.7, 0.3]^T$	0.6 dB	0.3 dB	-
$[0.8, 0.2]^T$	0 dB	0 dB	-
$[0.5, 0.3, 0.2]^T$	8.6 dB	0.8 dB	6.4 dB
$[0.6, 0.3, 0.1]^T$	5.7 dB	4.3 dB	4.6 dB
$[0.7, 0.2, 0.1]^T$	0.4 dB	0 dB	0.3 dB
$[0.85, 0.15, 0.05]^T$	0.7 dB	0.8 dB	0.1 dB

4.3. Experiment

In this part, we verify the improvement of BER performance of multi-user NOMA VLC systems by conducting an experiment. Due to the limitation of devices, the maximum SNR during signal detection is about 20 dB. Therefore, the verification of experiment is carried out in indoor environments under two-user and three-user NOMA VLC systems as examples.

The experimental setup of the NOMA VLC system is shown in Figure 8. The two users' 4-QAM data are generated offline using MATLAB in personal computer (PC). The two 4-QAM data with particular power allocation coefficient are superposed to generate a frequency-domain NOMA signal, which is converted to the time-domain OFDM signal, after passing through 64-point IFFT module. Note that Hermitian symmetry is used to the generate real-valued OFDM signal. Two thousand OFDM signals were uploaded to the arbitrary waveform generator (AWG) (RIGOL DG1062Z) with a sampling rate of 10 MSa/s. Subsequently, the signal generated by AWG drives the VLC transmitter module (HCCLS2021MOD01-TX), which converts the electrical signal into modulated light. After 1 m propagation, the modulated light reaches the VLC receiver module (HCCLS2021MOD01-RX), which converts it back to electrical signals. The converted electrical signal is captured by the oscilloscope (ROHDE&SCHWARZ RTE1022) and is downloaded to the PC for offline processing.

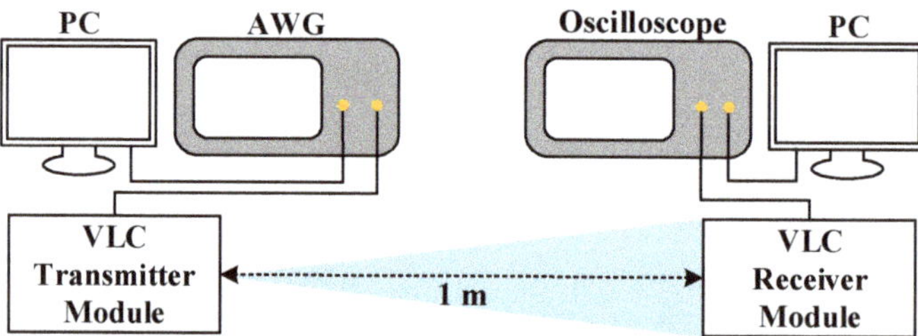

Figure 8. Experimental setup of the NOMA VLC system.

The experimental parameters are shown in Table 4, where the vectors of power allocation coefficient $\boldsymbol{\alpha}$ of two-user and three-user NOMA VLC systems are $[0.6, 0.4]^{\mathrm{T}}$ and $[0.6, 0.3, 0.1]^{\mathrm{T}}$, respectively. In the experiment, the values of γ_1 are changed under different powers of the transmitted signal. Therefore, we change the transmitted signal power by changing the voltage peak value of transmitted signal from 50 mV to 250 mV. BER performance is obtained under various γ_1.

BER performance when $K = 2$ is shown in Figure 9. Under the no adjustment scheme, user 1 and user 2 cannot achieve the BER of 10^{-3} with γ_1 in the range of 9 dB to 20 dB. Under the SC-A scheme, the values of γ_1 to achieve the BER of 10^{-3} are 14.7 dB and 15.8 dB for the two users, respectively. Under the SC-SIC-A scheme, the values of γ_1 to achieve the BER of 10^{-3} are 13.5 dB and 14.9 dB for the two users, respectively. Compared with the SC-A scheme, the corresponding reductions in γ_1 to achieve the same level of BER are 1.2 dB and 0.9 dB, respectively. In addition, from the Figures 4 and 9, we can see that the results of theoretical analysis and experiment match very well under different values of γ_1. BER performance when $K = 3$ is shown in Figure 10. Under the no adjustment scheme, when the value of γ_1 is between 9 dB and 20 dB, the error floors of all three users can be observed. Under the SC-A scheme, the BER of all the three users cannot reach 10^{-3} in the experiment. Under the SC-SIC-A scheme, although user 3 cannot achieve the BER of 10^{-3}, user 1 and user 2 can achieve the BER of 10^{-3} when the values of γ_1 are 18.7 dB and 20.0 dB, respectively. Note from Figures 6 and 10 that the results of theoretical analysis and experiment also match well when $K = 3$. The SCs under the three schemes obtained by experiment at the receiver are shown in the Figures 11 and 12. It can be found that the adjusted SCs under the SC-SIC-A scheme are better than the others, and the distances between adjacent points in the adjusted SCs under SC-SIC-A scheme are almost identical. Therefore, BER performance is improved. The joint transceiver optimization under the

SC-SIC-A scheme has an important impact on the improvement of BER performance of multi-user NOMA VLC systems.

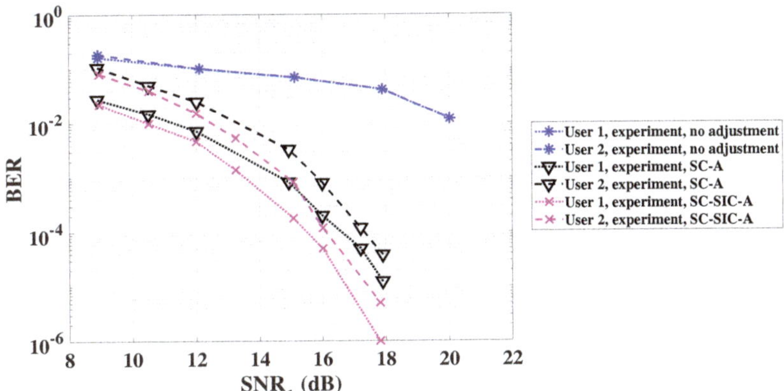

Figure 9. The experimental BER performance for each user of multi-user NOMA VLC system, when $K = 2$.

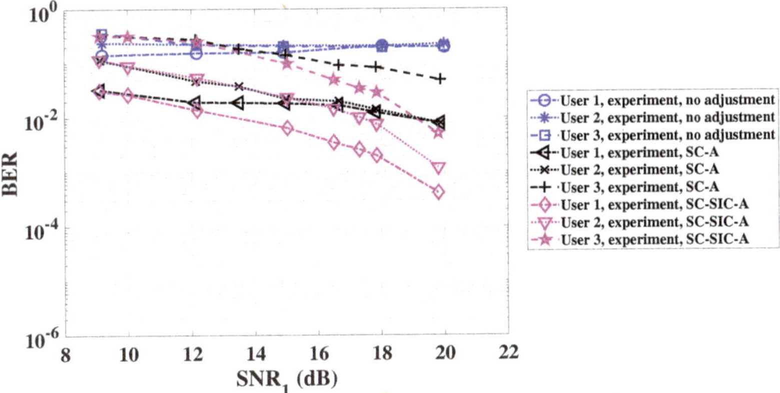

Figure 10. The experimental BER performance for each user of multi-user NOMA VLC system, when $K = 3$.

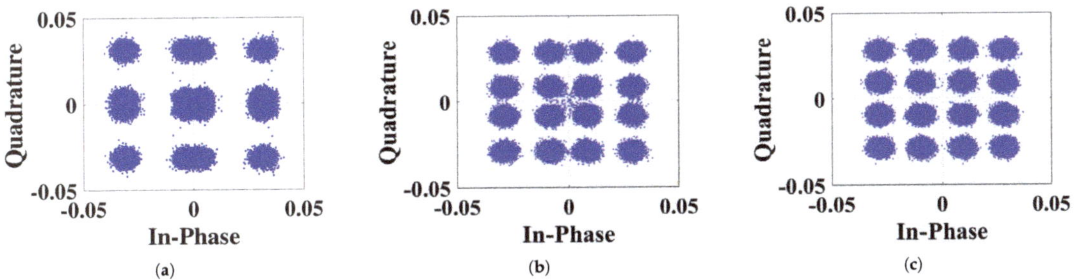

Figure 11. The SCs of the two-user NOMA VLC system obtained by experiment, when $\gamma_1 = 18$ dB. (**a**) No adjustment scheme; (**b**) SC-A scheme; (**c**) SC-SIC-A scheme.

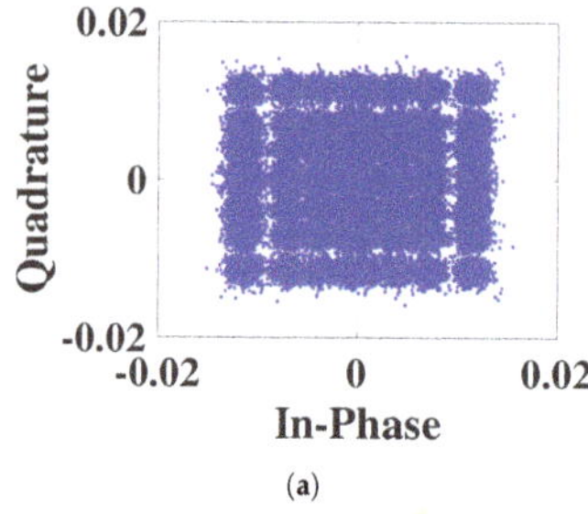

(a)

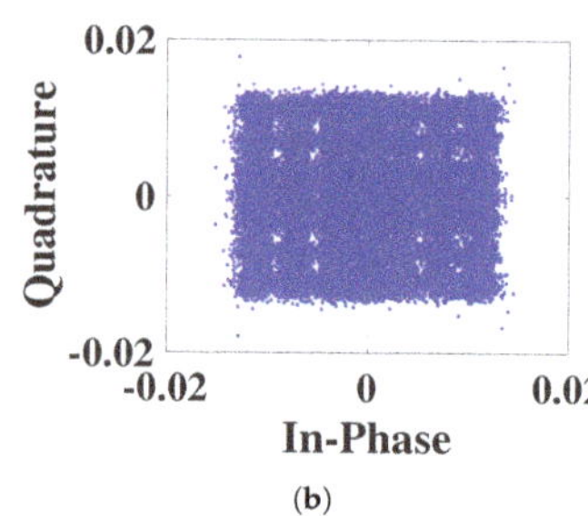

(b)

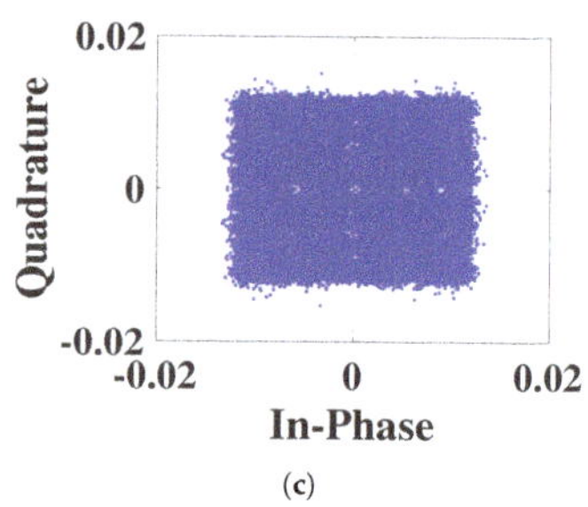

(c)

Figure 12. The SCs of the three-user NOMA VLC system obtained by experiment, when $\gamma_1 = 18$ dB. (**a**) No adjustment scheme; (**b**) SC-A scheme; (**c**) SC-SIC-A scheme.

5. Conclusions

In this paper, we propose a joint transceiver optimization scheme to improve BER and the capacity performance of multi-user NOMA VLC systems. This scheme is called SC-SIC-A scheme, where the parameters of SC at the transmitter and the SIC decoding at the receiver are jointly adjusted. We derive closed-form BER expressions under the proposed SC-SIC-A scheme for a multi-user NOMA VLC system. The rule for the number of terms for each user's BER expression is found, which is named as NOMA triangle regarding the terms of Q-function. Based on the derived BER expressions, we formulate the optimization problem to minimize the average BER for all users by adjusting the parameters of SC and SIC decoding, which is solved by the DE algorithm. We also derive the capacity expressions under the proposed SC-SIC-A scheme. In order to verify the effectiveness and feasibility of the proposed SC-SIC-A scheme, an indoor experiment was carried out. The results of theoretical analysis, MC simulation and experiment match very well, which verifies that the proposed SC-SIC-A scheme can achieve better performance compared with the SC-A scheme in terms of BER and sum capacity performance for multi-user NOMA VLC system. Specifically, for three-user NOMA VLC system, the results show that, compared with SC-A scheme, the SC-SIC-A scheme can reduce SNR to achieve BER of 10^{-3} by 5.7 dB, 4.3 dB and 4.6 dB for the three users, respectively, and can reduce SNR to achieve the maximum sum capacity by 1.0 dB. Thus, we recommend that the proposed SC-SIC-A scheme can be widely used for multi-user NOMA VLC systems.

Author Contributions: Conceptualization, T.W. and Z.W.; software, T.W.; validation, T.W., Z.W., S.H., J.Y. and Y.J.; writing—original draft preparation, T.W.; writing—review and editing, Z.W. and S.H.; supervision, Z.W., S.H., J.Y. and Y.J. All authors have read and agreed to the published version of the manuscript.

Funding: This research was supported by the National Natural Science Foundation of China under Grant 61835003.

Institutional Review Board Statement: Not applicable.

Informed Consent Statement: Not applicable.

Data Availability Statement: Not applicable.

Conflicts of Interest: The authors declare no conflict of interest.

Appendix A

Appendix A.1

We take the adjusted SC point $(x'_{2^{K-1}}, y'_{2^{K-1}}) = (m_{2^{K-1}}x_{2^{K-1}}, m_{2^{K-1}}y_{2^{K-1}})$ at the 2^{K-1}th column in the first quadrant of Figure 2 as an example to illustrate the derivation of BER_{1,x'_i}, where b_1 of the superposed signal at point $(x'_{2^{K-1}}, y'_{2^{K-1}})$ is denoted as '1'. The distance between point (x'_{2K-1}, y'_{2K-1}) and Q-axis is $\frac{m_{2K-1}\sum_{q=1}^{K} z_q \hat{\sigma}_N}{\sqrt{2}}$. When the amplitude of the noise's horizontal component $\hat{N}'_H$ is larger than $\frac{m_{2K-1}\sum_{k=1}^{K} z_k \hat{\sigma}_N}{\sqrt{2}}$, the noisy signal of point (x'_{2K-1}, y'_{2K-1}) in the horizontal direction falls into the region of b_1 ='0', which is the left side of the Q-axis. Consequently, b_1 is detected incorrectly as '0', where the bit error of b_1 occurs. Thus, the BER expression of b_1 for user 1 at the adjusted SC point (x'_{2K-1}, y'_{2K-1}) can be written as follows [29]:

$$\text{BER}_{1,x'_{2K-1}} = Q\left(m_{2K-1}\sum_{q=1}^{K} z_q\right) - Q(\infty), \tag{A1}$$

where $Q(\cdot)$ is the Q-function. Similarly, the BER for b_1 of user 1 at the adjusted SC point in other columns, i.e., $i = 1$ to $i = 2^{K-1} - 1$, follows the same principle.

Appendix A.2

We use Figure A1 to illustrate the derivation of user 2's BER explicitly. In Figure A1, we take the adjusted SC point (x'_{2K-1}, y'_{2K-1}) as an example, for which its bit sequence $\{b_1, b_2, b_3, b_4\}$ is denoted as '1010'. In the horizontal direction, the detection result of b_1 for user 1 affects its detection of b_3 for user 2. The condition that user 1's b_1 is detected correctly is shown in Figure A1a. When the scale of noise's horizontal component $\hat{N}'_H$ is smaller than $\frac{m_{2K-1}\sum_{q=1}^{K} z_q \hat{\sigma}_N}{\sqrt{2}}$, b_1 of user 1 is detected correctly. The corresponding decision threshold of b_1 is the Q-axis, and b_1 is detected as '1' when the noisy superposed NOMA signal falls into the right side of Q-axis. Similarly, when $\frac{m_{2K-1}\sum_{q=1}^{K} z_q \hat{\sigma}_N - z'_1 \hat{\sigma}_N}{\sqrt{2}} \leq \hat{N}'_H < \frac{m_{2K-1}\sum_{q=1}^{K} z_q \hat{\sigma}_N}{\sqrt{2}}$ or $\hat{N}'_H \geq \frac{m_{2K-1}\sum_{q=1}^{K} z_q \hat{\sigma}_N + z'_1 \hat{\sigma}_N}{\sqrt{2}}$, the noisy signal of adjusted SC point (x'_{2K-1}, y'_{2K-1}) in the horizontal direction falls into the region that b_3 = '0', which results in the erroneous detection of b_3. The decision thresholds of b_3 are denoted by the yellow dotted lines in Figure A1a, and the decision regions where b_3 is detected wrongly are the left side of the yellow dotted lines in the corresponding quadrants. The area of shaded region in Figure A1a is the error probability of b_3. Thus, the BER expression of b_3 for user 2 at adjusted SC point (x'_{2K-1}, y'_{2K-1}) under the condition that b_1 of user 1 is detected correctly is given by the following:

$$\text{BER}_{2,x'_{2K-1}\cap \text{U}_{1,0}} = Q\left(m_{2K-1}\sum_{q=1}^{K} z_q - z'_1\right) - Q\left(m_{2K-1}\sum_{q=1}^{K} z_q\right), \tag{A2}$$

where $z'_1 = \sqrt{\frac{\beta_1\gamma_1}{\alpha_1}}$. The condition that b_1 of user 1 is detected wrongly is shown in Figure A1b. Following the same principle as that depicted in Figure A1a, the area of shaded region in Figure A1b is the error probability of b_3 for user 2 at adjusted SC point (x'_{2K-1}, y'_{2K-1}) under the condition that user 1 is detected wrongly, and the corresponding BER expression is as follows.

$$\text{BER}_{2,x'_{2K-1}\cap \text{U}_{1,1}} = Q\left(m_{2K-1}\sum_{q=1}^{K} z_q + z'_1\right) - Q(\infty). \tag{A3}$$

Since the b_3 of user 2 at the adjusted SC point in the rest columns follows the same principle to derive its BER expression, the BER of user 2 can be obtained according to Equation (10).

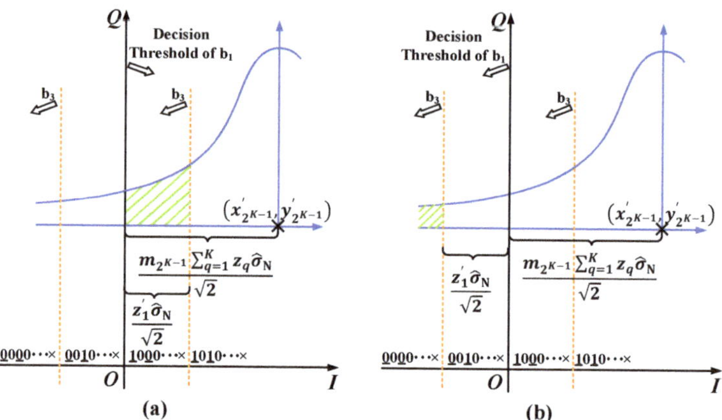

Figure A1. The constellations of the adjusted SC point (x'_{2K-1}, y'_{2K-1}) for deriving $\mathrm{BER}_{2,x'_{2K-1}}$ in K-user NOMA VLC system. (**a**) User 1 is detected correctly and (**b**) user 1 is detected incorrectly.

Appendix A.3

Under the condition that the bit of user 1 is detected correctly and the bits of user 2 and user 3 are detected incorrectly to derive the BER expression of user 4, we take the adjusted SC point (x'_8, y'_8) in first quadrant as an example to explain the special cases affected by SIC decoding parameters, which are shown in Figure A2. The bit sequence of the adjusted SC point (x'_8, y'_8) is denoted as $\{b_1, b_2, b_3, \ldots, b_8\}$. In the horizontal direction, the detection results of the first bit for user 1, user 2 and user 3 affect the detection of b_7 for user 4. Following the derivation principles of user 2's BER expression, the decision thresholds of b_1, b_3, b_5 and b_7 are the Q-axis, orange dotted line, purple dotted line and green dotted line in Figure A2, respectively. Regions that b_1 is detected correctly and b_3, b_5 as well as b_7 are detected incorrectly are the right side of Q-axis and the left side of all dotted lines in the corresponding quadrants, respectively. The area of the part where the decision regions of b_1, b_3, b_5 and b_7 overlap is affected by $z'_1 = \sqrt{\frac{\beta_1\gamma_1}{\alpha_1}}$, $z'_2 = \sqrt{\frac{\beta_2\gamma_1}{\alpha_1}}$ and $z'_3 = \sqrt{\frac{\beta_3\gamma_1}{\alpha_1}}$. The case that $z'_1 - z'_2 - z'_3 \geq 0$ is called Case I, which is illustrated in Figure A2a. From the constellation of the adjusted SC point (x'_8, y'_8) under Case I, we can find that the area of shaded region is the error probability of b_7 for user 4. Thus, the BER expression under Case I is given by the following.

$$\mathrm{BER}_{\text{Case I}} = Q\left[m_8 \sum_{q=1}^{4} z_q - (z'_1 - z'_2 - z'_3)\right] - Q\left(m_8 \sum_{q=1}^{4} z_q\right). \tag{A4}$$

The case where $z'_1 - z'_2 - z'_3 < 0$ is called Case II, which is illustrated in Figure A2b.We can find that the overlapping part of the decision regions of b_1, b_3, b_5 and b_7 do not exist. Thus, there is no bit error under the Case II, i.e., the BER expression under Case II is zero. Note that let $z'_1 - z'_2 - z'_3$ in Equation (A4) be zero, the BER expression Equation (A4) under Case I is changed to that under Case II. Thus, the BER expression under Case II can be written as follows.

$$\mathrm{BER}_{\text{Case II}} = Q\left(m_8 \sum_{q=1}^{4} z_q\right) - Q\left(m_8 \sum_{q=1}^{4} z_q\right) = 0. \tag{A5}$$

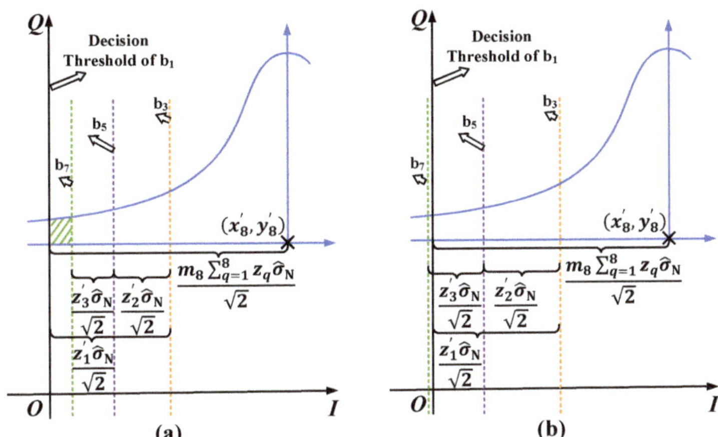

Figure A2. The constellations of the adjusted SC point (x_8', y_8') for deriving BER expression of user 4 in four-user NOMA VLC system. (**a**) The constellation of Case I and (**b**) the constellation of Case II.

References

1. Yang, H.; Alphones, A.; Zhong, W.; Chen, C.; Xie, X. Learning-Based Energy-Efficient Resource Management by Heterogeneous RF/VLC for Ultra-Reliable Low-Latency Industrial IoT Networks. *IEEE Trans. Ind. Inform.* **2020**, *16*, 5565–5576. [CrossRef]
2. Al Hammadi, A.; Sofotasios, P.C.; Muhaidat, S.; Al-Qutayri, M.; Elgala, H. Non-Orthogonal Multiple Access for Hybrid VLC-RF Networks with Imperfect Channel State Information. *IEEE Trans. Veh. Technol.* **2021**, *70*, 398–411. [CrossRef]
3. Karunatilaka, D.; Zafar, F.; Kalavally, V.; Parthiban, R. LED Based Indoor Visible Light Communications: State of the Art. *IEEE Commun. Surv. Tutor.* **2015**, *17*, 1649–1678. [CrossRef]
4. Liu, X.; Wang, Y.; Zhou, F.; Ma, S.; Hu, R.Q.; Ng, D.W.K. Beamforming Design for Secure MISO Visible Light Communication Networks with SLIPT. *IEEE Trans. Commun.* **2020**, *68*, 7795–7809. [CrossRef]
5. Obeed, M.; Dahrouj, H.; Salhab, A.M.; Zummo, S.A.; Alouini, M.S. User Pairing, Link Selection, and Power Allocation for Cooperative NOMA Hybrid VLC/RF Systems. *IEEE Trans. Wirel. Commun.* **2021**, *20*, 1785–1800. [CrossRef]
6. Pathak, P.H.; Feng, X.; Hu, P.; Mohapatra, P. Visible Light Communication, Networking, and Sensing: A Survey, Potential and Challenges. *IEEE Commun. Surv. Tutor.* **2015**, *17*, 2047–2077. [CrossRef]
7. Chen, C.; Zhong, W.D.; Yang, H.; Du, P. On the Performance of MIMO-NOMA-Based Visible Light Communication Systems. *IEEE Photon. Technol. Lett.* **2018**, *30*, 307–310. [CrossRef]
8. Shi, J.; He, J.; Wu, K.; Ma, J. Enhanced Performance of Asynchronous Multi-Cell VLC System Using OQAM/OFDM-NOMA. *J. Lightw. Technol.* **2019**, *37*, 5212–5220. [CrossRef]
9. Janjua, M.B.; da Costa, D.B.; Arslan, H. User Pairing and Power Allocation Strategies for 3D VLC-NOMA Systems. *IEEE Wirel. Commun. Lett.* **2020**, *9*, 866–870. [CrossRef]
10. Wang, G.; Shao, Y.; Chen, L.K.; Zhao, J. Subcarrier and Power Allocation in OFDM-NOMA VLC Systems. *IEEE Photon. Technol. Lett.* **2021**, *33*, 189–192. [CrossRef]
11. Dai, L.; Wang, B.; Ding, Z.; Wang, Z.; Chen, S.; Hanzo, L. A Survey of Non-Orthogonal Multiple Access for 5G. *IEEE Commun. Surv. Tutor.* **2018**, *20*, 2294–2323. [CrossRef]
12. Shi, J.; Hong, Y.; Deng, R.; He, J.; Chen, L.K.; Chang, G.K. Demonstration of Real-Time Software Reconfigurable Dynamic Power-and-Subcarrier Allocation Scheme for OFDM-NOMA-Based Multi-User Visible Light Communications. *J. Lightw. Technol.* **2019**, *37*, 4401–4409. [CrossRef]
13. Qiu, H.; Gao, S.; Tu, G. An Opportunistic NOMA Scheme for Multiuser Spatial Multiplexing VLC Systems. *IEEE Commun. Lett.* **2021**, *25*, 3017–3021. [CrossRef]
14. Ding, Z.; Lei, X.; Karagiannidis, G.; Schober, R.; Yuan, J.; Bhargava, V. A Survey on Non-Orthogonal Multiple Access for 5G Networks: Research Challenges and Future Trends. *IEEE J. Sel. Areas Commun.* **2017**, *35*, 2181–2195. [CrossRef]
15. Liu, Y.; Qin, Z.; Elkashlan, M.; Ding, Z.; Nallanathan, A.; Hanzo, L. Nonorthogonal Multiple Access for 5G and Beyond. *Proc. IEEE* **2017**, *105*, 2347–2381. [CrossRef]
16. Islam, S.M.R.; Avazov, N.; Dobre, O.A.; Kwak, K.S. Power-Domain Non-Orthogonal Multiple Access (NOMA) in 5G Systems: Potentials and Challenges. *IEEE Commun. Surv. Tutor.* **2017**, *19*, 721–742. [CrossRef]
17. Almohimmah, E.M.; Alresheedi, M.T. Error Analysis of NOMA-Based VLC Systems with Higher Order Modulation Schemes. *IEEE Access* **2020**, *8*, 2792–2803. [CrossRef]
18. Liu, X.; Chen, Z.; Wang, Y.; Zhou, F.; Luo, Y.; Hu, R.Q. BER Analysis of NOMA-Enabled Visible Light Communication Systems with Different Modulations. *IEEE Trans. Veh. Technol.* **2019**, *68*, 10807–10821. [CrossRef]

19. Guan, X.; Yang, Q.; Hong, Y.; Chan, C.C.K. Non-orthogonal multiple access with phase pre-distortion in visible light communication. *Opt. Express* **2016**, *24*, 25816–25823. [CrossRef]
20. Ren, H.; Wang, Z.; Du, S.; He, Y.; Chen, J.; Han, S.; Yu, C.; Xu, C.; Yu, J. Performance improvement of NOMA visible light communication system by adjusting superposition constellation: A convex optimization approach. *Opt. Express* **2018**, *26*, 29796–29806. [CrossRef]
21. Marshoud, H.; Sofotasios, P.; Muhaidat, S.; Karagiannidis, G.; Sharif, B. On the Performance of Visible Light Communication Systems with Non-Orthogonal Multiple Access. *IEEE Trans. Wirel. Commun.* **2017**, *16*, 6350–6364. [CrossRef]
22. Du, C.; Ma, S.; He, Y.; Lu, S.; Li, H.; Zhang, H.; Li, S. Nonorthogonal Multiple Access for Visible Light Communication IoT Networks. *Wirel. Commun. Mob. Com.* **2020**, *2020*, 1–10. [CrossRef]
23. Wu, T.; Wang, Z.; Yu, J.; Han, S.; Jiang, Y. Joint Transceiver Optimization for Performance Improvement of Multi-User NOMA VLC System. In Proceedings of the 2021 IEEE 21th International Conference on Communication Technology (ICCT), Tianjin, China, 13–16 October 2021; pp. 1477–1481.
24. Yin, L.; Popoola, W.; Wu, X.; Haas, H. Performance Evaluation of Non-Orthogonal Multiple Access in Visible Light Communication. *IEEE Trans. Commun.* **2016**, *64*, 5162–5175. [CrossRef]
25. Komine, T.; Nakagawa, M. Fundamental analysis for visible-light communication system using LED lights. *IEEE Trans. Consum. Electron.* **2004**, *50*, 100–107. [CrossRef]
26. Price, K.; Storn, R.; Lampinen, J. *Differential Evolution: A Practical Approach to Global Optimization*; Springer: Berlin/Heidelberg, Germany, 2005.
27. Baig, S.; Ali, U.; Asif, H.M.; Khan, A.A.; Mumtaz, S. Closed-Form BER Expression for Fourier and Wavelet Transform-Based Pulse-Shaped Data in Downlink NOMA. *IEEE Commun. Lett.* **2019**, *23*, 592–595. [CrossRef]
28. Assaf, T.; Al-Dweik, A.; Moursi, M.E.; Zeineldin, H. Exact BER Performance Analysis for Downlink NOMA Systems Over Nakagami-*m* Fading Channels. *IEEE Access* **2019**, *7*, 134539–134555. [CrossRef]
29. Wang, X.; Labeau, F.; Mei, L. Closed-Form BER Expressions of QPSK Constellation for Uplink Non-Orthogonal Multiple Access. *IEEE Commun. Lett.* **2017**, *21*, 2242–2245. [CrossRef]

MDPI
St. Alban-Anlage 66
4052 Basel
Switzerland
Tel. +41 61 683 77 34
Fax +41 61 302 89 18
www.mdpi.com

Photonics Editorial Office
E-mail: photonics@mdpi.com
www.mdpi.com/journal/photonics

www.ingramcontent.com/pod-product-compliance
Lightning Source LLC
LaVergne TN
LVHW070102170726
843515LV00007B/1593

9783036540870